Soil Quality, Soil Fertility and Land Management

Soil Quality, Soil Fertility and Land Management

Edited by **Henry Wang**

R CALLISTO REFERENCE

New York

Published by Callisto Reference,
106 Park Avenue, Suite 200,
New York, NY 10016, USA
www.callistoreference.com

Soil Quality, Soil Fertility and Land Management
Edited by Henry Wang

International Standard Book Number: 978-1-63239-566-5 (Hardback)

Printed in the United States of America.

Contents

Permissions

List of Contributors

Preface

This is comprehensive book which covers extensive topics like soil quality, soil fertility and land management. Soil serves as an essential resource for maintaining quality of life throughout the world. It performs multiple functions including facilitation of food production, storage of nutrients, waste disposal, storage of water, supporting our structures and environment etc. Maintenance of environmental quality and sustenance of biological productivity promotion of animal and plant health can be achieved by maintaining soil health. The effects of land management practices on soil properties and processes need to be understood to assess environmental and economic sustainability. Topics in this book elucidate multidisciplinary composition of current trends in soil health. It further elaborates the development of remediation strategies and feasible management to preserve and maintain soil health. The book talks about strategies to improve land management with the help of relevant case studies. The importance of characterizing soil properties in order to develop remediation and management strategies has also been emphasized. Introduction of new approaches for indicating soil pollution have been presented vividly in this book. It draws attention towards contemporary management of various environmental scenarios of high concern.

After months of intensive research and writing, this book is the end result of all who devoted their time and efforts in the initiation and progress of this book. It will surely be a source of reference in enhancing the required knowledge of the new developments in the area. During the course of developing this book, certain measures such as accuracy, authenticity and research focused analytical studies were given preference in order to produce a comprehensive book in the area of study.

This book would not have been possible without the efforts of the authors and the publisher. I extend my sincere thanks to them. Secondly, I express my gratitude to my family and well-wishers. And most importantly, I thank my students for constantly expressing their willingness and curiosity in enhancing their knowledge in the field, which encourages me to take up further research projects for the advancement of the area.

Editor

Part 1

Soil Characteristics
Control Biogeochemical Processes

The Role of Aluminum-Organo Complexes in Soil Organic Matter Dynamics

Maria C. Hernández-Soriano
Department of Soil Science, College of Agriculture and Life Sciences
North Carolina State University, Raleigh NC
USA

1. Introduction

The knowledge achieved during the last decades on the dynamics of organic matter (OM) and inorganic elements in soils has been essential to predict long-term effects of land management and develop sustainable practices that contribute to mitigate the decline in soil quality and the potential threats for human health.

Knowledge about soil organic matter (SOM) properties is essential to understand soil processes. The distribution and chemical speciation of organic carbon (OC) in soil have a major role on biogeochemical processes, e.g. its own chemical stability and the mobility and bioavailability of nutrient and contaminants (Eusterhues et al. 2005; von Lutzow et al. 2008). Therefore, SOM properties are directly related to essential environmental processes, e.g. plant production, carbon sequestration or water pollution. The decline in SOM quality results in land degradation, increasing flooding events or the rates of irrigation and fertilization necessary for agricultural activities.

Organo-mineral associations and complexation of SOM with metals ions largely determines the stability and degradability of OM (Kogel-Knabner et al. 2010). The relevance of inter-molecular interactions of OM with metal ions in solution on substrate degradation, e.g. complexation with aluminum (Al), has been already highlighted in the existing literature (Sollins et al. 1996) and prompted the research presented.

1.1 Stability of organic carbon in soils

Overall, this chapter addresses the relevance of understanding the mechanisms responsible for OC stabilization in soil. For instance, knowledge on SOM stability is essential to develop strategies for carbon sequestration in soil. Soil organic matter constitutes approximately 2/3 of the global terrestrial C pool (Batjes 1996), and OC dynamics in soils control a large part of the terrestrial carbon (C) cycle. In general, human activities cause a net release of CO_2 to the atmosphere of about 800 Gt C per year (Schlesinger 1984; Schlesinger and Andrews 2000) and more concretely, forest conversion to agriculture can release up to 75% of stored soil OC as CO_2 (Lal 2004). Besides, world soils have constituted a major source of enrichment of atmospheric concentration of carbon dioxide (CO_2) ever since the dawn of settled agriculture, about 10,000 years ago.

Nowadays carbon sequestration in soils is a main area of research because of the importance of soils for food production and its role in the global carbon cycle. Depending on environmental conditions and land use, soils may act as sources or sinks for C. Therefore, understanding the mechanisms that control stabilization and release of C is essential for the prediction of the effects of global climate change and for the development of management strategies to increase carbon sequestration in soils, which constituted a major demand at the Kyoto Protocol on climate change in 1992.

In general, it can be assumed that the pool of stable SOM in the soil solid phase is in equilibrium with the soluble organic matter (dissolved organic carbon, DOC) in the soil solution (Fig. 1). The term SOM refers to all organic substances in the soil. Organic carbon in soil can originate from natural or anthropogenic inputs, i.e. plant and animal litter decomposition, substances synthesized through microbial and chemical reactions and biomass of soil micro-organisms, but also soil addition with organic amendments. The turnover of C in soils is controlled mainly by water regimes and temperature, but is modified by factors such as size and physicochemical properties of C additions in litter or root systems or distribution of C within the soil matrix and its interactions with clay surfaces (Oades 1988).

The stabilization of organic materials in soils by the soil matrix is a function of the chemical nature of the soil mineral fraction and the presence of multivalent cations in the soil solution (Fig. 1), the presence of mineral surfaces capable of adsorbing organic materials, and the architecture of the soil matrix. The degree and amount of protection offered by each mechanism depends on the chemical and physical properties of the mineral matrix and the morphology and chemical structure of the organic matter. Thus, each mineral matrix will have a unique and finite capacity to stabilize organic matter (Baldock and Skjemstad 2000).

Three types of pathways are commonly considered in the formation of stable OM in soils (Christensen 1996; Sollins et al. 1996): Selective enrichment of organic compounds, which refers to the inherent recalcitrance of specific organic molecules against degradation by microorganisms and enzymes (Fig. 1); Chemical stabilization, involving all intermolecular interactions between organic substances and inorganic substances leading to a decrease in availability of the organic substrate due to surface condensation and changes in conformation, i.e., sorption to soil minerals and precipitation; Physical stabilization, related to the decrease in the accessibility of the organic substrates to microorganisms caused by occlusion within aggregates. According to Kogel-Knabner et al. (2008), the protection against decomposition imparted to soil organic carbon (SOC) by these mechanisms decreases in the order: chemically protected > physically protected > biochemically protected > non-protected. Hence, the relationship between soil structure and the ability of soil to stabilize SOM is a key element in soil C dynamics.

The sorption of OM is assumed as a chemisorptive process that occurs concomitantly with changes in OM conformation. Organomineral interactions lead to aggregations of clay particles and organic materials, which stabilizes both soil structure and the C compounds within aggregates, but due to the heterogeneity of natural soil systems different adsorption mechanism(s) may operate. Besides, different studies have evidenced the different depth distribution of OC in soils (Kaiser and Guggenberger 2000; Gillabel et al. 2010) and the dissimilarity in controls on C dynamics and decomposition of soil OC with depth in topsoil and subsoil (Fontaine et al. 2007; Salomé et al. 2009). For instance, according to Guggenberger and Kaiser (2003) sorptive preservation by location of OM in small pores rarely occur in topsoil horizons but primarily in subsoil horizons. Furthermore, the authors indicated that Fe oxides may be the most important sorbents for the formation of organo-mineral associations in the subsoils, which was later corroborated by Kogel-Knabner (2008).

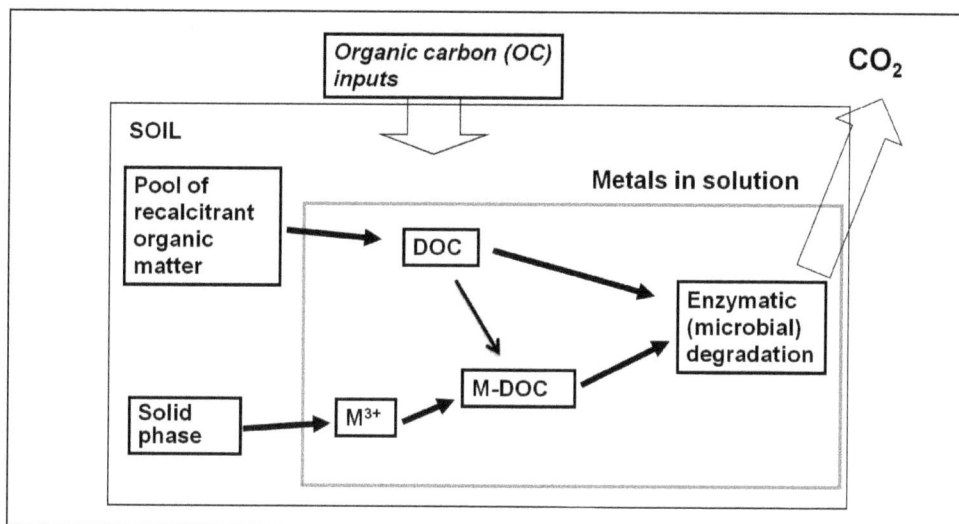

Fig. 1. Conceptual scheme of carbon cycling in soil.

Chemical protection involves interactions of OM with minerals; physical protection makes OC inaccessible to microbes and enzymes; and biochemical protection results from differential degradability of organic structures.

Chemical stabilization may result from association of OC compounds with mineral surfaces primarily in silt- and clay-sized particles (Sollins et al. 1996), but the mechanisms by which the different OM fractions adsorb onto mineral surfaces and the relationship between mineralogy and the chemistry of OM bound in organo-mineral associations are not yet fully understood.

1.2 Chemical protection of organic matter

For the purposes of the research discussed, this chapter focuses on the mechanisms involved in chemical stabilization. The primary scope is to provide a comprehensive understanding of chemical stabilization of OC in subsoil, considering the following statements:

- Chemical protection of OM might be predictable from soil clay mineralogy and extractable forms of Fe and Al (Kögel-Knabner et al. 2008).
- Complexation of fresh organic matter with Fe or Al oxides might constitute a major mechanism for OM stabilization (Eusterhues et al. 2005), forming aggregates protected from microbial degradation. Moreover, organic matter has been described to be more protected in subsoils than in topsoils (Gillabel et al. 2010). The total amount of C in subsoils is in general larger than in topsoils, but concentrations are in general lower in subsoil. The specific surface area (SSA) available in the subsoil for adsorbing OM is determined by the subsoil mineralogy and therefore the amount of OM that can be stabilized before reaching saturation (Eusterhues et al. 2005).

1.3 Complexation of aluminum by organic compounds

Aluminium is the third most abundant element in the Earth's crust, occurring at about 8%, and is a main or secondary component of numerous minerals, especially silicates. The only stable ion, Al^{3+}, is known to coordinate with oxygen-bearing ligands (Kabata-Pendias 2011). Metal ions in aqueous solution exist as aqua ions, where water molecules act as ligands, and coordinate to the metal ion via the oxygen donor atoms. Solution properties of Al are complex (Fig. 2), it is present as Al^{3+}, $Al(OH)^{2+}$ $Al(OH_2)^+$ at pH<5 but above pH 7.5 the dominant specie might be $Al(OH)_4^-$ (McBride 1994). Aluminum is a strongly hydrolyzing metal and relatively insoluble in the neutral pH range (6.0 to 8.0) (May et al. 1979). Under acidic (pH <6.0) or alkaline (pH >8.0) conditions, and/or in the presence of complexing ligands, the solubility of Al is enhanced, making it more available for biogeochemical transformations.

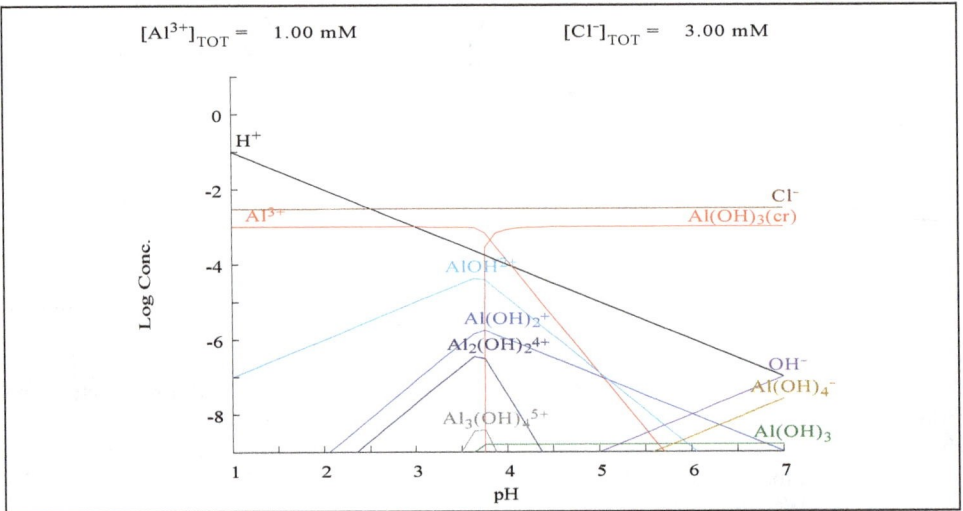

Fig. 2. Species distribution diagram (logarithmic) for Al(III) in aqueous solution.

The chemistry of Al in soils has been thoroughly evaluated (Sposito 1996) and highlighted as a major scientific problem, due to the worldwide concern about deforestation and acidic deposition and the resulting environmental impact.

The bioavailability and potential toxicity of Al in soils and waters are highly dependent on chemical interactions with natural organic matter. Solution Al (M^{3+}, Fig. 1) is the most chemically and biologically available form, although this pool represents an extremely small fraction of total Al in the environment. Within the aqueous phase, Al may be associated with a variety of inorganic or organic ligands. The extent of complexation depends on the availability of soil/sediment Al, solution pH, concentrations of complexing ligands, ionic strength and temperature (Driscoll and Schecher 1990). Aqueous Al may be redeposited to free soil/sediment pools, assimilated by living biomass or transported from the system (Figure 1).

Boudot et al. (1989) showed the direct protective effect of amorphous Al compounds against the mineralization of various associated organics. The results indicated that insoluble metallic hydroxides were responsible of capturing organic molecules, and therefore either preserving them from the access by soluble soil enzymes or preventing their movement to immobile enzymatic constituents associated with microbial cells, which would better account for a protective effect than chemical binding.

Organic matter has been described to floculate with Al salts. Maison et al. (2000) described the existence of specific binding sites for Al for given structures or ligands within the OM composition. Those results suggested that the organic ligands present in the OM are responsible for the distribution of metals.

1.4 Methodological approaches to characterize aluminum-organic interactions

The chemical complexity of SOM, the heterogeneity of its distribution and the variations in size and decomposition rate create significant analytical problems and partly explain the current deficiency in our understanding of SOM chemistry and dynamics (Lehmann et al. 2010).

Spectroscopic techniques are powerful tools in environmental research and a growing field of research (Schulp et al. 2008; Scheckel et al. 2010). The momentum of synchrotron research is leading to a continuous success of synchrotron studies addressing complex environmental issues and moreover, it can be expected that regulations and policy decisions will increasingly rely on such techniques.

Fluorescence spectroscopy has the required sensitivity to characterize metal ion binding properties of organic matter and determine its micromolar complexing capacities, i.e. to allow differentiating the binding sites or the ligand types involved in the formation of metal-organo complexes (Ryan and Weber 1982). Besides, the fact that fluorescence differentiates free from bound ligand provides an excellent complement to other complexing capacity techniques which measure free metal ion, like anodic stripping voltammetry or ion selective electrode potentiometry. Thus, fluorescence spectroscopy has been probed as valuable for the analysis of complexes between humic materials and several metal ions. For instance, fulvic acid exhibits fluorescence that is quenched upon binding to a paramagnetic metal ion.

2. Spectroscopic analysis to determine the formation of aluminum-organo complexes

Several methods are available for examining aluminium interactions with natural organic ligand solutions. Spectroscopic approaches allow an accurate characterization of metal-organic complexes. Thus, ultraviolet, infrared, fluorescence and ^{13}C NMR spectroscopy are used to evaluate the functional groups involved in binding, i.e. the aluminium-organic complexes. Fluorescence spectroscopy is well-known as a powerful technique useful for chemical characterization of humic acids while Fourier transform infrared spectroscopy (FTIR) can provide an insight into structural characteristics of complex organic macromolecules and allow typing organic molecules at the micrometer resolution. For instance, FTIR mapping have shown the location of organic C forms in relation to mineral surfaces, and relevant information on microaggregate formation.

2.1 Ultraviolet spectrophotometric determination of aluminum-organic complexes

Spectrophotometric titrations with Al(III) were carried out for gallic acid (Figure 3a) and salicylic acid (Figure 3b) to demonstrate the formation of the metal-organo complexes. UV–visible spectra were recorded for increasing concentrations of Al(III). A proportional decrease was observed for the absorbance at 220 and 270 nm for gallic acid (GA) and at 240 and 300 nm for salicylic acid (SA). The two isosbestic points (IUPAC 2007) confirmed the formation of the metal-organo complexes and are indicative of a transition between two light absorbing species in all recorded spectra (Harrington et al. 2010).

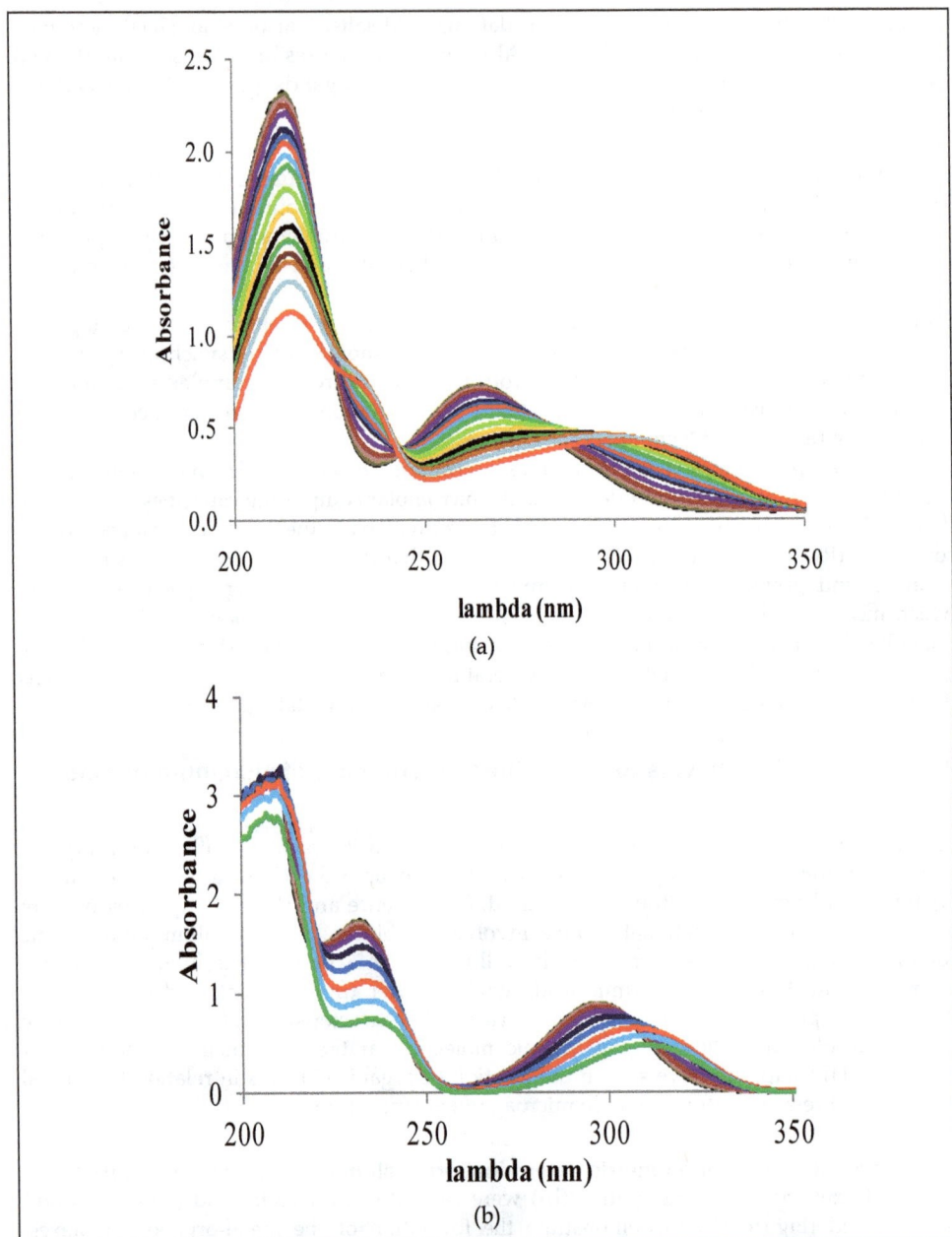

Fig. 3. Spectrophotometric titration of a) gallic acid 50 µM and b) salicylic acid 50 µM with Al(III).

Titrations were carried out at a pH range 3.91–4.25, which corresponds with a fraction of GA dissociated of 86-92%, 90-94% for SA and 7–13% hydrolization of Al(III). Therefore the decrease in absorbance can be mostly related with the formation of an inner-sphere complex. The shift of the absorbance maximum at the highest concentrations of Al(III) can be only partially explained by the pH variation and might be largely related to the formation of outer-sphere complexes.

2.2 Luminiscence spectrophotometric determination of aluminum-organic complexes

Fluorescence excitation-emission spectra were collected to characterize the formation of aluminium:gallic acid (Al:GA) complexes. Multidimensional fluorescence spectroscopy has been previously demonstrated to provide better characterization of metal binding to organic compounds than traditional quenching at a single excitation-emission wavelength.

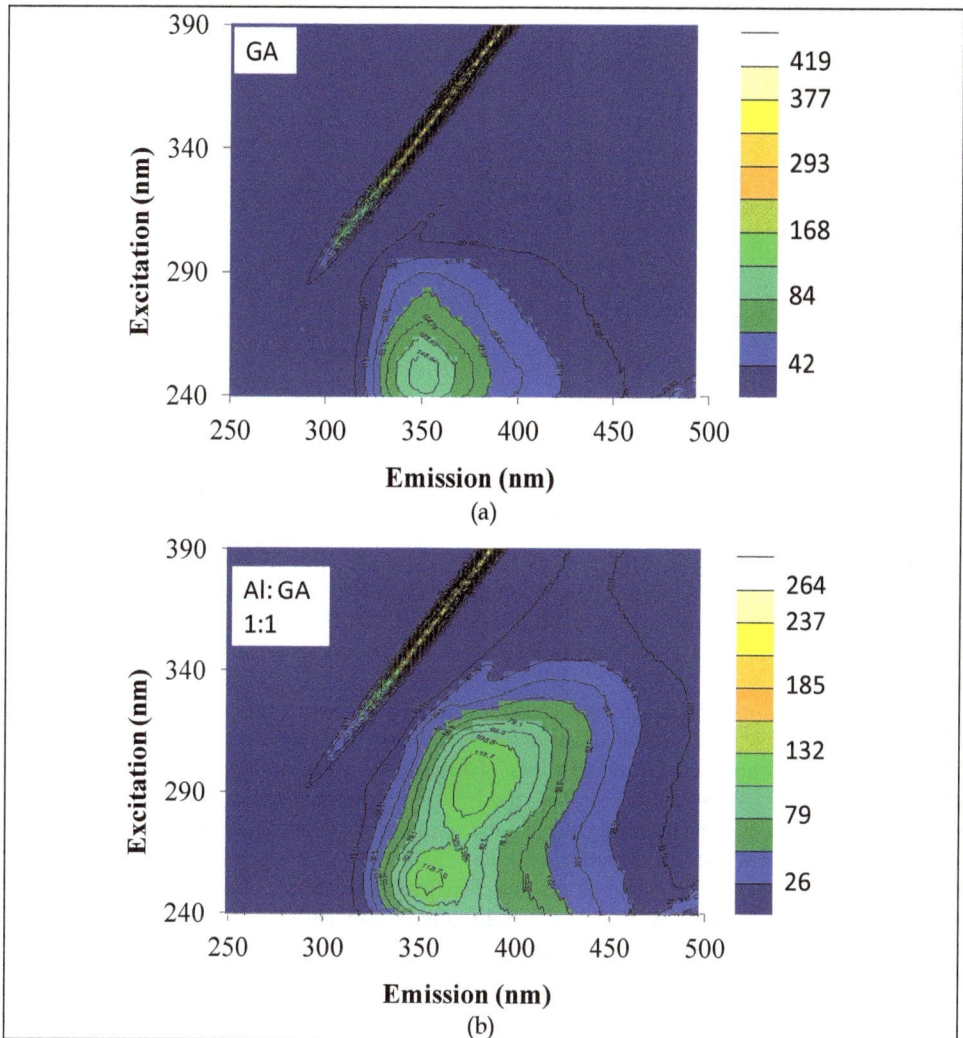

Fig. 4. Excitation-Emission spectra for gallic acid 50 µM (a) and Al:GA 1:1 (b).

Aluminum complexation with GA varied the fluorescence fingerprinting and therefore provided information on the types of metal-organic complexes, which depended on Al(III) concentration in solution (Ohno et al. 2008). Spectral variations confirmed the formation of metal-organo complexes (Figure 4) but furthermore suggested the formation of at least two different types of complexes (Figure 4b).

Similarly, for the titration of glucose (Glu) with Al(III) a shift in the maximum of fluorescence was determined, as well as an increase in the intensity of the signal for Al:Glu ratios up to 1:1, as depicted for the excitation-emission at 240 nm (Figure 5).

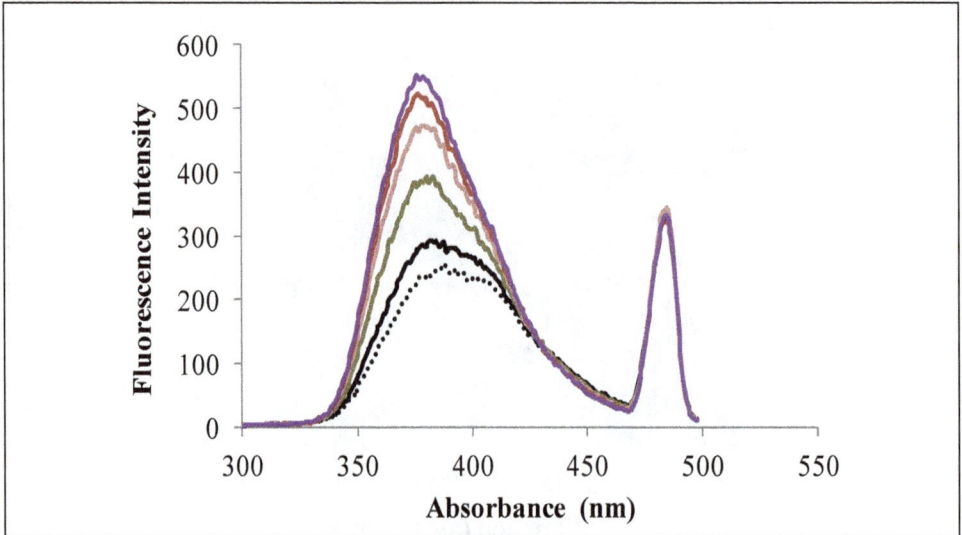

Fig. 5. Excitation-Emission spectra for glucose 1mM (a) and Al:Glu 1:1 (b).

The multidimensional spectra of Glu and Al:Glu 1:1 (Figure 6) provides a better portrayal of the complexation of Al with Glu. The tendency of Al(III) ions to displace aliphatic hydroxyl protons from various ligands in which they are in positions favourable for metal ion coordination, as might be the anomeric carbon, was already described by Motekaitis and Martell (1984).

Those results are particularly relevant to explain the stability of organic matter. Bartoli and Philippy (1990) described association of exchangeable Al with organic matter rich in polysaccharides as one of two major types of organo-mineral associations responsible for aggregate stability. Moreover, they described such aggregates to be easily disrupted by Al/Na exchange, a process that might partially explain the effect of sodicity in the increase of organic matter solubility in salt-affected soils.

The turnover of soil organic matter pools is a slow process, essential for soil structure and to promote soil biodiversity and support phytoremediation (Burke et al. 1995). Thus, soil addition with organic materials might decrease water run-off and erosion potential of topsoil.

Artificial soils were created according to the OECD guidelines (2010): 20% kaolinite, 70% sand, ≤ 1% CaCO3 and 10% Pahokee peat were mixed, moistened with deionised water at field capacity and autoclaved. Soils were equilibrated for 1 week in order to

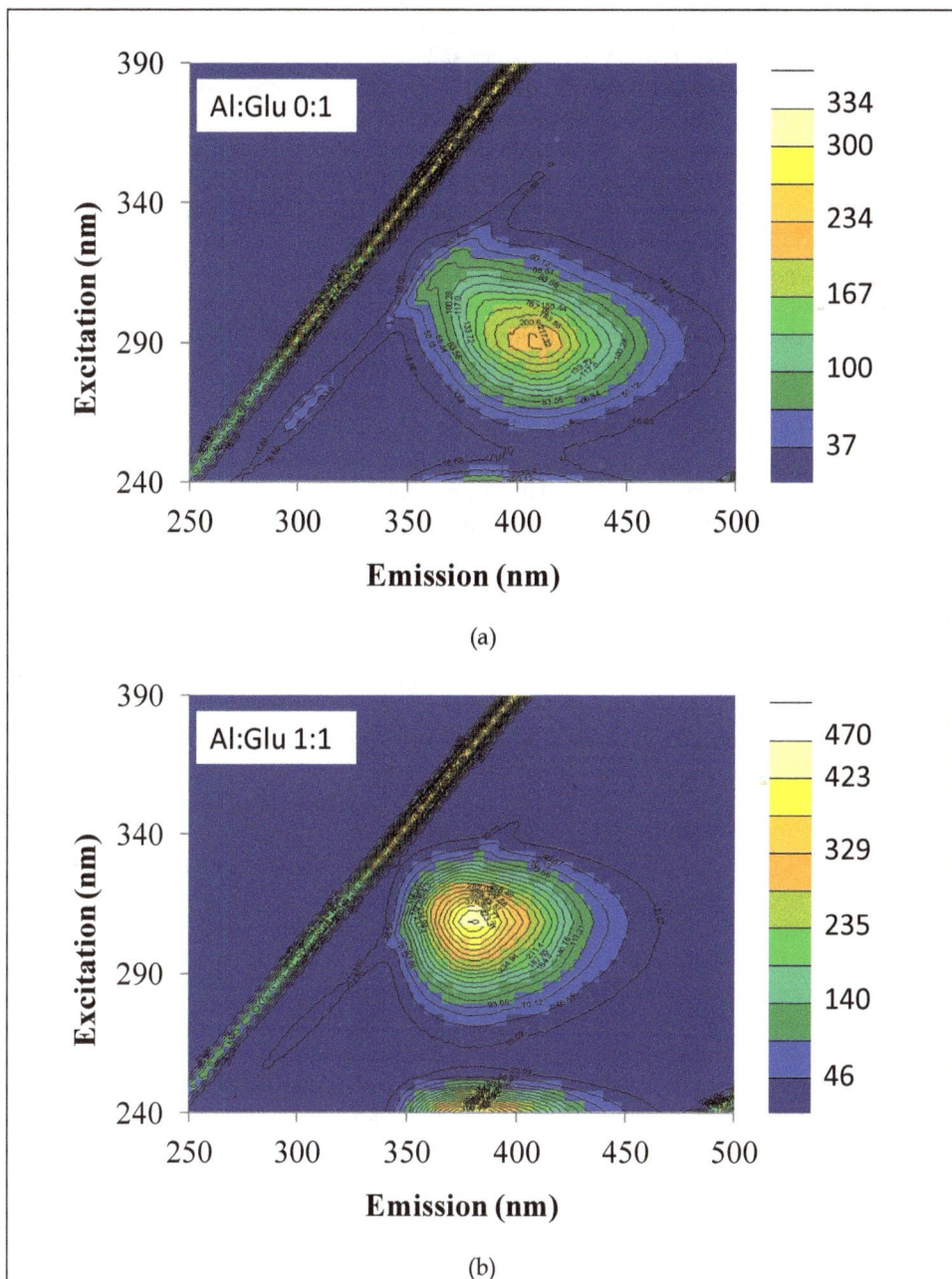

Fig. 6. Excitation-Emission spectra for glucose 1mM (a) and Al:Glu 1:1 (b).

equilibrate/stabilise the acidity. For the determination of pH a mixture of soil and 1 M KCl solution in a 1:5 ratio was used. The pH value was 6.0 ± 0.5. Subsamples of the artificial soil were added with a solution of glucose 1 mM or an Al(III) complex Al:Glu 1:1. The soil solution was normalized to a UV-Absorbance 0.2 at 254 nm by adequate dilutions and analyzed by luminescence spectroscopy. Emission spectra were recorded over the range of 250 to 500 nm at a constant excitation wavelength of 240 nm. Relative fluorescence intensity was based on a unitless reciprocal to the gain used to normalize each emission spectrum and was expressed in arbitrary units (Plaza et al. 2006).

Despite the chemical diversity of dissolved organic matter (DOM), similar steady state fluorescence spectra has been observed for DOM excitation –emission matrices (EEM), indicating the presence of common pool of fluorophores (Cory and McKnight 2005). Thus, EEM spectroscopy provides the sensitivity to examine subtle changes in dissolved organic matter (DOM) fluorescence and provide insight into alteration of DOM pool composition (Coble et al. 1998). Thus, fingerprinting of soil organic pools allows a thorough characterization of the organic matter in a given soil and provides a relevant tool to evaluate the alterations due to a particular soil addition with organic materials.

Otherwise, metal ion mobility and bioavailability in soil is extensively controlled by soil organic matter (Sposito 1996), and especially humic substances, of which humic acids (HAs) and fulvic acids (FAs) represent two major fractions. These materials are the most important soil organic ligands in terms of metal binding capacity due to their large content of oxygenated functional groups, including various carbonyl, carboxyl, phenolic, alcoholic and enolic hydroxyl groups (Tipping et al. 2002).

Soil addition with Al:Glu 1:1 resulted in a significant increase of the visible humic-like organic matter pool (Ex/Em 320-360/400-460) (Figure 7) compared to soil added with Glu.

This substantial increase in the fraction of UV and visible humic-like organic matter suggest that glucose complexation with Al(III) hampers the degradation of such labile compound, increasing the pool of highly stable, low degradation rate organic matter. Moreover, the increase in fluorescence intensity confirms the presence of a metal-organo complex in the solution Al:Glu 1:1. These results are consistent with previous results describing the preferential binding of aluminium to polysaccharides (Masion et al. 2000). The authors attempted to describe a model structure where cellulose is partially responsible for the locally ordered arrangement of Al atoms. Because of the complex chemical nature of humic and fulvic materials, which represent up to 70% of DOC and are major components of natural organic matter, these results are of major relevance for better understanding organic matter structure at a molecular level. Currently, little is known about stabilization of organic matter by formation of insoluble Al–OM complexes, which has been already described as a major pathway for the formation of stable soil OM (Scheel et al. 2007).

3. Future perspectives: The role of aluminum-organo complexes on carbon speciation, a health benchmark in salt-affected soils

Soil health is the balance of inherent soil properties (physical, chemical and biological), environmental conditions and management practices. Soil health is measured in terms of individual ecosystem services provided relative to a specific benchmark: e.g. microbial activity, CO_2 release, or humus level.

(a)

(b)

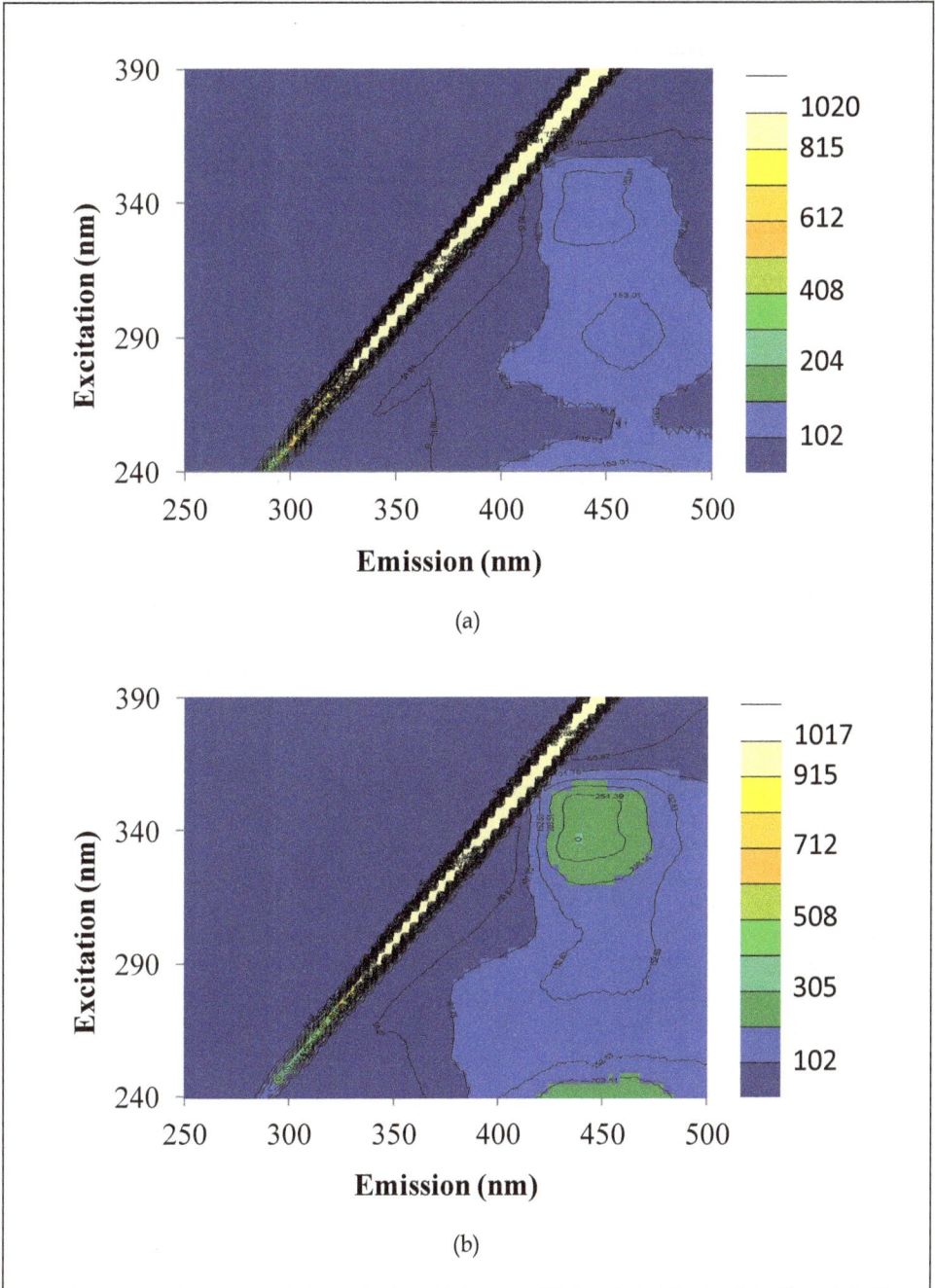

Fig. 7. Excitation-Emission spectra for pore water extractions from artificial soils added with glucose 1mM (a) or Al(III) complexes Al:Glu 1:1 (b).

A healthy soil is productive, sustainable and profitable. In general, a healthy soil presents the following characteristics which promote the health of plants, animals and humans while also maintaining environmental quality:

1. Soil organic matter equilibrium maintained
2. Soil fertility is balanced
3. Water entry, storage & supply optimised
4. Enhanced soil biological function
5. Supports productive land uses
6. Enhances environmental & community health and well-being

Among these properties, knowledge about soil organic matter (SOM) properties is central to soil health. The optimal level of SOM for any given soil is one which supports the functional capacity of the soil to hold and supply plant available water, store plant nutrients, provide energy for soil fauna, improve crop/biomass yields, and moderate net greenhouse gas emissions. Decomposition of organic matter (OM) regulates the flow of energy and nutrients in soil. It plays a key role in C, N, S and P cycling and also acts to improve soil structure. Agricultural practices and plant inputs influence both the quantity and quality of SOM, which in turn directly impacts on soil productivity and the ability of soil to recover from stress (soil resilience). For instance, the increasing use of wastewater for irrigation purposes in areas of southern Europe introduces surfactants in the soil system. Surfactants present in wastewater can have side-effects such as the increase of soil sodicity. A study conducted for a collection of calcareous soils correlated surfactants effect with soil properties, and additionally the effect of some amendments commonly used in agriculture was evaluated. Increasing sodicity and calcium sequestration were the mechanism driving trace metal release from soil treated with anionic surfactants (Hernandez-Soriano et al. 2011).

The amount of OM in a soil is used as an indicator of the potential sustainability of a system. Salt-affected soils present low OM contents due to poor plant growth, dispersion, erosion and leaching. Altered physical, biological and chemical properties directly impact SOM dynamic, particularly the active pool (rapid turnover). Thus, disruption of soil aggregates, changes in OM distribution and increased solubility of OM in the presence of Na increase SOM mineralization (Nelson et al. 1996; Oster and Shainberg 2001), while salinity alters soil microbial biomass (Rietz and Haynes 2003; Wong et al. 2008). Moreover, the opposing saline and sodic processes and the effects of OM result on conflicting effects of OM addition to saline and sodic soils. Currently, our understanding about carbon stocks and fluxes in saline and sodic soils is still limited while data related to carbon dynamics in such soils is contradictory (Wong et al. 2010).

A better understanding of the mechanisms of chemical protection of organic matter will contribute to protect and ameliorate soils health. This knowledge can be achieved by the accomplishment of the following goals:

• To characterize the specific binding mechanisms in organomineral associations using high-resolution spectroscopic techniques;
• To characterize the composition of the organic matter associated with minerals and/or ions and determine molecular-level changes in soil Fe and Al species for relevant scenarios.

The organomineral associations provide an alteration of the molecular structure of organic matter (organic compounds) such that enzymes for decomposing specific functional groups will be inactive. Thus, another major goal for future research is to relate soil respiration rates and biological transformation of organic matter to organomineral associations

4. Conclusions

Aluminium complexation with organic compounds plays a fundamental role in the dynamics of organic matter. Spectrometric analysis can help demonstrating the formation of complexes of Al (III) with organic compounds, but also identifying the types of metal-organic complexes in aqueous solution, which largely depends on metal concentration. Spectra alteration for the complexation of Al with gallic acid, salicylic acid and glucose, confirmed the formation of metal-organo complexes. Structural changes to organic molecules due to metal complexation might alter their biological transformation and similar processes can be expected to control soil organic matter turnover.

A comprehensive characterization of soil organic matter properties is central to soil health. The increasing concern about salt-affected soils and the necessity of developing feasible preventive and remediation strategies demand a better knowledge on carbon dynamics, particularly for the management of agricultural soils. Therefore, characterization of metal-organo complexes in the soil system constitutes a major research goal for soils scientist. Moreover, a multidisciplinary approach is required for such knowledge to truly contribute to the preservation of organic matter equilibrium in soil.

5. Acknowledgment

Maria C. Hernandez-Soriano thanks the Fulbright program and the Spanish Ministry of Education for a postdoctoral fellowship (FMECD-2010).

6. References

Baldock JA, Skjemstad JO (2000) Role of the soil matrix and minerals in protecting natural organic materials against biological attack. Organic Geochemistry 31 (7-8):697-710

Bartoli F, Philippy R (1990) Al-organic matter associations as cementing substances of ochreous brown soil aggregates: Preliminary examination. Soil Science 150 (4):745-751

Batjes NH (1996) Total carbon and nitrogen in the soils of the world. Eur J Soil Sci 47 (2):151-163

Boudot JP, Bel H, Brahim A, Steiman R, Seigle-Murandi F (1989) Biodegradation of synthetic organo-metallic complexes of iron and aluminium with selected metal to carbon ratios. Soil Biology and Biochemistry 21 (7):961-966

Burke IC, Lauenroth WK, Coffin DP (1995) Soil organic matter recovery in semiarid grasslands: implications for the conservation reserve program. Ecological Applications 5 (3):793-801

Coble PG, Del Castillo CE, Avril. B (1998) Distribution and optical properties of CDOM in the Arabian Sea during the 1995 Southwest Monsoon. Deep-sea research Part 2 Topical studies in oceanography 45:2195-2223

Cory RM, McKnight DM (2005) Fluorescence spectroscopy reveals ubiquitous presence of oxidized and reduced quinones in dissolved organic matter. Environ Sci Technol 39 (21):8142-8149. doi:Doi 10.1021/Es0506962

Christensen B (1996) Carbon in primary and secondary organomineral complexes. In: Advances in Soil Science - Structure and Organic Matter Storage in Agricultural Soils. CRC Lewis Publishers,, Boca Raton, FL.

Driscoll CT, Schecher WD (1990) The chemistry of aluminum in the environment. Environmental Geochemistry and Health 12 (1):28-49. doi:10.1007/bf01734046

Eusterhues K, Rumpel C, Kögel-Knabner I (2005) Organo-mineral associations in sandy acid forest soils: importance of specific surface area, iron oxides and micropores. Eur J Soil Sci 56 (6):753-763

Fontaine S, Barot S, Barre P, Bdioui N, Mary B, Rumpel C (2007) Stability of organic carbon in deep soil layers controlled by fresh carbon supply. Nature 450 (7167):277-280

Gillabel J, Cebrian-Lopez B, Six J, Merckx R (2010) Experimental evidence for the attenuating effect of SOM protection on temperature sensitivity of SOM decomposition. Global Change Biology 16 (10):2789-2798

Guggenberger G, Kaiser K (2003) Dissolved organic matter in soil: challenging the paradigm of sorptive preservation. Geoderma 113 (3-4):293-310

Harrington JM, Chittamuru S, Dhungana S, Jacobs HK, Gopalan AS, Crumbliss AL (2010) Synthesis and Iron Sequestration Equilibria of Novel Exocyclic 3-Hydroxy-2-pyridinone Donor Group Siderophore Mimics. Inorganic Chemistry 49 (18):8208-8221

Hernandez-Soriano MC, Degryse F, Smolders E (2011) Mechanisms of enhanced mobilisation of trace metals by anionic surfactants in soil. Environ Pollut 159 (3):809-816

IUPAC (2007) Glossary of terms used in photochemistry, 3rd edition.

Kabata-Pendias A (2011) Trace Elements in Soils and Plants 4edn. CRC Press, Boca Raton, FL

Kaiser K, Guggenberger G (2000) The role of DOM sorption to mineral surfaces in the preservation of organic matter in soils. Organic Geochemistry 31 (7-8):711-725

Kogel-Knabner I, Amelung W, Cao ZH, Fiedler S, Frenzel P, Jahn R, Kalbitz K, Kolbl A, Schloter M (2010) Biogeochemistry of paddy soils. Geoderma 157 (1-2):1-14

Kögel-Knabner I, Guggenberger G, Kleber M, Kandeler E, Kalbitz K, Scheu S, Eusterhues K, Leinweber P (2008) Organo-mineral associations in temperate soils: Integrating biology, mineralogy, and organic matter chemistry. Journal of Plant Nutrition and Soil Science 171 (1):61-82

Lal R (2004) Soil Carbon Sequestration Impacts on Global Climate Change and Food Security. Science 304 (5677):1623-1627. doi:10.1126/science.1097396

Lehmann J, Solomon D, Balwant S, Markus G (2010) Organic Carbon Chemistry in Soils Observed by Synchrotron-Based Spectroscopy. In: Developments in Soil Science, vol Volume 34. Elsevier, pp 289-312

Masion A, Vilge-Ritter A, Rose J, Stone WEE, Teppen BJ, Rybacki D, Bottero JY (2000) Coagulation-flocculation of natural organic matter with Al salts: Speciation and structure of the aggregates. Environmental Science & Technology 34 (15):3242-3246. doi:10.1021/es9911418

May HM, Helmke PA, Jackson ML (1979) Gibbsite solubility and thermodynamic properties of hydroxyaluminum ions in aqueous solutions at 25°. Geochim Cosmochim Acta 43:861- 868

McBride MB (1994) Environmental Chemistry of Soils. Oxford University Press, Oxford

Motekaitis RJ, Martell AE (1984) Complexes of Aluminum(III) with hydroxy carboxylic acids. Inorg Chem 23:18-23

Nelson PN, Ladd JN, Oades JM (1996) Decomposition of C-14-labelled plant material in a salt-affected soil. Soil Biology & Biochemistry 28 (4-5):433-441. doi:10.1016/0038-0717(96)00002-8

Oades J (1988) The retention of organic matter in soils. Biogeochemistry 5 (1):35-70

OECD/OCDE (2010) OECD guidelines for the testing of chemicals.

Ohno T, Amirbahman A, Bro R (2008) Parallel factor analysis of excitation-emission matrix fluorescence spectra of water soluble soil organic matter as basis for the determination of conditional metal binding parameters. Environ Sci Technol 42 (1):186-192

Oster JD, Shainberg I (2001) Soil responses to sodicity and salinity: challenges and opportunities. Australian Journal of Soil Research 39 (6):1219-1224. doi:10.1071/sr00051

Plaza C, Brunetti G, Senesi N, Polo A (2006) Fluorescence characterization of metal ion-humic acid interactions in soils amended with composted municipal solid wastes. Analytical and Bioanalytical Chemistry 386 (7-8):2133-2140. doi:10.1007/s00216-006-0844-0

Rietz DN, Haynes RJ (2003) Effects of irrigation-induced salinity and sodicity on soil microbial activity. Soil Biology & Biochemistry 35 (6):845-854. doi:10.1016/s0038-0717(03)00125-1

Ryan DK, Weber JH (1982) Fluorescence quenching titration for determination of complexing capacities and stability constants of fulvic acid. Anal Chem 54:986-990

Salomé C, Nunan N, Pouteau V, Lerch TZ, Chenu C (2009) Carbon dynamics in topsoil and in subsoil may be controlled by different regulatory mechanisms. Global Change Biology 16 (1):416-426

Scheckel KG, Ford RG, Balwant S, Markus G (2010) Role of Synchrotron Techniques in USEPA Regulatory and Remediation Decisions. In: Developments in Soil Science, vol Volume 34. Elsevier, pp 147-169

Scheel T, Dörfler C, Kalbitz K. (2007) Precipitation of Dissolved Organic Matter by Aluminum Stabilizes Carbon in Acidic Forest Soils. Soil Sci. Soc. Am. 71(1): 64-74

Schlesinger W (1984) Soil organic matter: A source of atmospheric CO_2. The role of the terrestrial vegetation in the global carbon cycle. Wiley, New York, USA.

Schlesinger WH, Andrews JA (2000) Soil respiration and the global carbon cycle. Biogeochemistry 48 (1):7-20. doi:10.1023/a:1006247623877

Schulp CJE, Nabuurs G-J, Verburg PH (2008) Future carbon sequestration in Europe - Effects of land use change. Agr Ecosyst Environ 127 (3-4):251-264

Sollins P, Homann P, Caldwell BA (1996) Stabilization and destabilization of soil organic matter: mechanisms and controls. Geoderma 74 (1-2):65-105

Sposito G (1996) The Environmental Chemistry of Aluminum 2nd edn. Lewis Publishers, Boca Raton, FL

Tipping E, Rey-Castro C, Bryan SE, Hamilton-Taylor J (2002) Al(III) and Fe(III) binding by humic substances in freshwaters, and implications for trace metal speciation. Geochimica et Cosmochimica Acta 66 (18):3211-3224

von Lutzow M, Kogel-Knabner I, Ludwig B, Matzner E, Flessa H, Ekschmitt K, Guggenberger G, Marschner B, Kalbitz K (2008) Stabilization mechanisms of organic matter in four temperate soils: Development and application of a conceptual model. J Plant Nutr Soil Sci 171 (1):111-124

Wong VNL, Dalal RC, Greene RSB (2008) Salinity and sodicity effects on respiration and microbial biomass of soil. Biology and Fertility of Soils 44 (7):943-953. doi:10.1007/s00374-008-0279-1

Wong VNL, Greene RSB, Dalal RC, Murphy BW (2010) Soil carbon dynamics in saline and sodic soils: a review. Soil Use Manage 26 (1):2-11

2

Soil Fertility Status and Its Determining Factors in Tanzania

Shinya Funakawa[1], Hiroshi Yoshida[1], Tetsuhiro Watanabe[1],
Soh Sugihara[1], Method Kilasara[2] and Takashi Kosaki[3]
[1]*Graduate School of Agriculture, Kyoto University,*
[2]*Faculty of Agriculture, Sokoine Agricultural University,*
[3]*Graduate School of Urban Environmental Sciences, Tokyo Metropolitan University*
[1,3]*Japan*
[2]*Tanzania*

1. Introduction

The pedogenetic conditions in Tanzania vary widely. In particular, the country has a wide variety of parent materials of soils because of the presence of volcanic mountains, the Great Rift Valley, and several plains and mountains with different elevations (hence, different temperatures). In addition, the amount and seasonal distribution pattern of the annual precipitation vary, from less than 500 mm to more than 2500 mm. The potential land use and agricultural production differ greatly among regions, due to the presence of different soils.

There have been several reports on the distribution patterns of soils and their physicochemical and mineralogical properties. According to a review of the history of soil surveys in Tanzania by Msanya *et al.* (2002), the major soil types described in the country are Ferric, Chromic, and Eutric Cambisols (39.7%); followed by Rhodic and Haplic Ferralsols (13.4%) and Humic and Ferric Acrisols (9.6%). To obtain basic information on soil mineralogy, Araki *et al.* (1998) investigated soil samples collected from regions at different altitudes in the Southern Highland and reported that the cation exchange capacity (CEC) per unit amount of clay content showed a negative correlation with elevation, which was accompanied by clay mineralogical transformation from mica to kaolinite. The authors suggested that soil formation on different planation surfaces is mainly controlled by the geological time factor whereby the lower surfaces are formed at the expense of the higher surfaces. Szilas *et al.* (2005) analyzed the mineralogy of well-drained upland soil samples collected from important agricultural areas in different ecological zones in the sub-humid and humid areas of Tanzania. They concluded that all soils were severely weathered and had limited but variable capacities to hold and release nutrients in plant-available form and to sustain low-input subsistence agriculture. Generally, there seems to be a consensus that the soils in Tanzania and the neighboring countries are not very fertile. The relevance of soil organic carbon management and appropriate fallowing systems such as agroforestry have been pointed out since as critical for sustaining agricultural production (Kimaro *et al.*, 2008; Nandwa, 2001).

In the present study, the regional trend in soil fertility with respect to the soil mineralogical and chemical properties was investigated. Soil properties were correlated with different pedogenetic factors such as geology and climate. A comprehensive understanding of the distribution of some soil properties as influenced by soil-forming factors is essential for planning an appropriate land-use strategy. Besides, this knowledge will allow developing and sustaining agricultural production, while preserving natural resources such as forest and woodland ecosystems.

2. Materials and methods

2.1 Soil samples

Ninety-five topsoil samples were collected from different regions of Tanzania. All the sampling points were located on slopes or plains, covering regions with different parent materials and with a wide variety of annual precipitation (less than 250 to more than 1500 mm) (Fig. 1; prepared based on Atlas of Tanzania [1967]). Apparent lowland soils were excluded from the analysis. The parent materials of the soils were broadly classified according to the following categories: (1) volcanic rocks (mostly basic), (2) granite and other plutonic rocks, (3) sedimentary and metamorphic rocks, and (4) Cenozoic rocks and recent deposits. The sampling plots corresponded to croplands or areas covered by either seminatural vegetation (forest or woodland) or secondary vegetation that had grown after human disturbance.

2.2 Analytical methods

The soil samples collected were air-dried and passed through a 2-mm mesh sieve. Soil pH in water or 1 mol L^{-1} KCl solution was measured with a glass electrode with a 1:5 soil:solution ratio. The pH(NaF) was measured with a glass electrode in 1 mol L^{-1} NaF solution after stirring for 2 min; the soil to solution ratio was 1:50. The CEC and the amount of exchangeable bases were measured after extracting with 1 mol L^{-1} NH_4OAc at pH 7.0 and then with a 10% NaCl solution (Thomas, 1982). The NH_4^+ extracted with 10% NaCl solution was distilled after the addition of concentrated NaOH solution, and collected into a 2% H_3BO_4 solution. Subsequently, the NH_4 content was determined by HCl titration (0.01 mol L^{-1}). The exchangeable base (Na, K, Mg, and Ca) content in the NH_4OAc solution was

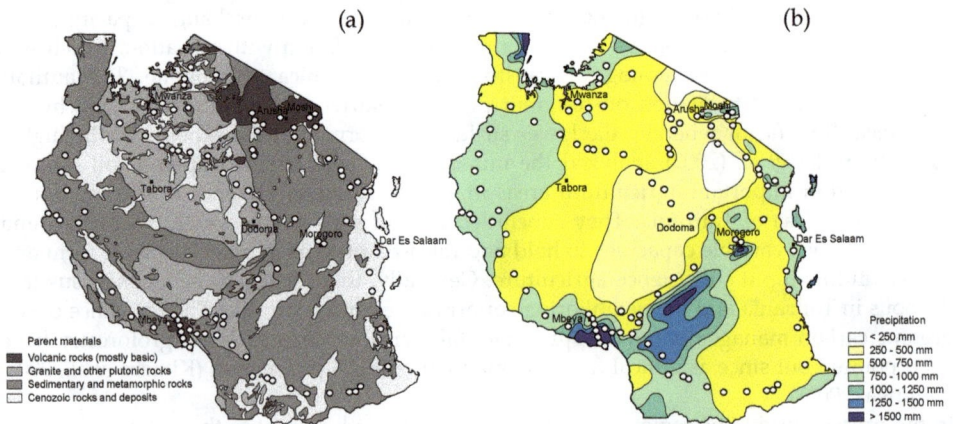

Fig. 1. Geological (a) and climatic (b) conditions of the sampling plots

determined by atomic absorption spectrophotometry (AAS) (Shimadzu, AA-840-01). The exchangeable Al and H were extracted using 1 mol L^{-1} KCl. The exchange acidity (Al + H) was determined to pH 8.3 by titration with 0.01 mol L^{-1} NaOH wherein phenolphthalein was used as an indicator. Then, after the addition of 4% NaF solution to liberate OH^- from the $Al(OH)_3$ precipitates, the exchangeable Al was determined by back titration to obtain the same pH (8.3) using 0.01 mol L^{-1} HCl. The exchangeable H content was determined as the difference between the exchange acidity and the exchangeable Al. The total C and total N content were measured with an NC analyzer (Sumigraph NC-800; Sumika Chem. Anal. Service, Ltd., Tokyo, Japan). The available phosphate was determined by the modified Bray-II method (soil:solution = 1:20; shaking time 60s; Bray & Kurz, 1945; Olsen & Sommers, 1982). The particle size distribution was determined using a combination of sieving and pipette methods, in which a complete dispersion of silt and clay particles was achieved by adjusting the pH to 9–10 and supersonication, after pretreatment with H_2O_2 at 80°C to remove organic matter (Gee & Bauder, 1986). The clay mineral composition was semiquantified by the relative peak areas corresponding to mica (1.0 nm), kaolin minerals (1.0 and 0.7 nm), and expandable 2:1 minerals (1.4 nm) in the X-ray diffractograms obtained by using Cu–Kα radiation (RAD–2RS; Rigaku, Tokyo, Japan). The free oxides (Fe, Al, and Si) were extracted by the following two methods: (1) extraction in the dark with acid (pH 3) 0.2 mol L^{-1} ammonium oxalate (McKeague & Day, 1966) to obtain Fe_o, Al_o, and Si_o and (2) extraction with a citrate-bicarbonate mixed solution buffered at pH 7.3 by the addition of sodium dithionite (DCB) at 80°C (Mehra & Jackson, 1960) to obtain Fe_d and Al_d. The Fe, Al, and Si content in each extract were determined by multi-channel inductively coupled argon plasma atomic emission spectroscopy (ICP-AES) (SPS-1500; Seiko, Chiba, Japan) after filtration of the extracts by 0.45 μm Millipore filters.
The data analysis was performed with the software SYSTAT version 8.0 (SPSS, 1998).

3. Results and discussion

3.1 Physicochemical and mineralogical properties of the soils

Selected physicochemical and mineralogical properties for the soils studied, and the corresponding statistical analysis, are summarized in Table 1. The surface soils studied were, in general, slightly acidic, with the average values of $pH(H_2O)$ and pH(KCl) being 6.17 and 5.37, respectively. The exchangeable Al content was low and the base saturation was high, exceeding 95% on average; hence, soil acidity was not considered a serious constraint for agricultural production. Although the average soil texture was sandy clay loam to clay loam, the particle size distribution varied widely. The average C content was 20.7 g kg^{-1}, and the dominant clay mineral was kaolinite, followed by clay mica. However, the values obtained for most of the listed properties varied significantly over the regions under study. The coefficients of variation often exceeded 100%; which indicates a significant variability among the soil characteristics for the different regions across Tanzania.

Table 2 summarizes the data obtained, categorized according to the parent materials and land use. In terms of soil parent materials, the physicochemical and mineralogical properties of the volcanic-derived soils (n = 12) were significantly different from the other soil groups in terms of CEC, total C content, available P, and free oxide-related properties. Moreover, the proportion of 1.4-nm minerals was significantly higher for the soils originated from Cenozoic rocks or deposits. On the other hand, these soil properties generally did not significantly differ for different land uses.

Variable	Number of samples	Ave.(STD)	Min. –Max.	CV (%)
pH(H$_2$O)	95	6.17 (0.80)	4.36–8.66	13.0
pH(KCl)	95	5.37 (0.89)	3.71–7.96	16.5
pH(NaF)	95	8.15 (0.66)	7.12–11.01	8.0
EC (μS dm^{-1})	95	74.3 (59.4)	10.0–325	79.9
CEC (cmol$_c$ kg^{-1})	95	14.0 (11.0)	1.61–59.5	78.6
Exch. Na (cmol$_c$ kg^{-1})	95	0.18 (0.29)	0.00–1.92	161
Exch. K (cmol$_c$ kg^{-1})	95	1.10 (1.12)	0.10–5.62	102
Exch. Mg (cmol$_c$ kg^{-1})	95	2.78 (2.14)	0.18–11.4	77.0
Exch. Ca (cmol$_c$ kg^{-1})	95	6.98 (8.76)	0.00–49.5	126
Exch. Al (cmol$_c$ kg^{-1})	95	0.21 (0.51)	0.00–2.99	241
Exch. bases (cmol$_c$ kg^{-1})	95	11.0 (11.4)	0.43–60.7	103
Base satur. (%)	95	95.4 (10.5)	49.0–101	11.0
Sand (%)	95	63.6 (23.3)	3.4–96.7	36.7
Silt (%)	95	11.2 (11.3)	0.2–48.1	101
Clay (%)	95	25.2 (17.6)	1.5–81.4	69.8
Total C (g kg^{-1})	95	20.7 (24.4)	2.13–152	124
Total N (g kg^{-1})	95	1.49 (1.84)	0.21–13.7	129
Available P (gP$_2$O$_5$ kg^{-1})	95	0.15 (0.24)	0.01–1.0	161
Feo (g kg^{-1})	95	2.46 (3.28)	0.02–14.7	133
Alo (g kg^{-1})	95	3.61 (9.34)	0.08–64.3	259
Sio (g kg^{-1})	95	1.10 (3.32)	0.00–21.7	303
Fed (g kg^{-1})	95	23.7 (25.8)	0.19–159	109
Ald (g kg^{-1})	95	4.55 (7.48)	0.01–50.9	164
0.7 nm minerals (%)	90	72.5 (27.0)	5.4–100	37.3
1.0 nm minerals (%)	90	19.6 (21.4)	0.0–91.2	109
1.4 nm minerals (%)	90	7.9 (17.9)	0.0–94.6	227

Table 1. Physicochemical and mineralogical properties of the soils studied

3.2 Principal component analysis for summarizing soil properties

A principal component analysis was performed to evaluate soil parameters related to soil fertility. The variables selected were pH(H$_2$O); pH(KCl); pH(NaF); CEC; amounts of exchangeable Na$^+$, K$^+$, Mg^{2+}, Ca^{2+}, and Al^{3+}; sand, silt, and clay content; total C and total N content; available P content; and Fe$_o$, Al$_o$, Si$_o$, Fe$_d$, and Al$_d$ content. Table 3 summarizes the factor pattern for the first five principal components after varimax rotation. The analysis resulted in the soil parameters categorized into five principal components, which explained 85.4% of the total variance.

Highly positive coefficients were obtained for pH(NaF), total C and total N, Fe_o, Al_o, Si_o, and Al_d for the first component (Table 3). These variables correspond to the soil properties related to the presence of organic materials that are bound to amorphous compounds, which might be originated on recent volcanic activity. Hence, the first component is referred to as the "soil organic matter (SOM) and amorphous compounds" factor. The second component presents strongly negative coefficients for sand content and highly positive coefficients for clay content, exchangeable Mg, and Fe_d. These soil characteristics can be related to parent materials and clay formation, i.e. soils derived from mafic and/or clayey parent materials tend to exhibit fine-textured properties with high concentrations of exchangeable Mg and Fe_d through rapid mineral weathering and clay formation. Hence, the second component is denominated as the "texture" factor. The coefficients corresponding to the third component have highly positive or negative values for pH(H_2O), pH(KCl), and exchangeable Ca and Al, indicating that a close relationship exists between this component and soil acidity. This relationship can be denominated as the "acidity" factor. The fourth and fifth components are denominated "available P and K" and the "sodicity" factors, respectively, on the basis of the coefficients correlating each of the components and the soil variables (exchangeable K and available P, and exchangeable Na, respectively).

Variable	Averages for soils from different parent materials[1]				Averages for soils under different land uses[1]		
	Volcanic rocks	Granite and other plutonic rocks	Sedimentary and metamorphic rocks	Cenozoic rocks and deposits	Natural and matured secondary vegetation	Incipient fallow vegetation	Cropland
Number of samples	12(9)[2]	14	50(48)[2]	19	37(35)[2]	16	42(39)[2]
pH(H_2O)	5.91 ab	5.70 a	6.28 ab	6.43 b	6.34 a	6.14 a	6.04 a
pH(KCl)	5.18 a	4.86 a	5.49 a	5.56 a	5.53 a	5.32 a	5.26 a
pH(NaF)	9.04 b	7.95 a	8.04 a	8.05 a	8.12 a	7.99 a	8.22 a
EC (μS dm^{-1})	103.2 b	48.1 a	78.7 ab	63.9 a	87.2 b	35.9 a	77.6 b
CEC ($cmol_c$ kg^{-1})	29.5 b	6.93 a	12.5 a	13.3 a	14.3 a	8.6 a	15.7 a
Exch. Na ($cmol_c$ kg^{-1})	0.26 ab	0.07 a	0.11 a	0.39 b	0.11 a	0.12 ab	0.27 b
Exch. K ($cmol_c$ kg^{-1})	2.49 b	0.45 a	1.11 a	0.68 a	1.17 a	0.70 a	1.19 a
Exch. Mg ($cmol_c$ kg^{-1})	4.34 b	1.25 a	2.70 ab	3.12 b	2.95 a	2.53 a	2.72 a

Variable	Averages for soils from different parent materials[1]				Averages for soils under different land uses[1]		
	Volcanic rocks	Granite and other plutonic rocks	Sedimentary and metamorphic rocks	Cenozoic rocks and deposits	Natural and matured secondary vegetation	Incipient fallow vegetation	Cropland
Exch. Ca ($cmol_c$ kg^{-1})	11.60 b	1.68 a	5.87 ab	10.85 b	8.02 a	4.17 a	7.13 a
Exch. Al ($cmol_c$ kg^{-1})	0.18 a	0.28 a	0.25 a	0.07 a	0.22 a	0.28 a	0.18 a
Exch. bases ($cmol_c$ kg^{-1})	18.7 b	3.4 a	9.8 ab	15.0 b	12.2 a	7.5 a	11.3 a
Base satur. (%)	97.4 ab	88.5 a	95.4 ab	99.0 b	96.5 a	92.3 a	95.5 a
Sand (%)	36.2 a	73.9 b	64.7 b	70.2 b	66.2 a	69.0 a	59.1 a
Silt (%)	28.7 b	6.6 a	9.2 a	9.0 a	9.5 a	7.0 a	14.3 a
Clay (%)	35.1 a	19.5 a	26.1 a	20.8 a	24.3 a	23.9 a	26.5 a
Total C (g kg^{-1})	43.3 b	12.5 a	20.1 a	13.9 a	28.4 a	11.8 a	17.2 a
Total N (g kg^{-1})	3.40 b	0.98 a	1.42 a	0.87 a	1.98 a	0.80 a	1.33 a
Available P (gP_2O_5 kg^{-1})	0.431 b	0.044 a	0.128 a	0.112 a	0.154 a	0.067 a	0.180 a
Fe_o (g kg^{-1})	8.35 b	0.73 a	2.04 a	1.12 a	2.26 a	1.11 a	3.16 a
Al_o (g kg^{-1})	13.89 b	1.50 a	2.76 a	0.91 a	4.22 a	1.11 a	4.02 a
Si_o (g kg^{-1})	4.83 b	0.16 a	0.75 a	0.34 a	1.18 a	0.29 a	1.33 a
Fe_d (g kg^{-1})	40.2 b	13.2 a	26.5 ab	11.4 a	23.2 a	26.2 a	23.1 a
Al_d (g kg^{-1})	11.34 b	3.50 a	4.37 a	1.50 a	5.58 a	2.98 a	4.24 a
0.7 nm minerals (%)	75.8 a	80.4 a	73.3 a	63.0 a	73.9 a	79.6 a	68.3 a
1.0 nm minerals (%)	20.1 a	13.8 a	22.0 a	17.8 a	20.9 a	16.1 a	19.9 a
1.4 nm minerals (%)	4.2 a	5.8 a	4.7 a	19.2 b	5.2 a	4.3 a	11.8 a

[1] The values with the same letters are not significantly different by Tukey test ($p < 0.05$).

[2] Parenthesis denotes the number of samples considered for XRD analysis (i.e. the percentage of 0.7, 1.0 and 1.4 nm minerals). Some samples were excluded from the analysis because of their X-ray amorphous natures.

Table 2. Average values of measured soil variables in terms of parent materials or land uses

Variable	PC1	PC2	PC3	PC4	PC5
pH(H$_2$O)	-0.19	-0.07	-0.94	0.10	0.04
pH(KCl)	-0.05	-0.11	-0.95	0.10	-0.02
pH(NaF)	0.84	0.00	-0.05	0.29	0.05
CEC	0.57	0.51	-0.09	0.39	0.43
Exch. Na	-0.01	-0.01	0.04	0.08	0.93
Exch. K	0.01	0.32	-0.29	0.84	0.07
Exch. Mg	-0.03	0.62	-0.42	0.36	0.39
Exch. Ca	0.05	0.28	-0.64	0.38	0.47
Exch. Al	0.20	0.12	0.64	-0.08	0.07
Sand	-0.30	-0.83	-0.11	-0.35	-0.17
Silt	0.45	0.29	0.05	0.65	0.21
Clay	0.11	0.91	0.11	0.04	0.09
Total C	0.89	0.20	0.08	0.02	0.01
Total N	0.88	0.19	0.14	0.03	0.01
Avail. P	0.06	0.02	-0.26	0.87	0.04
Fe$_o$	0.62	0.32	0.18	0.57	-0.01
Al$_o$	0.97	0.00	0.13	0.05	0.01
Si$_o$	0.94	-0.07	0.07	0.10	0.04
Fe$_d$	0.09	0.86	0.15	0.13	-0.27
Al$_d$	0.87	0.24	0.26	-0.06	-0.08
Eigenvalue	5.98	3.43	3.14	2.92	1.60
Proportion (%)	29.9	17.2	15.7	14.6	8.0
	"SOM and amorphous compounds" factor	"Texture" factor	"Acidity" factor	"Available P and K" factor	"Sodicity" factor

Table 3. Factor pattern for the first four principal components (n = 95)

3.3 Pedogenetic conditions determining the distribution patterns of factor scores for each of the principal components

Figure 2 shows a scattergram of the factor scores of SOM and amorphous compounds and those of available P and K. Both factor scores were significantly higher in soils derived from volcanic rocks than in other soils, but no significant correlation was observed. The factor scores are plotted on the geological map, as shown in Figure 3. There are two representative volcanic areas in Tanzania, namely, Mount. Kilimanjaro and the surrounding region and the southern mountain ranging between the east of Mbeya and Lake Malawi. Generally, the scores of the factor for SOM and amorphous compounds were highest in the region of the southern volcanic mountain ranges, followed by some plots around Mt. Kilimanjaro (Fig. 3a), whereas the scores of the factor for available P and K tended to be high in both volcanic regions (Fig. 3b). Msanya *et al.* (2007) indicated that the volcanic soils in the southern mountain ranges were rich in K, compared to several Japanese volcanic soils, most likely reflecting lithological differences among the parent materials. The predominantly high scores of the factor for SOM and amorphous compounds in the southern volcanic mountain ranges indicate a relatively incipient feature of soils after recent active volcanic events and potentially high soil fertility relating to SOM in these regions. In addition, soils located in those volcanic regions could be more fertile in terms of P and K nutrients supply from soils.

Figure 3c represents the distribution pattern of the factor scores of texture in terms of the geological conditions. There is a certain regional trend in these factor scores, though no statistical difference was observed in terms of the geological condition as a whole. Among the soils of volcanic origin, those in the northern volcanic regions exhibited higher scores in the texture factor, consistent with a previous report by Mizota *et al.* (1988), in which they postulated that these soils were in the advanced stages of weathering of volcanic materials. The scores were high for some soils originated from sedimentary and metamorphic rocks, which are mostly distributed in the western region around Kigoma and the hill slopes near Tanga. Otherwise, scores were in general low for soils originated from granite, except for those of the southern highland.

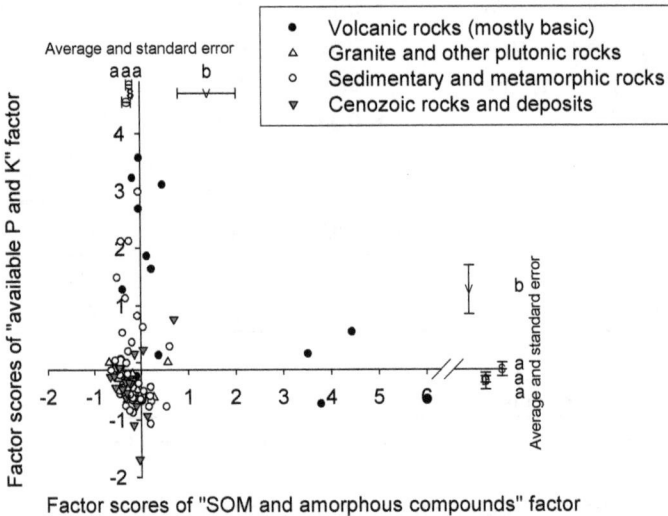

Fig. 2. Relationship between the scores of the "SOM and amorphous" and "available P and K" factors

Figure 4 shows the influence of the amount of precipitation on the scores of selected factors. There was no significant relationship between the amount of precipitation and the factor scores of acidity or texture. Although the positive contribution of precipitation on mineral weathering might accompany soil acidification or the formation of clays and secondary Fe oxides, there was no correlation between those processes, which indirectly suggests that the influence of parent materials on soil properties is stronger than climatic factors among the soils studied.

(a)

(b)

(c)

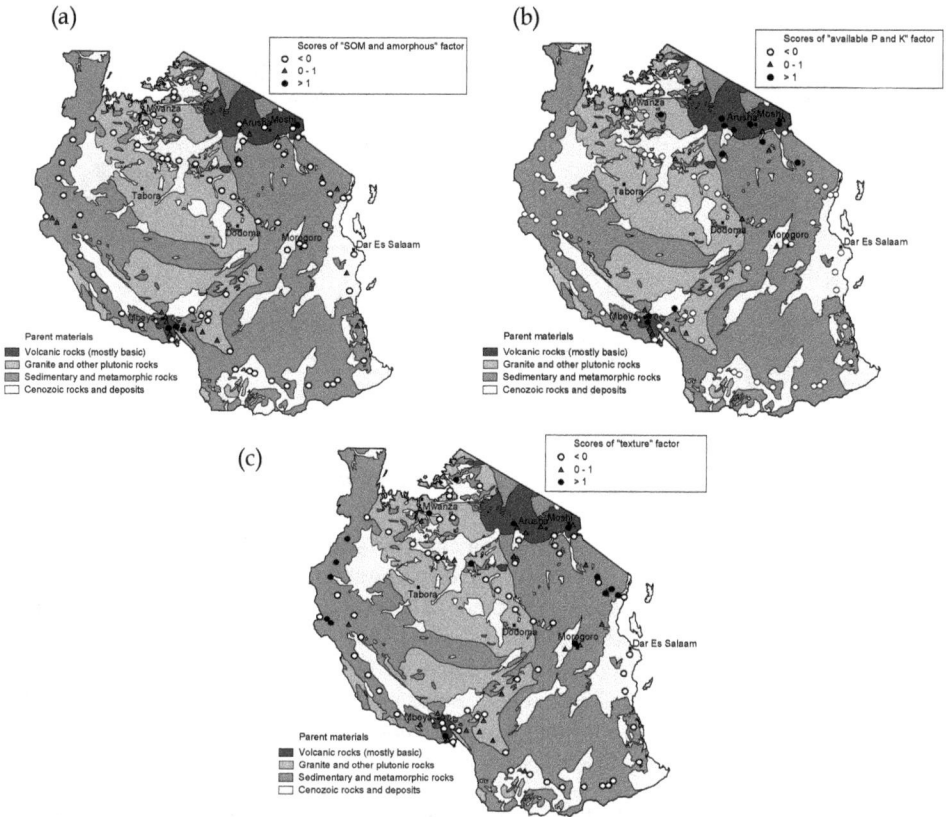

Fig. 3. Distribution patterns of scores of (a) "SOM and amorphous," (b) "available P and K," and (c) "texture" factors in relation to geological conditions

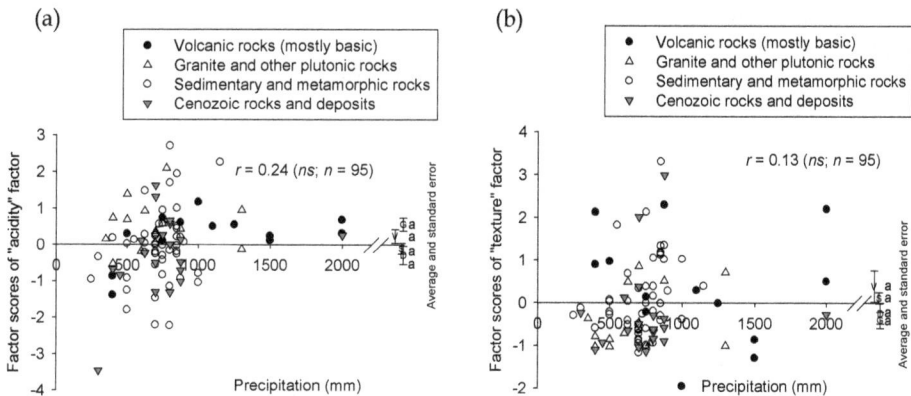

(a)

(b)

Fig. 4. Relationships between precipitation and scores of (a) "acidity" and (b) "texture" factor

(a)   (b)

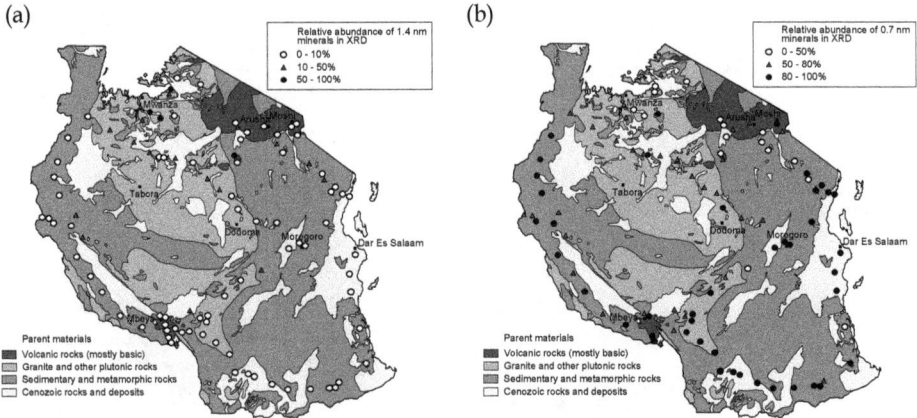

Fig. 5. Distribution patterns of clay mineralogy in relation to geological or climatic conditions. Abundances of (a) 1.4 nm and (b) 0.7 nm minerals

3.4 Pedogenetic conditions determining the clay mineralogy of the soils

Figure 5 shows the distribution patterns of the clay mineralogy in relation to the geological and climatic conditions. The relative abundance of 1.4-nm minerals was often higher in the northern region of the Great Rift Valley and around Lake Victoria. On the other hand, the abundance of 0.7-nm minerals tended to be lower in the central steppe, which has lower precipitation than other regions. These relationships are more clearly presented in Figure 6. Stepwise multiple regression indicated that the abundances of 1.4-nm minerals (mostly smectite) could be expressed by the following equation:

$$1.4 - \text{nm minerals } (\%) = 6.38 + 13.4 \text{ (sodicity factor)} - 9.78 \text{ (SOM / amorphous factor)}$$
$$+ 3.17 \text{ (P / K factor)}; \quad r^2 = 0.58 \; (p < 0.01, n = 90) \tag{1}$$

The 1.4-nm minerals were probably formed under the strong influence of the high sodicity of the parent materials around the Great Rift Valley, and were often observed in the soils in the flat plains near Lake Victoria.

On the other hand, the abundances of the 0.7-nm minerals (kaolin minerals) can be expressed by the following equation:

$$0.7 - \text{nm minerals } (\%) = - 56.2 + 19.5 \; \ln(\text{precipitation in mm}) + 5.92(\text{acidity factor})$$
$$+ 4.82 \text{ (texture factor)} - 11.2 \text{ (sodicity factor)} - 7.70 \text{ (P / K factor)}; \tag{2}$$
$$r^2 = 0.45 \; (p < 0.01, n = 90)$$

From this equation, it can be stated that the kaolin formation is promoted under highly humid conditions with the positive influence of soil acidity and texture (or clayey parent materials) as well as the negative influence of sodicity. Hence, it can be inferred that the clay mineralogical properties of the soils studied herein were formed under the strong influence of the present climatic conditions as well as the parent materials on a countrywide scale in Tanzania.

(a)

(b)

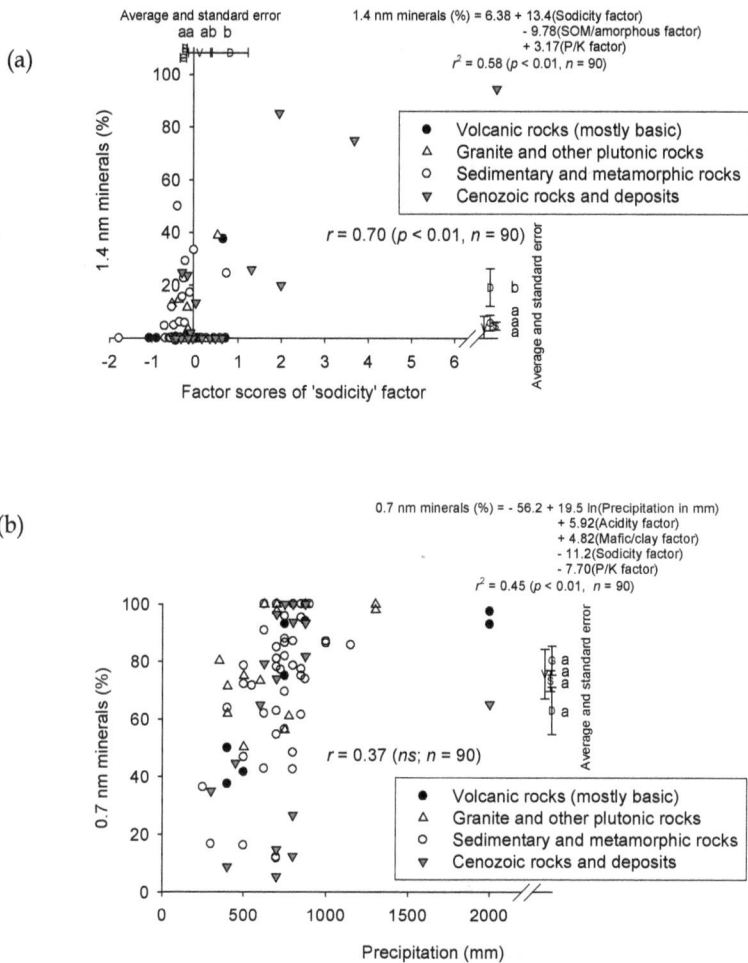

Fig. 6. Relationships between clay mineralogy and soil and climatic factors. Abundances of (a) 1.4 nm and (b) 0.7 nm minerals

3.5 General discussion on the soil conditions in Tanzania with specific reference to potential agricultural development

As previously stated, soils can be considered as significantly fertile in the volcanic regions and areas around, due to the high SOM contents and the high P and K nutrient status. In addition, the soils around Lake Victoria are fertile due to the strong influence of the 1.4-nm minerals, which contributes to the retention of base cations. Both regions, namely, the volcanic regions and the regions around Lake Victoria, are included in the Great Rift Valley, which is the center of intensive agricultural activities of the country. However, in other areas of Tanzania, soils are generally low in SOM-related parameters and the 1.4-nm minerals are virtually absent, presumably due to consecutive mineral weathering under ustic soil moisture regime (Watanabe *et al.*, 2006). The proportion of kaolin minerals increases with the precipitation; hence, soil fertility decreases in regions of high humidity. Soil fertility in terms of clay mineralogy is comparatively higher in dry regions than in humid regions because of the greater abundance of mica minerals. However, water availability decreases in such dry regions. Thus, the semiarid regions in Tanzania suffer from water scarcity, while the relatively humid areas have less fertile soil that predominantly contains kaolin minerals. In summary, high scores in SOM-related properties and the 1.4-nm minerals contribute to relatively high soil fertility in Great Lift Valley regions, whereas either water scarcity or low soil fertility are not favorable for agricultural production in the other regions of Tanzania. These conditions should be considered when studying the feasibility of agricultural development in different areas in the future.

4. Conclusion

From the principal component analysis of the collected soil samples, five individual factors—SOM and amorphous compounds, texture, acidity, available P and K, and sodicity—were determined which explained 85.4% of total variance. From the clay mineralogical composition and the relation between the geological conditions (or parent materials) and the annual precipitation and the scores of the five factors, the following conclusions can be summarized:

1. The maximum scores of "SOM and amorphous compounds" were found at the volcanic center of the southern mountain ranges from the east of Mbeya to Lake Malawi.
2. The scores of the "available P and K" were high in the volcanic regions around Mt. Kilimanjaro and in the southern volcanic mountain ranges.
3. The abundance of 1.4-nm minerals (mostly smectite) can be expressed by the following equation (Equation 1):

$$1.4 - \text{nm minerals } (\%) = 6.38 + 13.4 \text{ (sodicity factor)} - 9.78 \text{ (SOM / amorphous factor)}$$
$$+ 3.17 \text{ (P / K factor)}; \quad r^2 = 0.58 \ (p < 0.01, n = 90)$$

    The 1.4-nm minerals were probably formed under conditions of high sodicity and were often observed in the soils near Lake Victoria.
4. The abundance of 0.7-nm minerals (kaolin minerals) can be expressed by the following equation (Equation 2):

$$0.7 - \text{nm minerals } (\%) = -56.2 + 19.5 \ \ln(\text{precipitation in mm}) + 5.92 \text{ (acidity factor)}$$
$$+ 4.82 \text{ (texture factor)} - 11.2 \text{ (sodicity factor)} - 7.70 \text{ (P / K factor)};$$
$$r^2 = 0.45 \ (p < 0.01, n = 90)$$

    Equation 2 suggests that kaolin formation is promoted under highly humid conditions, which is also controlled by the acidity and texture of the soil (or parent materials). Hence, the results indicate that the formation of the soils studied in the present study was strongly influenced by climatic conditions and parent materials.
5. In Tanzania, the volcanic regions and the Great Rift Valley region, where soil is generally more fertile than in other regions, are favorable to modernized agriculture. The semiarid regions in Tanzania suffer from water scarcity, while the relatively humid areas have less fertile soil that predominantly contains kaolin minerals. These

conditions are not favorable for agricultural production and must be strongly considered when studying the feasibility of agricultural development in different areas in the future.

5. Acknowledgements

This study was supported by a Grant-in-Aid for Scientific Research (No. 17208028) from the Ministry of Education, Culture, Sports, Science and Technology, Japan.

6. References

Araki, S., Msanya, B.M., Magoggo, J.P., Kimaro, D.N. & Kitagawa, Y. 1998. Characterization of soils on various planation surfaces in Tanzania. In: *Summaries of 16th World Congress of Soil Science*, Vol. I, pp. 310, Montpellier, France.

Bray, R.H. & Kurz, L.T. 1945. Determination of total organic and available forms of phosphorus in soils. *Soil Science*, 59, 39–45

Gee, G.W. & Bauder, J.W. 1986. Particle size analysis. In: *Methods of Soil Analysis* (ed. A. Klute), pp. 383–411. Soil Science Society of America, Madison, WI.

Kimaro, A.A., Timmer, V.R., Chamshama, S.A.O., Mugasha, A.G. & Kimaro, D.A. 2008. Differential response to tree fallows in rotational woodlot systems in semi-arid Tanzania: Post-fallow maize yield, nutrient uptake, and soil nutrients. *Agriculture, Ecosystems & Environment*, 125, 73–83.

McKeague, J.A. & Day, J.H. 1966. Dithionite- and oxalate-extractable Fe and Al as aids in differentiating various classes of soils. *Canadian Journal of Soil Science*, 46, 13–22.

Mehra, O.P. & Jackson, M.L. 1960. Iron oxide removal from soils and clays by a dithionite-citrate system buffered with sodium bicarbonate. *Clays and Clay Minerals*, 7, 317–327.

Mizota, C., Kawasaki, I. & Wakatsuki, T. 1988. Clay mineralogy and chemistry of seven pedons formed in volcanic ash, Tanzania. *Geoderma*, 43, 131–141.

Msanya, B.M., Magoggo, J.P. & Otsuka, H. 2002. Development of soil surveys in Tanzania (Review). *Pedologist*, 46, 79–88.

Msanya, B.M., Otsuka, H. Araki, S. & Fujitake, N. 2007. Characterization of volcanic ash soils in southwestern Tanzania: Morphology, physicochemical properties, and classification. *African study monographs (Supplementary issue)*, 34, 39–55.

Nandwa, S.M. 2001. Soil organic carbon (SOC) management for sustainable productivity of cropping and agro-forestry systems in Eastern and Southern Africa. *Nutrient Cycling in Agroecosystems*, 61, 143–158.

Olsen, S.R. & Sommers, L.E. 1982. 24. Phosphorus. In: *Methods of Soil Analysis*, Part 2, Chemical and Microbiological Properties, Second Edition (eds. A.L. Page, R.H. Miller & D.R. Keeny), pp. 403–430. American Society of Agronomy & Soil Science Society of America, Madison, WI.

SPSS 1998. SYSTAT 8.0. Statistics. SPSS, Chicago.

Surveys and Mapping Division, Tanzania 1967. Atlas of Tanzania. Dar es Salaam, Tanzania.

Szilas, C., Møberg, J.P., Borggaard, O.K. & Semoka, J.M.R. 2005. Mineralogy of characteristic well-drained soils of sub-humid to humid Tanzania. *Acta Agriculturae Scandinavica, Section B - Plant Soil Science*, 55, 241–251.

Thomas, G.W. 1982. Exchangeable cations. In: *Methods of Soil Analysis*, Part 2, Chemical and Mineralogical Properties (eds. A.L. Page, R.H. Miller & D.R. Keeney), pp. 159–165. American Society of Agronomy & Soil Science Society of America, Madison, WI.

Watanabe, T., Funakawa, S. & Kosaki, T. 2006. Clay mineralogy and its relationship to soil solution composition in soils from different weathering environment of humid Asia: Japan, Thailand and Indonesia. *Geoderma*, 136, 51–63.

Part 2

Land Use Impact on Soil Quality

3

Quantifying Soil Moisture Distribution at a Watershed Scale

Manoj K. Jha
North Carolina A&T State University
USA

1. Introduction

Soil moisture content is a very vital component of the hydrological cycle. It is a key variable controlling water and energy fluxes in soils (Vereecken et al. 2007). It provides the plant-available transpirable pool of water for vegetative life. In addition, the availability or retention of moisture in the soil controls the rainfall-runoff process. Despite its importance to vital lives and ecosystem, the distribution of soil moisture varies tremendously over the time and space. Spatial patterns of soil moisture are determined by a number of pysiographic factors that affect vertical and lateral redistribution of water in the unsaturated zone. These include topography and landscape position, slope aspect, vegetation, and texture. Temporal patterns depend on meteorological factors and their variation over the time. During the dry period (nonrainly periods), spatial variation in soil moisture is controlled by vegetation (Seyfried and Wilcox 1995). Different vegetation will have different impacts on soil moisture as their uptake will vary widely. Moisture content also exerts a strong control on soil biogeochemistry including microbial activity, nitrogen mineralization, and biogeochemical cycling of nitrogen and carbon (Turcu et al. 2005). Therefore, understanding the spatio-temporal distribution and quantity of available soil moisture that can be used without damaging the natural ecosystem are keys to sustainable development and prevention of ecosystem decline.

Soil moisture has been traditionally measured through point measurements, which is useful to understand field-scale soil water dynamics (Topp and Ferre 2002), and predominantly developed for applications in agriculture. Recent advancements in remote sensing technologies has developed capabilities that contribute to understanding of soil moisture distribution at very large scales such as large basins or continental or global scales; however, these prediction needs to be validated through a large number of ground based point measurements. It would be difficult to provide such information on a larger scale. Several techniques used in the past to represent spatial variation of soil moisture on a large scale using geostatistical anslyses tools such as kringing and semivariogram analysis, but these require a dense sampling character of the soil moisture field. The concept of temporal stability was able to capture spatial variation but limited to smaller scales (Brocca et al. 2010). Robinson et al. (2008) have extensively reviewed and summarized the challenges and opportunities for soil water content measurement in terms of laboratory, equipment, monitoring, remote sensing, and modelling challenges.

Recent advancement in watershed scale hydrology models have increasingly been adopted for soil and water management (Jha et al. 2007, 2010a, 2010b). These models provide a more holistic approach of modelling complex interconnected and nonlinear hydro-geological movement of water across all physical processes. This study used a watershed scale hydrologic model, called Soil and Water Assessment Tool (SWAT) (Arnold et al. 1998), to quantify long-term variation in spatial distribution of soil moisture on a medium-size watershed located in Midwestern USA. SWAT has been shown to perform well on both large river basins and small watersheds in terms of annual water and sediment yield (Arabi et al. 2006, Gitau et al. 2004, Spruill et al. 2000, and Jha et al. 2011, among may other studies). Gassman et al. (2007) has reviewed over a hundred of peer-reviewed SWAT related peer-reviewed publications, which speaks of the magnitude and reliability of model use for hydrology and water quality analyses.

The combination of favourable climate and fertile soil makes the Midwest one of the most productive agricultural areas in the world. However, this brings an enormous application of fertilizers and manures on the cropland, unmanaged and overapplication, which led water quality problems in the local rivers and ultimately to larger ecosystems, e.g. hypoxia problem in the Gulf of Mexico (Rabalais et al. 1996). Many conservation practices have been proposed and implemented over decades. One such practice is the inclusion of winter cover crops in the traditional corn-soybean rotation. Winter cover crops can reduce nitrogen (N) leaching by extending the growing season and the uptake of N beyond that for corn and soybean (Shepherd and Webb 1999). These crops take up residual N, released by mineralization during fall and spring, and N released from fall-applied anhydrous ammonia. The cover crops then release this N as their residue decays the next spring or summer. While this practice was shown to have a tremendous potential for N reduction (Kaspar et al. 2005, Singer et al. 2011), it might have implication in soil moisture dynamics over a long period of time. This study analyzed the impacts of this conservation practice on spatial distribution of soil moisture.

The main objective of this present study is to use SWAT model to quantify soil moisture distribution on a watershed scale and evaluate the impact of applying cover crop conservation practice on soil moisture content.

2. Methods and materials

2.1 Watershed description

The Raccoon River Watershed (RRW) covers nearly 3,630 mi^2 area in portions of 17 Iowa counties in west central Iowa (Figure 1). The North and Middle Raccoon Rivers flow through the recently glaciated (< 12,000 years old) Des Moines Lobe landform region, a region dominated by low relief and poor surface drainage. In contrast, the South Raccoon River drains an older (> 500,000 years old) Southern Iowa Drift Plain landscape region characterized by higher relief, steeply rolling hills, and well-developed drainage. The RRW is dominated by agricultural row crop production, with over 70% of the areas planted primarily in corn and soybeans. Other main land use includes grassland (16.3%), woodland (4.4%), and urban (4.0%). The grasses and trees generally are scattered throughout the South Raccoon basin on terrain difficult to cultivate. Figure 2 show the land use ypes in the watershed. As explained by the landorm region, north Raccoon is mostly tiled due to inadequate soil drainge property. Figure 3 depicts the tile drainage densitiy in the watershed that was very extenstively done in North Raccoon. The RRW stream system has been impacted by elevated levels of nitrogen, phosphorus, sediment, and bacteria pollutants during recent decades, primarily from nonpoint sources (Hatfield et al., 2009; Jha et al., 2010; Schilling et al., 2008).

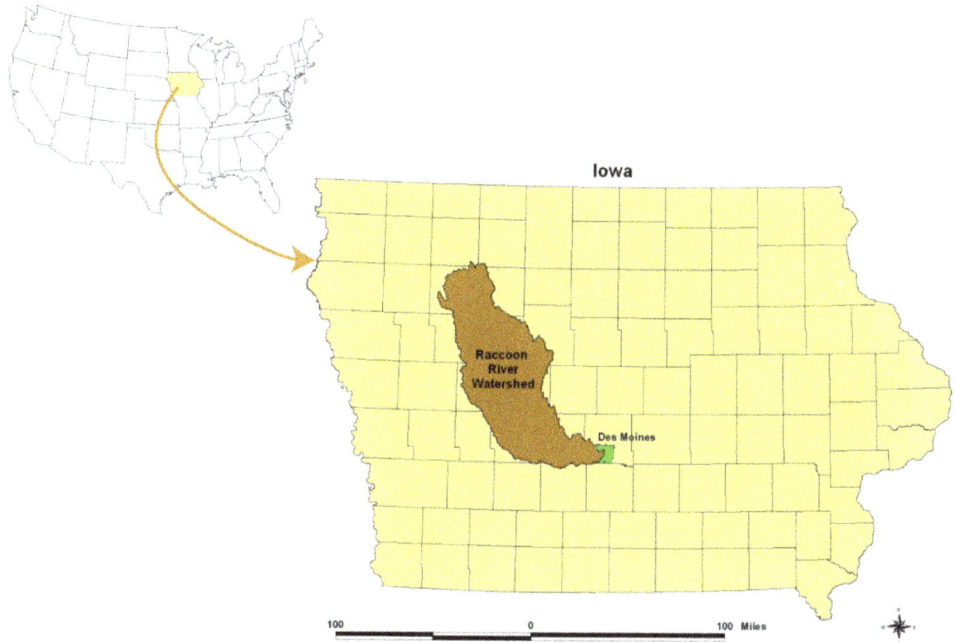

Fig. 1. Location of the study watershed

The modeling framework of the SWAT model for RRW was adapted from Jha et al. 2010. It has used SWAT vesion 2005 and relied on standard 12-digit watersheds (USGS 2009) as a basis for the subwateshed delineation. The process of watershed delineation and HRU creation was performed using the ArcView SWAT interface (AVSWATX). The resulting watershed configuration consisted of 112 subwatersheds. The hydrologica response unites (HRUs) were then created by overlaying Soil Survey Geographic (SSURGO) data (USDA-NRCS, 2008) and 2002 land cover data obtained from IDNR (2008). All together, a total of 3640 HRUs were created for modeling. Daily weather data was obtained from the National Weather Service COOP monitoring sites available through the Iowa Environmental Mesonet (www.mesonet.agron.iastate.edu). AVSWATX assigned the appropriate weather station information to each subwatershed based on the proximity of the station to the centroid of the subwatershed. Ten weather stations were used to provide the temperature and precipitation data for the entire simulation time frame. The SWAT model was run on a daily time step for the 1986 to 2004 period, with the first ten years (1986 to 1995) consisting of a model calibration period and a second nine year period (1996 to 2004) comprising a model validation period. The Penman-Monteith method was selected to estimate potential evapotranspiration and the Muskingum method was selected for channel flow routing simulation. Model calibration required varying model parameters within their ranges for

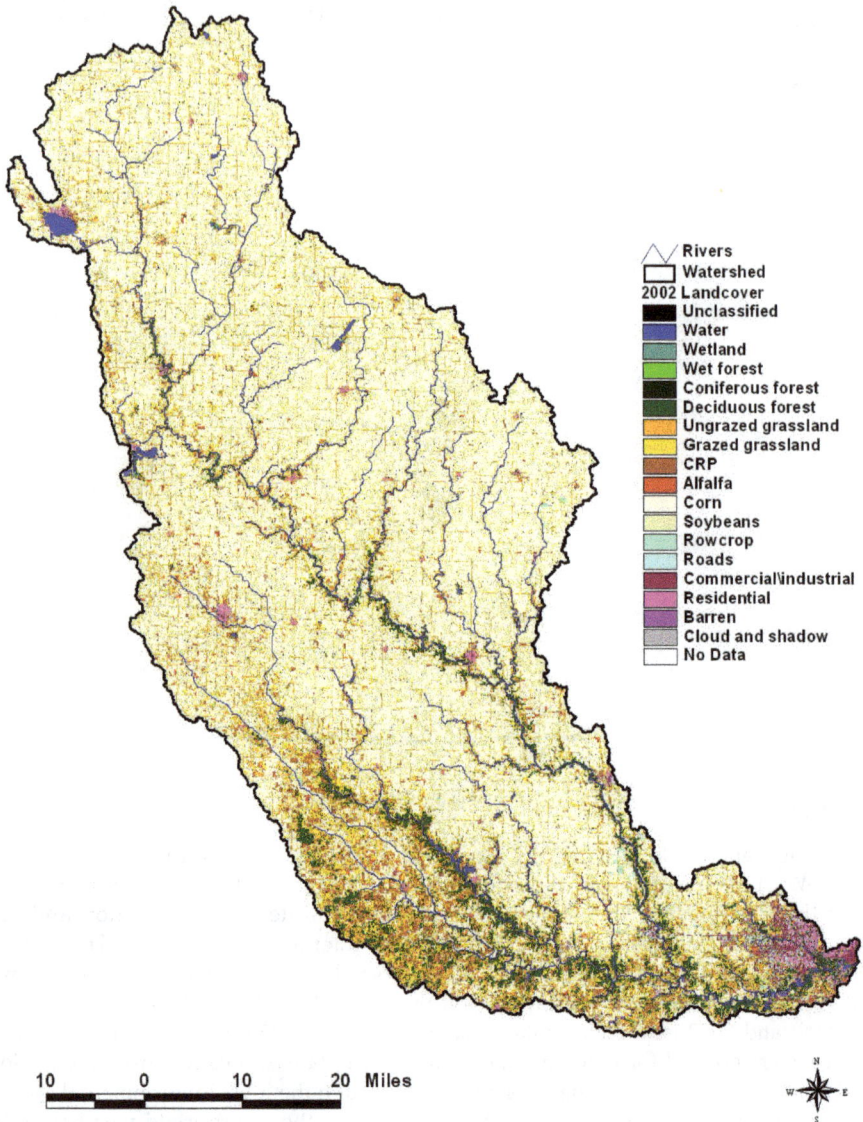

Fig. 2. Land use pattern in the Raccoon River Watershed

match observed variables with the simulated variables. Figure 4 shows the monthly comparison of flow at the watershed outlet for both calibration and validation periods. Details on modeling setup can be found in Jha et al. 2010. Over the entire simulation period, the modeled average annual streamflow at the outlet (220 mm) was very close to the measured value (215 mm). Comparison of monthly values resulted in R² and E (Nash-Sutcliffe's coefficient) values of 0.86 and 0.86 for calibraiton and 0.88 and 0.87 for validation.

The modeled average monthly streamflow (18.4 mm) closely matched the measured monthly average (17.9 mm) over the 228 months (19 years) simulation period. These statistical results can be viewed as quite strong for the resutls when viewed in the context of the suggested criteria by Moriasi et al. (2007).

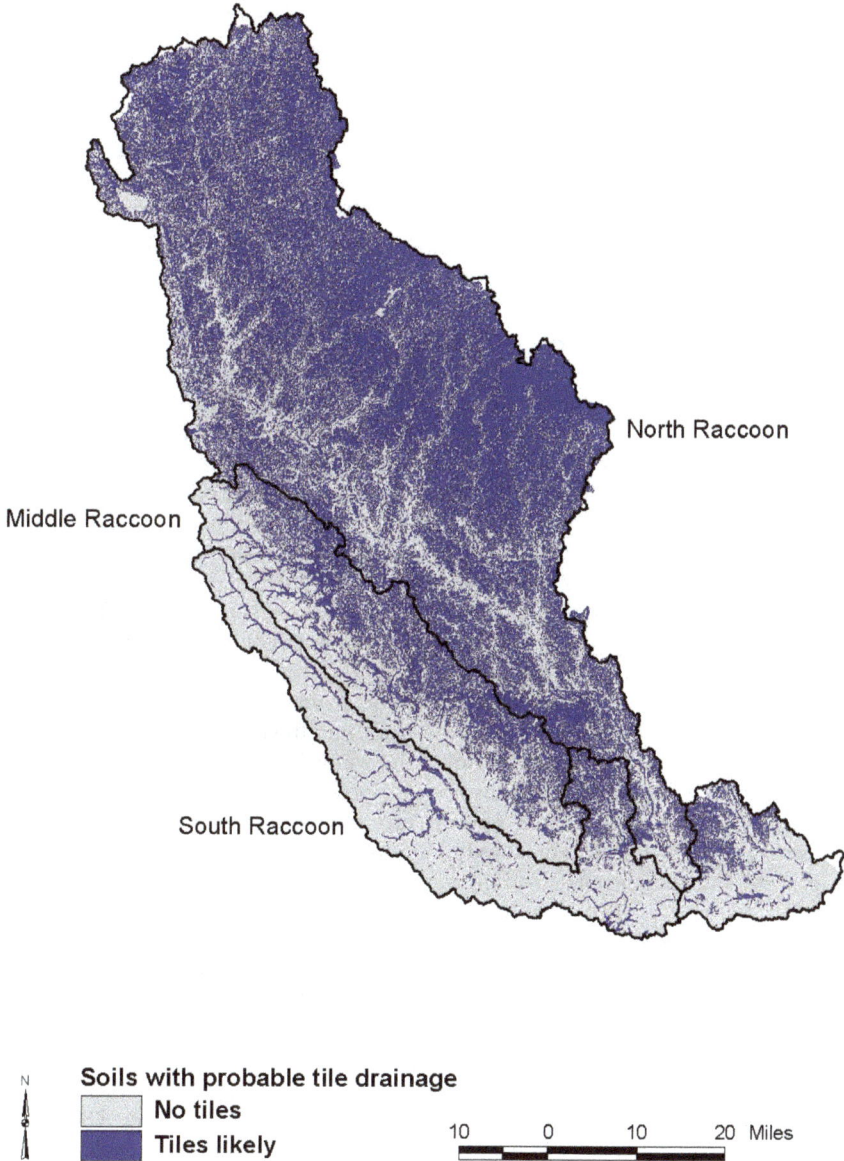

Fig. 3. Soils with probable tile drainage in the watershed (adapted from Schilling et al. 2008)

Monthly Streamflow (mm)

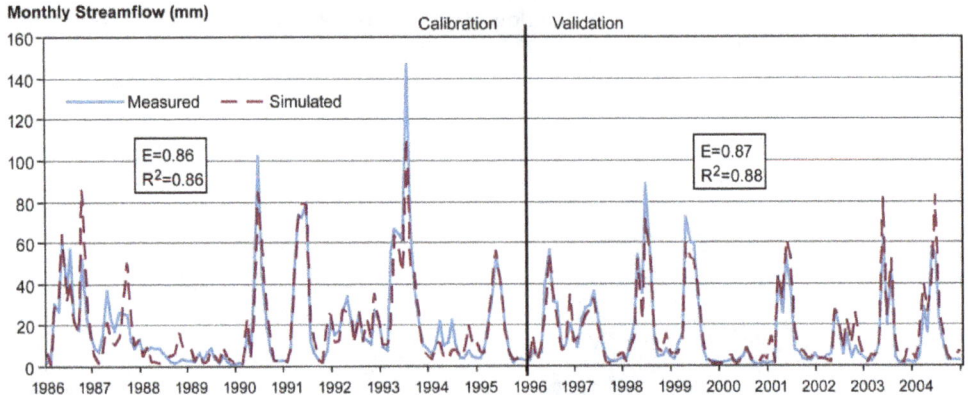

Fig. 4. Long-term average (1986-2004) streamflow comparison at the watershed outlet (adapted from Jha et al. 2010)

2.2 Description of the watershed model, SWAT

The Soil and Water Assessment Tool (SWAT; Arnold et al. 1998) model is a watershed-based hydrologic and water quality model that operates on a daily times step and is capable of modeling the impact of different land use and management practices on hydrology and water quality of the watershed . It was developed by the U.S. Department of Agriculture (USDA) Agricultural Research Service (ARS) and has experienced continuous evolution since the first releases in the early 1990s. Major model components include hydrology, weather, soil temperature, crop growth, nutrient, bacteria, and land management. In SWAT, watersheds are divided into subwatersheds, which are further delineated by HRUs that consist of homogeneous soil, land use and management characteristics. The HRUs represent percentages of a subwatershed area and thus are not spatially defined in the model. The water balance of each HRU is represented by four storage volumes: snow, soil profile, shallow aquifer, and deep aquifer. Flow generation, sediment yield and pollutant loadings are summed across all HRUs within a subwatershed, and the resulting alues are then routed through channels, ponds, and/or reservoirs to the watershed outlet. The model has several options to estimate potential evapotranspiration including Hargreaves method, Penman-Monteith method, and others. Two options are available to simulate channel routing: variable storage method and Muskingum method. SWAT simulates a complete plant growth process and model nutrient dynamics throughout several interconnected nutrient pools.

Water that enters the soil profile may move along one of several different pathways. The water may be removed from the soil by plant uptake or evaporation. It can percolate past the bottom of the soil profile and ultimately become aquifer recharge. A final option is that water may move laterally in the profile and contribute to streamflow. Of these different pathways, plant uptake of water removes the majority of water that enters the soil profile. Two stages of water content are recognized: field capacity (water held at a tension of 0.033 MPa) and permanent wilting point (water held at a tension of 1.5 MPa). The amount of water held in the soil between field capacity and permanent wilting point is considered to be the water available for plant extraction. SWAT directly simulates saturated flow only. The model records the water contents of the different soil layers but assumes that the water is uniformly distributed within a given layer. This assumption eliminates the need to model unsaturated flow in the horizontal direction. Unsaturated flow between layers is indirectly modelled with the depth distribution of plant water uptake (Equation 1) and depth distribution of soil water evaporation (Equation 2).

Depth distribution of plant water uptake:

$$w_{up,z} = \frac{E_t}{[1-\exp(-\beta_w)]} \cdot [1 - \exp(-\beta_w \cdot \frac{z}{z_{root}})_0] \tag{1}$$

Where $w_{up,z}$ is the potential water uptake from the soil profile to a specified depth, z, on a given day (mm), E_t is the maximum plant transpiration on a given day (m), β_w is the water-use distribution parameter, z is the depth from the soil surface (mm), and z_{root} is the depth of root development in the soil (mm). The potential water uptake from any soil layer can be calculated by solving above equation for the depth at the top and bottom of the soil layer and taking the difference.

Depth distribution of soil water evaporation:

$$E_{soil,ly} = E_{soil,zl} - E_{soil,zu} \tag{2}$$

Where $E_{soil,ly}$ is the evaporative demand for layer ly (mm), $E_{soil,zl}$ is the evaporative demand at a lower boundary of the soil layer (mm), and $E_{soil,zl}$ is the evaporative demand at the upper boundary of the soil layer (mm).

2.3 Design experiment for soil moisture analyses

The calibrated SWAT model was examined for predicting the hydrological response at a subwatershed level. The level of spatial detail framed in this study is the size of the subwatershed (total number of which is 112 in the Raccoon River watershed with an average area of about 83.5 km^2). Various hydrological processes including precipitation, water yield, evapotranspiration, and soil water content were looked at from the perspective of spatial distribution across the watershed on a long-term average annual basis. While the spatial distribution of precipitation was derived from historical climatic observation from 10 weather stations located in and around the watershed, other parameters are simulated outcomes from the calibrated SWAT model.

It is hypothesized that the total water yield (surface runoff and baseflow) is very close (if not equal) to the difference between precipitation and evapotranspiration, while soil moisture content remains unaffected over a long-period of time. This hypothesis was tested at a subwatershed level to evaluate the model's ability to predict hydrological processes at smaller spatial scales. There is no set specific criterion to evaluate the hypothesis, but it was assumed that the model performance would be considered acceptable if the bias was found to be less than or equal to 10%. Model prediction of soil moisture was not directly validated by comparing with actual measurement due to the lack of available data on such a large scale (a motivation of this study). However, the reasonable prediction of other hydrological parameters by the model satisfied the validity of the model's ability to replicate hydrological response of the watershed through prediction of hydrological processes.

After the model validation, it was used to evaluate the effect of incorporating winter cover crops into standard corn soybean rotation in the watershed. In this scenario, rye was planted after the corn and soybean harvest. Harvest of the rye crop was not simulated but was simply plowed in prior to corn or soybean planting. This scenario provided an opportunity to assess the impact of adoption of this practice on soil moisture content on a long-term basis. Winter cover crops provide ground cover on cultivated cropland after the growing season. Rye, oats, and alfalfa have been used as cover crops in cropland areas in the Midwest for number of years, and continuously increasing. It has shown a promise of significant reduction in N losses from agricultural lands (Kaspar et al. 2004) thereby protecting local streams from nonpoint source pollution, and contributing positively to regional ecosystems. Implementation of this practice into vast majority of traditional corn and soybean rotation in the Midwest has potential to reduce N loss significantly, and ultimately reducing the concern of delivering significant nutrient loadings from Iowa and Illinois watersheds into the Mississippi and ultimately to the Gulf of Mexico.

3. Results and discussion

Meteorological input to the modelling system was from 10 weather stations located in and around the watershed. Spatial distribution of the most important hydrological driver precipitation is shown in Figure 5. It can be seen that the distribution does not vary

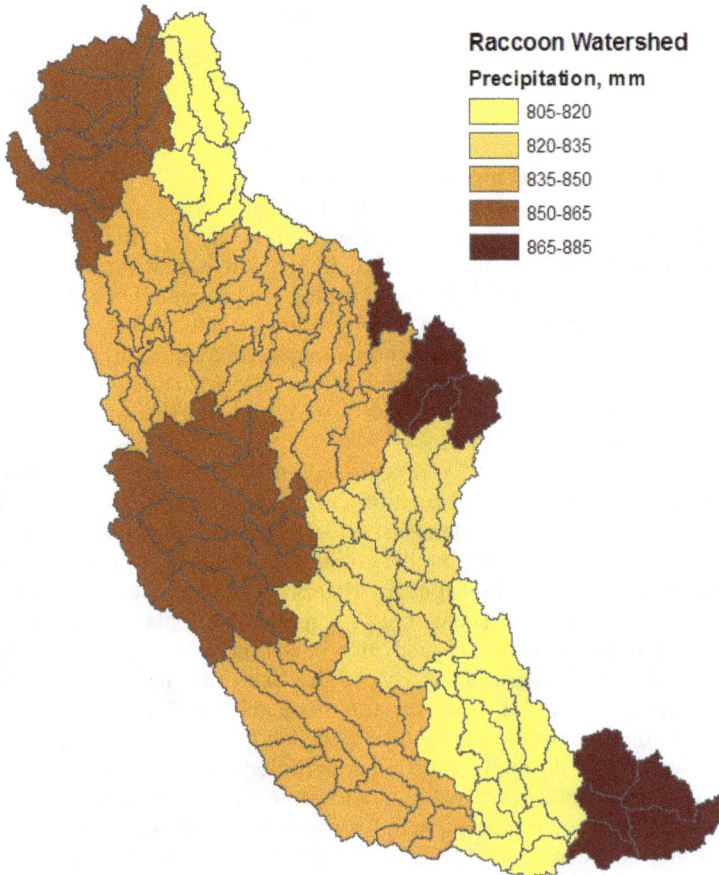

Fig. 5. Spatial representation of precipitation on a long-term annual basis

significantly over the watershed spatially, and values range from 805 to 885 mm on a long-term average annual basis over the period of 19 years (1986-2004). Based on the input on temperature, other meteorological data, and information on land cover, SWAT estimated evapotranspiration (ET) using Penman-Monteith method (Figure 6). Spatial distribution of ET ranged from 470 to 660 mm with higher values in north and central portion of the watershed. Average ET among subwatersheds was found to be 564 mm with standard deviation of 36.

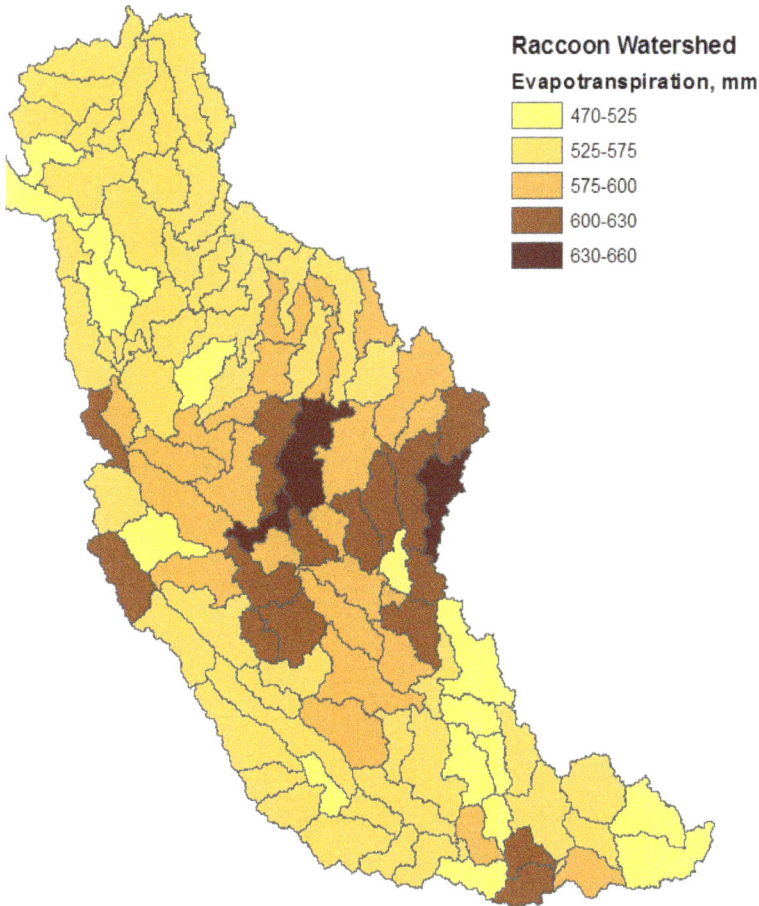

Fig. 6. Estimated evapotranspiration (ET) over a subwatershed scale on a long-term annual basis

Hydrological model performed daily water balance on scale much finer than subwatershed (at HRU or response unit level). The total water yield (sum of surface runoff and baseflow) calculated at each response unit were aggregated at subwatershed level. The distribution of water yield at the subwatershed level is show in Figure 7. This was achieved after the model was calibrated for overall watershed hydrology and then for time-series data of streamflow at the watershed outlet. Our hypothesis about water yield be equal to precipitation minus evapotranspiration on a long term basis, was tested for each subwatershed individually for the calibrated model. It was found that the absolute deviation of water yield values as compared with the difference in precipitation and evapotranspiration values were very small (mean = 3 mm, standard deviation = 3 mm, and values range from +6 to -10 mm) over the entire watershed. This is the error of less than 1% in predicting water yield on a long-term basis on such a large scale. This validates the accuracy of model prediction on a long-term average annual basis. The resulting soil water content and its spatial distribution are shown in Figure 8. Its value ranges from 164 to 300 mm with an average value of 250 mm and standard deviation of 25mm. Higher moisture content was seem to exist mostly in the eastern portion of the watershed.

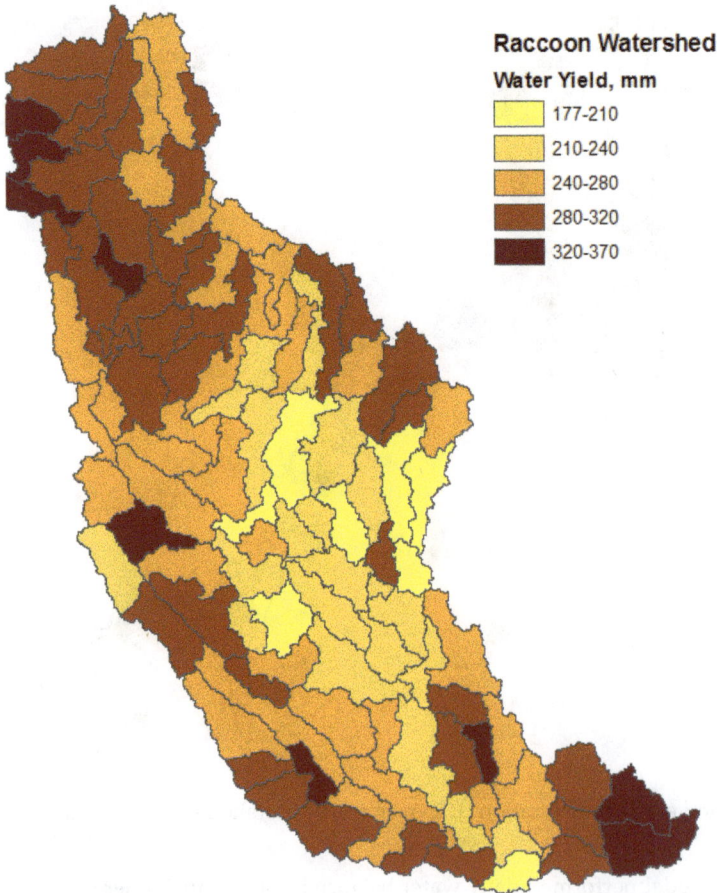

Fig. 7. Total water yield distribution as predicted by SWAT on a long-term basis

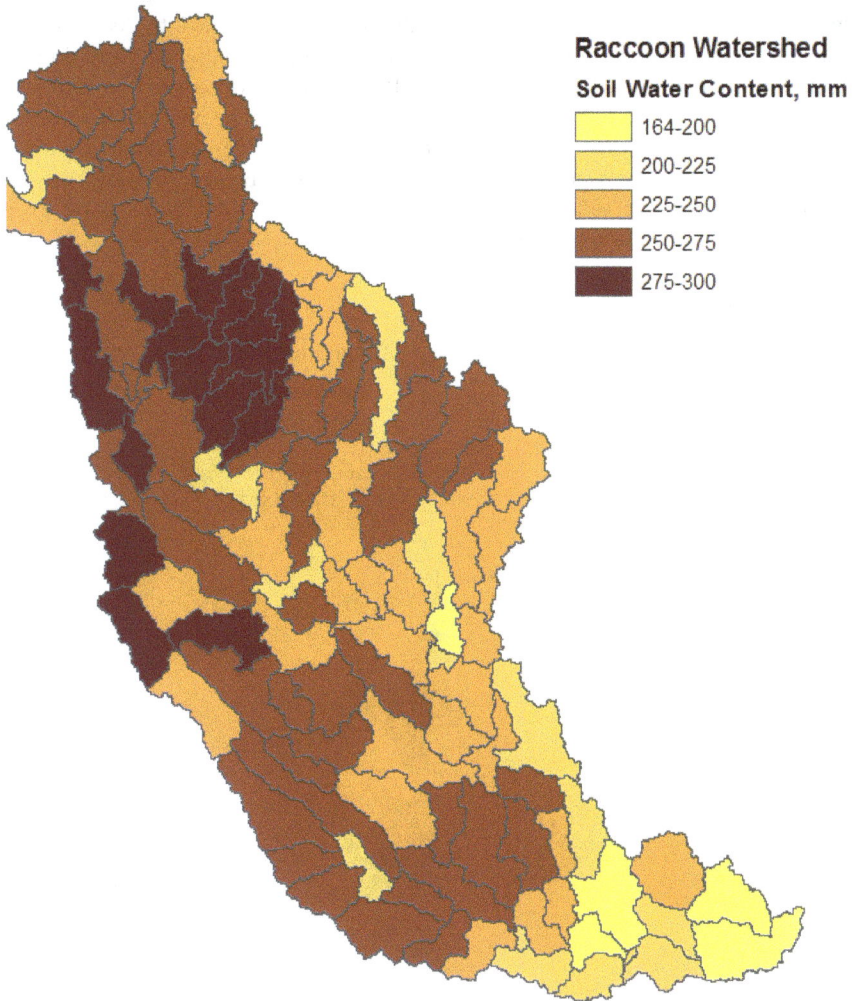

Fig. 8. Soil moisture content predicted by SWAT at subwatershed scale on a long-term basis

Once the model was successfully tested to predict soil moisture content, a scenario was conducted to examine the impact on soil moisture content for a promising land management practice: inclusion of winter cover crops into cropland (corn and soybean in this case). A winter cover crop, rye, was simulated to be planted after corn and soybean harvest each year. While this practice is well known for both soil and water quality and conservation, this study attempts to quantify its impact on soil moisture content. The modelling setup was run with cover crop simulation included into the original baseline condition, and soil moisture content was predicted at each subwatershed. The long-term impact of this management practice on soil moisture content is reflected as shown in Figure 9. Soil moisture content was found to reduce significantly across the watershed with a new mean of 167 mm and standard deviation of 21. The range of values across subwatershed was found to be 116 to 207 mm, while compared to the baseline condition which was 164 to 300 mm. Spatial distribution of soil moisture was consistent with the original baseline condition where Eastern part of the watershed had higher moisture content. Moreover, the reduction in moisture content was found to be consistent on a spatial scale. The magnitude of reduction was found significant as evident by reduction in mean by 67%. Even though it is an outcome of a simulation model, the signal of impact is very high. Figure 10 show the spatial distribution of reduction in soil water content due to inclusion of winter cover crops in standard corn-soybean rotation on a long-term basis.

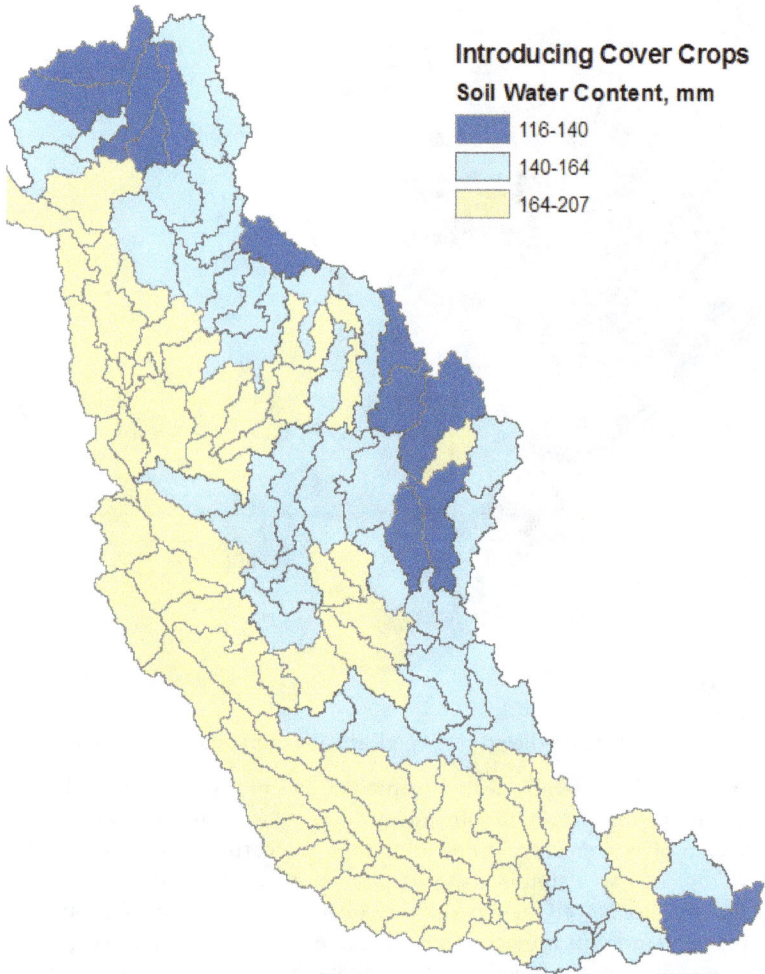

Fig. 9. Soil moisture content (after introducing winter cover crop) as predicted by SWAT at subwatershed scale on a long-term basis

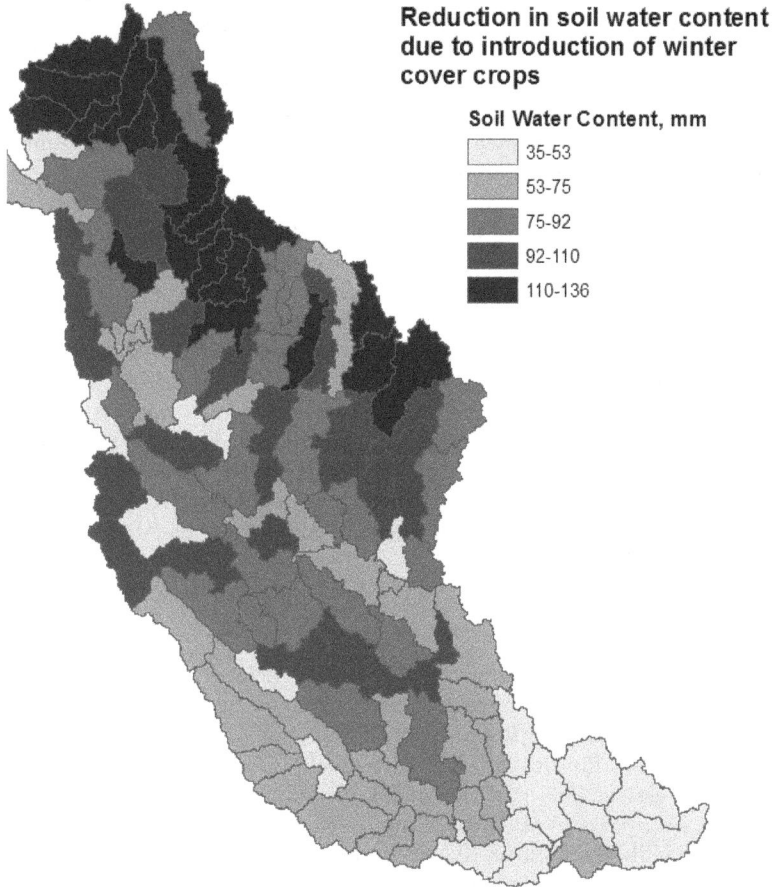

Reduction in soil water content due to introduction of winter cover crops

Soil Water Content, mm

- 35-53
- 53-75
- 75-92
- 92-110
- 110-136

Fig. 10. Reduction in soil moisture content due to inclusion of winter cover crops in standard corn-soybean rotation on a long-term basis

Significant reduction in soil water content raises the sustainability concern of the future crop production and regional ecosystem. As soil water content is very vital for crop growth and other ecosystem variables, it is imperative that it needs to be conserved. Added to that, the uncertainties in climate change with a certain increase of temperature and uncertain changes (may increase or decrease) in the amount of precpitation pose more threat to the sustainable agriculture system. It is warranted that the large scale implementation of winter cover crops should be examined with caution for changes in soil moisture content and its impact on future use of the land for agricultural production.

4. Conclusion

Understanding the spatio-temporal distribution and quantity of available soil moisture that can be used without damaging the natural ecosystem are keys to sustainable development and prevention of ecosystem decline. This study attempted to quantify the distribution of soil moisture content on a 3,630 mi^2 Raccoon River Watershed located in the Midwest United States through the use of a watershed scale hydrologic model SWAT. After a successful test of SWAT's ability to predict soil moisture content, it was used to quantify the impact of introducing winter cover crops in standard corn-soybean rotation in the Midwest. The unit of analyses was at a subwatershed scale; a finer unit with total number of 112 comprise the entire watershed. Successful calibration of the SWAT modelling setup for the watershed input parameters and databases was found to produce total water yield very accurately (less than 1% error) which lead to the accurate estimation of soil moisture content at a subwatershed scale. While introducing winter cover crops has shown to be effective positively for both soil quality as well as water quality, this modelling study on the impact of this change in soil moisture found to have an adverse impact on a long-term basis. Soil moisture content was found to reduce significantly across the watershed with a mean of 167 mm and standard deviation of 21. The range of values across subwatershed was found to be 116 to 207 mm, while compared to the baseline condition which was 164 to 300 mm. The magnitude of reduction was found significant as evident by reduction in mean by 67%. Even though it is predicted by simulating a well calibrated model, signal of the impact is very high. It is warranted that the large scale implementation of winter cover crops should be examined with caution for changes in soil moisture content and its impact on future use of the land for agricultural production.

5. References

Arabi, M.; Govindraju, R.S. & Hantush, M.M. (2006). Role of watershed subdivision on evaluation of long-term impact of best management practices on water quality, *Journal of American Water Resources Association*, Vol. 42, No. 2, pp. 513-528.

Arnold, J.G.; Srinivasan, R.; Muttiah, R.S. & Williams, J.R. (1998). Large area hydrologic modeling assessment: Part 1. Model development, *Journal of American Water Resources Association*, Vol. 34, No. 1, pp. 73-89.

Brocca, L.; Morbidelli, R.; Melone, F. & Moramarco, T. (2007). Soil moisture spatial variability in experimental areas of central Italy, *Journal of Hydrology*, Vol. 333, No. 2-4, pp. 356-373.

Gassman, P.W.; Reyes, M.; Green, C.H. & Arnold, J.G. (2007). The Soil and Water Assessment Tool: Historical development, applications, and future directiosn, *Transactions of the ASABE*, Vol. 50, pp. 1211-1250.

Gitau, M.W.; Veith, T.L. & Gburek, W.J. (2004). Farm-level optimization of BMP placement for cost-effective pollution reduction, *Transactions of the ASABE*, Vol. 47, No. 6, pp. 1923-1931.

Hatfield, J.L.; McMullen, L.D. & Jones, C.S. (2009). Nitrate-nitrogen patterns in the Racoon River Basin related to agricultural practices, *Journal of Soil and Water Conservation*, Vol. 64, No. 3, pp. 190-199. Doi: 10.2489/jswc.64.3190.

IDNR (2008). Natural resources geographic information systems library. Iowa City, Iowa: Iowa Department of Natural Resources, Geological Survey. http://www.igsb.uiowa.edu/nrgislibx/, accessed July 2008.

Kaspar, T.C.; Jaynes, D.B.; Parkin, T.B. & Moorman, T.B. (2007). Rye Cover Crop and Gamagrass Strip Effects on NO3 Concentration and Load in Tile Drainage, *Journal of Environmental Quality*, Vol. 36, No. 5, pp. 1503-1511.

Jha, M.K.; Arnold, J.G. & Gassman, P.W. (2007). Water quality modeling for the Raccoon River Watershed using SWAT, *Transactions of the ASABE*, Vol. 50, No. 2, pp. 479-493.

Jha, M.K.; Wolter, C.F.; Schilling, K.E. & Gassman, P.W. (2010a). Assessment of total maximum daily load implementation strategies for nitrate impairment of the Raccoon River, Iowa, *Journal of Environmental Quality*, Vol. 39, pp. 1317-1327. Doi: 10.2134/jeq2009.0392.

Jha, M.K.; Schilling, K.E.; Gassman, P.W. & Wolter, C.F. (2010b). Targeting land-use change for nitrate-nitrogen load reductions in an agricultural watershed, *Journal of Soil and Water Conservation*, Vol. 65, No. 6, pp. 342-352. Doi: 10.2489/jswc.65.6.342.

Jha, M.K. (2011). Evaluating hydrologic response of an agricultural watershed for watershed analysis. *Water*, Vol. 3, No. 2, pp. 604-617.

Moriasi, D.N.; Arnold, J.G.; Van Liew, M.W.; Binger, R.L.; Harmel, R.D. & Veith, T. (2007). Model evaluating guidelines for systematic quantification of accuracy in water simulations. *Transactions of the ASABE*, Vol. 50, No. 3, pp. 885-900.

Rabalais, N. N.; Turner, R.E.; Justic, D.; Dortch, Q.; Wiseman, Jr., J. W. & Sen Gupta, B.K. (1996). Nutrient changes in the Mississippi River, *Estuaries*, Vol. 19, No. 2B, pp. 385-407.

Robinson, D.A.; Campbell, C.S.; Hopmans, J.W.; Hornbuckle, B.K.; Jones, S.B.; Knight, R.; Ogden, F.; Selker, J. & Wendroth, O. (2008). Soil moisture measurement for ecological and hydrological watershed-scale observations: A review, *Vadose Zone Journal*, Vol. 7, pp. 358-389. Doi: 10.2136/vzj2007.0143.

Schilling, K.E.; Wolter, C.F.; Christiansen, D.E.; Schnoebelen, D.J. & Jha,M.K. (2008). Water quality improvement plan for Raccoon River, Iowa. TMDL Report. Watershed Improvement Section, Iowa Department of Natural Resources, pp. 202. Available at:
http://www.iowadnr.gov/water/watershed/tmdl/files/final/raccoon08tmdl.pdf

Seyfried, M.S. & Wilcox, B.P. (1995). Scale and the nature of spatial variability: Field examples having implications for hydrologic modeling, *Water Resources Research*, Vol. 31, pp. 173–184.

Shepherd, M. A. & Webb, J. (1999). Effects of overwinter cover on nitrate loss and drainage from a sandy soil: Consequences for water management? *Soil Use Management*, Vol. 15, No. 2, pp. 109-116.

Singer, J.W.; Malone, R.W.; Jaynes, D.B. & Ma, L. (2011). Cover crop effects on nitrogen load in tile drainage from Walnut Creek, Iowa, using root zone water quality model (RZWQM), *Agricultural Water Management*, Vol. 98, No. 10, pp. 1622-1628.

Spruill, C.A.; Workman, S.R. & Taraba, J.L. (2000). Simulation of daily and monthly stream discharge from small watersheds using the SWAT model, *Transactions of the ASBE*, Vol. 43, No. 6, pp. 1431-1349.

Topp, G.C. & Ferre, P.A. (2002). Th ermogravimetric method using convective oven-drying. p. 422–424. *In* J.H. Dane and G.C. Topp (ed.) Methods of Soil Analysis: Part 4. Physical methods. SSSA, Madison, WI.

Turcu, V.E.; Jones, S.B. & Or, D. (2005). Continuous soil carbon dioxide and oxygen measurements and estimation of gradient-based gaseous flux, *Vadose Zone Journal*, Vol. 4, pp. 1161–1169.

USDA-NRCS. (2008). Overview and history of hydrologic units and the watershed boundary dataset (WBD). Washington, D.C.: USDA Natural Resources Conservation Service.
http://www.ncgc.nrcs.usda.gov/products/datasets/watershed/history.html, accessed July 2008.

USGS. (2009). Federal guidelines, requirements, and procedures for the National Watershed Boundary Dataset. U.S. Geological Survey, U.S. Department of the Interior, Reston, VA and Natural Resources Conservation Service, U.S. Department of Agriculture, Washington, D.C. Available at: http://pubs.usgs.gov/tm/tm11a3/.

Vereecken, H.; Kamai, T.; Harter, T.; Kasteel, R.; Hopmans, J. & Vanderborght, J. (2007). Explaining soil moisture variability as a function of mean soil moisture: a stochastic unsaturated flow perspective, *Geophysical Research Letters*, Vol. 34, L22402. Doi: 10.1029/2007GL031813.

4

Fire Impact on Several Chemical and Physicochemical Parameters in a Forest Soil

Andrea Rubenacker, Paola Campitelli, Manuel Velasco and Silvia Ceppi
Departamento de Recursos Naturales, Facultad de Ciencias Agropecuarias,
Universidad Nacional de Córdoba, Córdoba
Argentina

1. Introduction

Cordoba is a Mediterranean State, with semiarid climate, dry autumn and winter, in which the wild fire can take place, especially at the end of dry season. Forest fires happen frequently in the mountain zones of the province of Córdoba, Argentina, which are located at west and south-west region. The vegetation, in the south-west zone, are principally *Pinus halepensis Mill.; Pinus elliottii* implanted and the native vegetation cover is *Stipa caudata, Piptochaetium hackelii, P. napostaense y* Briza *subaristata,* between others. Taxonomically the soil corresponds to an Ustorthent.

In the west area the native vegetation is principally *Acacia caven, Festuca hieronymi, Stipa, Poa stukerti,* between others. The soil is an Argiustoll.

Forest wild fires constitute a serious environmental problem, not only due to the destruction of vegetation but also because the degradation that may be induced in a soil as a consequence of the change produced in its properties. Wild fire can strongly modify the abiotic and biotic characteristics of soil, altering its structure, chemical and physicochemical properties, carbon content and macronutrient levels. The degree of the alteration produced depends on the frequency and intensity of fire, all these modifications being particularly important in the surface horizons.

Organic matter is a key factor for forest soil. It has a direct and /or an indirect influence on all physical and chemical characteristics of the soil. While low severity fires, such as those prescribed for forest management, have been reported to have transient but positive effect on soil fertility, severe wildfire result in significant losses of soil organic matter, and nutrient, and deterioration of the overall physical-chemical properties of soil that determine its fertility, such as porosity, structure among others (Certini, 2005).

Fire may directly consume part or all of the standing plant material and litter as well as the organic matter in the upper layer of the soil. One of the most important soil change, during the burning is the alteration in the organic matter content therefore, the nutrient contained in the organic matter are either more available or can be volatilized and lost from the site. The soluble nutrient would be loss for erosion or leaching if they are not immediately absorbed by plants or retained by soil.

Humic substances are one of the most important fractions of the organic matter and are considered the most abundant organic component in nature and largely contribute to soil structuring and stability, to its permeability for water and gases, to its water holding capacity, to the nutrient availability, to the pH buffering and to the interaction with metal ions (Schnitzer 2000; Hayes & Malcom, 2001, Campitelli & Ceppi 2008)

Depending on the fire severity the organic matter may change not only their level, but also, their different fractions, i.e. the humic substances (humic and fulvic acids) content and their principal characteristics (Vergnoux et al., 2011a, 2011b; Duguy & Rovira 2010).

The fire could induce transformation in the solubility of humic substances in different media, alkali or acid, and of their fraction (humic and fulvic acids). Thus, it could means that the humic and fulvic acids suffer structural modification, provably of the peripheral chains and the oxygenated moieties.

The study of the burned soil is necessary to analyze the soil degradation, not only to estimate the modification of the nutrient content but either the physicochemical characteristics of the organic matter, specially analyzing the humic substances and their fractions.

Humic acids have an important role in soil structure and nutrient capacity due to their surface charge development which may change due to the fire event. Some researchers have mainly shown changes in the aromaticity and in the oxygen-containing functional groups (Almedros et al., 2003; Gonzalez-Vila & Almendros 2009; Kniker et al., 2005; Vergnoux et al., 2007).

Most of the researches are focusing in the nutrient content, carbon content and their fraction, the fate of nutrient after fire, the effect of erosion, in general, the effect of fire disturbance of the soil properties. Moreover, a single parameter is insufficient to give an accurate evaluation of soil alteration. That is why several parameters need to be taken into consideration (Vergnoux et al., 2009).

Because the abundance and importance in soil of the organic matter and their fraction, mainly the humic acids, is necessary to focus the study not only onto the principal nutrient content, but also, on the characteristics of the humic acids and compare it with those extracted from the same unburned soil.

The objective of this research was to study, through different analytical techniques, i) several soil parameters related to chemical soil fertility and ii) chemical and physicochemical properties of organic carbon and their different fractions, focusing mainly, on the humic acids extracted from the forest soil after the fire event in order to compare it with the humic acids from the unburned soil.

2. Materials

2.1 Study site and sampling

The area selected was the south-west area of the province of Cordoba, named as San Agustín (Departamento de Calamuchita).

The annual average rainfall is about 600-800 mm. The mean temperature in the area fluctuates from 9 ^{0}C in winter to 20 ^{0}C in summer.

The tree stratum are principally *Pinus halepensis Mill.*; *Pinus elliottii* implanted and native vegetation cover is *Stipa caudata, Piptochaetium hackelii, P. napostaense y* Briza *subaristata*, between others. Three composite samples (10 subsamples) were taken from the upper layer (0-10 cm) of soil, some days after the wildfire occurrence and before any rainfall event. The samples were taken from the burned (BS) and adjacent unburned soil (UBS), at the same sampling moment. Litter and the ash were not removed from the soil surface before

sampling in the unburned and burned soil, respectively. The soil was taxonomically characterized as Ustorthent. The samples were air-dried, crushed and passed through a 2 mm sieves before all the analytical analysis
The humic acids (HA) analyzed were extracted from the burned (HA-BS) and unburned soil (HA-UBS)

2.2 Methods
The samples of burned and unburned soil were analyzed for pH at a rate 1:2.5 (w:v), electric conductivity (EC), total nitrogen content (TN) by the Kjeldahl method, phosphorus available (P) by Bray & Kurtz method (1945), total organic carbon (TOC) by combustion at 540 ^{0}C for 4 h (Abad, et al., 2002) and oxidable carbon (Cox) by the methodology proposed by de Richter & Von Wistinghausen (1981). Organic light fraction (OLF) were determined according the method proposed by Janzen et al., 1992, the C and N content of the OLF by dry combustion using a Perkin Elmer CN Elemental Analizer.
The carbon content of humic substances (CHS), humic acids (CHA) and fulvic acids (CFA) were determined according to the technique proposed by Syms & Haby 1971. The carbon content of each fraction (CHS, CHA and CFA) were calculated as percentage of the TOC, therefore, the % CHA correspond to the Humification Index (HI) (Roletto et al., 1985; Ciavatta et al., 1988).
The apolar or free lipidic fraction (FLF) were extracted with petroleum ether (40-60 ^{0}C) in 250 ml Soxhtel loaded with 50 g of soil; the extraction phase was renewed every 12h. The total extract was dehydrated with anhydrous Na_2SO_4 evaporated under reduced pressure to approximately 50 ml, dried under N_2 stream at room temperature (20-25 ^{0}C), and finally weighted, following the methodology proposed by Zancada et al., 2004.
The spectroscopy characteristics of the alkaline extract of both soil samples, the absorbance at different wavelength (280, 470 y 664 nm), were determined by the methodology proposed by Sapeck & Sapeck (1999). The ratio E2/E6, E4/E6 and E4/E6 were calculated from the corresponding absorbance value of the alkaline extract. The measures were determined using Spectronic 20 Genesys Spectrophotometer.

2.2.1 Humic Acids isolation
HA from burned and unburned soil were extracted with NaOH 0.1 mol L^{-1}, purified with HCl:HF (1:3) and dried at low temperature until constant weight, according to the procedure recommended by Chen et al. (1978). All solutions were prepared with tridistilled water and all the reagents were ACS reagent grade.

2.2.2 Humic Acids analyses
HA ash content was measured by heating it at 550 ^{0}C for 24 h. The elemental composition for C, H, N, S was determined by an analyzer instrument Carlo Erba 1108, using isothiourea as standard. Oxygen was calculated by difference: O%= 100 - (%C+%H+%N+%S) (ash and moisture-free basis).

2.2.3 Spectroscopic characteristics
The absorbance of the extracted HA were measured on a solution containing 3.0 mg of each HA in 10 mL of 0.05 mol L^{-1} $NaHCO_3$ at different   wavelength (280, 470 y 664 nm)

according the methodology proposed by Kononova, 1982; Zbytniewski & Buszewski, 2005; Sellami et al., 2008. From the absorbance value then were calculated the E2/E4, E2/E6 and E4/E6 ratio

2.2.4 Potentiometric titration

Potentiometric titrations were carried out according to the technique proposed by Campitelli et al. (2003), which is briefly: HA solution of each samples were prepared by dissolving HA ($\approx$ 50 mg) with minimum volume of NaOH solution (0.1 mol L^{-1}) and adding water up to the final volume (50 ml). An aliquot containing the desired amount of HA ($\approx$7–8 mg) was transferred to the titration flask containing 10 mL of tridistilled water. The titrant (HCl =0.05 mol L^{-1}) was added from an automatic burette (Schott Geräte T80/20) at a titrant rate of 0.1 ml/40 s. This rate was chosen taking into consideration that the variation of pH values should range between 0.02 and 0.04 pH units. The pH values were measured with an Orion Research 901 pH meter equipped with a glass-combined electrode (Orion 9103 BN). All titrations were performed in KCl 0.01 mol L^{-1} as background electrolyte. The same titration was followed in absence of HA (reference or blank titration) for each titration curve, in order to subtracts it from the raw data titration, and thus obtain the charge developed by the HA sample. Each HA solution, with the corresponding blank solution, was titrated by triplicate and the reported data representing the average values. All the reagents were ACS reagent grade.

2.2.5 Capillary zone electrophoresis

Capillary zone electrophoresis (CZE) experiments were performed on an Agilent Technology Capillary electrophoresis system equipped with a diode array. Operation of the instrument, data collection and analysis were controlled by Agilent ChemStation software. The polarity was negative, voltage of -30kV, temperature 25 ^{0}C, total run time 30 min (for time migration higher than 30 min no significant peak were observed). The samples were injected hydrodynamically using pressure of 5000 Pa for 20 s. The absorbance was monitored at four different wavelengths (210 nm, 230 nm, 260 nm and 450nm) and 260 nm was selected to report.

Each HA electropherogram was carried out by triplicate and the reported data representing the average values. The dimensions of the fused-silica capillary were 75 µm internal diameter; 73.0 cm total length and 64.5 cm effective length. All the solutions and background electrolyte (BGE) were prepared from analytical (p.a. or HPLC) chemicals and ultra pure water. BGE was buffer borate 20 mmol L^{-1} at pH=9.3, the concentration of the HA solutions were 1000 ppm. At the beginning of daily work, the capillary was washed for 5 min with 0.1 mol L^{-1} NaOH solutions, followed by 5 min washing with ultra pure water and 20 min with BGE at 25ºC and 10^4 Pa. At the end of the daily work, the capillary was rinsed with BGE for 5 min and water for 10 min, at the same temperature and pressure condition.

The capillary was treated before each sampling injection, as following, pre-condition: 2 min with NaOH 0.1 mol L^{-1} at 10^4 Pa, followed by washing with BGE for 3 min at 10^4 Pa, and finally waiting for 1 min. Post-run conditions were: 1 min with NaOH 0.1 mol L^{-1} at 10^4 Pa, followed by 5 min with water at the same pressure.

3. Results and discussion

3.1 Soil characterization

3.1.1 Main properties

The results of the principal chemical parameters are shown in table 1. The concentration of the cations, such as Na+ and K+ were not altered by fire, Ca^{2+} content slightly increase after fire, probably due to their release from the litter layer and Mg^{2+} decrease. The increase observed in the availability of Ca^{2+}, may be remarkably in a fire event, but ephemerally (Certini, 2005).

Sample	pH	EC	TN	P	TOC	Cox	TOC/TN	Na	K	Ca	Mg	CIC
Unburned soil (UBS)	6.20a	0.60a	6.6a	23.0a	105a	24.2a	15.9a	0.22a	1.15a	10.25a	2.25b	23.5a
Burned soil (BS)	6.53a	1.19b	7.1a	52.4b	128b	24.9a	18.0b	0.22a	1.03a	11.1a	1.5a	23.6a

EC: dSm^{-1}; P: mg kg^{-1}; TOC, Cox: g kg^{-1}; CIC, Na, K, Ca Mg: cmol kg^{-1} Different letters (a–b) in the same column indicate significant differences (p<0.05) according to Tukey test.

Table 1. Principal chemical characteristic of burned and unburned soil

The effect of burning onto soil Total Nitrogen (TN) content present a paradox, which have been debated for years (Neary et al., 1999; Knicker & Skjemstad,, 2000). Fisher & Binkley (2000) found that the immediate response of soil N to heating is a decrement because some loss through volatilization; Certini (2005), suggested that organic N could be volatilizes and in part mineralized to ammonium. Santin et al. (2008) found that the TN after fire increase and González-Vila et al. (2009) suggest that wildfire promote the accumulation of recalcitrant organic-N forms. The N, would be as NH_4^+ or NO_3^- , the NH_4^+ could be adsorbed onto negative charge of mineral and/or organic surface, but with time transformed to NO_3^-. Nitrate, without any plant uptake, will be lost from the ecosystem either by denitrification or leaching (Certini, 2005; Knicker, 2007).

The TN content increased slightly in the burned soil, but this change is not statistically significant; this behavior may be due to the nitrogen supplied by the burned litter and/or the ash contained in the sample.

The forest fires have not necessarily the same impact on soil P as on N, because losses of P through volatilization or leaching are small. The combustion of vegetation and litter causes modification on biogeochemical cycle of P. Burning convert the organic pool of soil P to orthophosphate, which is the form of P available to biota. Furthermore, the peak of P bioavailability is around pH 6.5. These could be the reason for which an enrichment of P is observed in the studied burned soil, but this enrichment will decline soon, because it precipitates as slightly available mineral forms (Certini, 2005; Cade-Menun et al., 2000). In agreement with this suggestion, the increase in the available P content in this burned soil could be due to the soil pH value (table 1).

Cation Exchange Capacity (CIC), on average, decrease after a fire event due to the loss of organic matter (Certini 2005; Badía & Martí, 2003), in this soil CIC was not changed, probably because the Cox content is the same before and after fire.

In general the soil pH increase by soil heating as a result of organic acids denaturalization, this increase take place when the temperatures are higher than 450 or 500ºC, in coincidence with the complete combustion of fuel and the bases release (Arocena & Opio 2003; Knicker,

2007; Certini 2005). For the soil analyzed, the increasing observed in soil pH after the fire event was slight (around 5%), this is in agreement with the cation (Na, K, Ca, Mg) content which were not largely modified by the heating soil, suggesting that the temperature did not raise up to 450°C or greater.

The electric conductivity (EC) increase in the burned soil, it could be assigned to the release of inorganic ions from the combusted organic matter present as litter or ash; this increase could be temporary (Kutiel & Imbar 1993; Hernandez, et al.,1997; Certini, 2005).

3.1.2 Organic matter

The most intuitive expected change in the soils during a fire event is the loss of organic matter. This change depends on the fire severity, vegetation type, soil texture and even slope. The impact on the organic matter consist of slight distillation (volatilization of minor constituents), charring or complete oxidation. Substantial consumption of organic matter begins in the 200-250 °C range to complete at around 450-500 °C (Fernandez et al., 1997; Giovannini et al., 1988; Certini, 2005; Knicker, 2007).

The influence of fire on the organic matter content have been reported a wide range of effects, showing even contrasting results (Gonzalez-Perez et al., 2004; Czimczik et al., 2005, Dai et al., 2005; Knicker et al., 2005; Alexis et al., 2007)

The oxidable organic carbon (Cox) content was not altered by fire, but, total organic carbon (TOC) increase around 21%, this behavior could be attributed to the accumulation of recalcitrant hydrophobic fraction of organic matter (Gonzalez-Perez et al., 2004; Santin et al., 2008).

The organic fraction extracted with petroleum ether, the soil free lipids, represents a diverse group of hydrophobic substances ranging from simple compounds such as fatty acids, to more complex substances as sterols, terpenes, polynuclear hydrocarbons, chlorophylls, fats, waxes and resins. The hydrophobic fraction extracted (FLF) from the sample after fire was greater ($\approx$ 38%) than that quantified for the sample of the control soil (table 2), in agreement with those found by Almendros et al. (1988), for a soil under *Pinus pinea*. Although, such compounds occur in fire unaffected soil, their abundance is increased by fire due to greater stability of lipids and lignin derivatives but also due to the neoformation of aromatic polymers (Almendros et al., 2003; Fernandez et al., 2004; Knicker et al., 2005a).

The high TOC content before and after fire event could be due to the sampling methodology, taking the soil sample with all the litter and grass in soil before fire and litter from decaying fire affected vegetation. The increase in the TOC content suggests that this fire event contributes to an enhancement of the organic matter, through the incomplete combust vegetation and thus contributes to a soil TOC increase. With the time residence in the soil this unstable organic matter could be incorporated to the stable pool of organic matter, this behavior is related to the process of accumulation of organic compounds in soil controlled by their chemical affinity with the native organic matter. The randomness of the process and the heterogeneity of the organic molecules, probable produced by fire, lead to the accumulation of organic matter in which hydrophilic association may be contiguous with hydrophobic domains or contained in one other, and thus the native organic matter pool could behave as sink of the decaying fire affected vegetation. (Spaccini et al., 2000; Santin et al., 2008; Gonzales-Perez et al., 2004; Knicker et al., 2005).

The increase in the TOC/TN ratio (Table 1) after the fire event is due, principally, to the TOC increase more than to the TN change after fire. This could confirm the accumulation of incompletely burnt plants necromass, or a post-fire enhancement of the litter from decaying fire-affected vegetation production (Knicker et al., 2005a; 2007; Gonzalez-Perez et al., 2004; Santin et al., 2008)

The light fraction (LF) content (Table 2), which represents all residues, with a density value lower than 1.7 g ml-1, on the top soil before and after the wildfire event, could be the reason for the high value of the TOC observed. This fraction (LF) increases after fire in the same way as the TOC, around 28%; which represent one possible source of organic material (incomplete burnt plants) that would be incorporated to the native pool of soil organic matter and thus a way to a progressive stabilization of the different organic compounds produced by the fire effect, such as, aliphatic compounds, polysaccharides, peptides of plant and microbial origin and other organic compound generated by fire. The carbon content slightly increases and nitrogen content decreases significantly ($\approx$ 27%) after fire in the LF. The C/N ratio indicate that this fraction is formed by an unstable organic fraction, composed by debris with incomplete combustion, thus, it could produce a nitrogen immobilization during the stabilization and the incorporation to the native soil organic matter.

3.2 Organic matter fractions analysis

The carbon content of each fraction (CHS, CHA and CFA) were calculated as a percentage of the TOC, therefore, the % CHA correspond to the Humification Index (HI) (Roletto et al., 1985; Ciavatta et al., 1988).

Vergnoux et al.(2011a, 2011b), found that the different fraction of the humic substances decrease after fire, in agreement with Almendros et al.(1990); Fernandez et al.(1997); Gonzalez-Perez et al.(2004); Kincker et al.(2005). Other studies suggest that during the wildfire a humic-like fraction can be produced from burned plant biomass and thus it would be extractable in alkaline solution. In general, medium heating, i.e. temperatures not higher than 250°C, leads to increase complexity of the organic matter: newly formed compounds, oxidation and thermal fixation of alkyl moieties, etc. (Almendros et al., 1992; Gonzalez-Perez et al., 2004).

The organic carbon content of each fraction (CHS, CHA, and CFA) of the burned and unburned soil is shown in table 2.

Sample	CHS	CFA	CHA(HI)	CHA/CFA	LF	N%	C%	C/N	FLF
Unburned soil (UBS)	1.92a	0.68a	1.23b	1.80b	56.7a	1.54b	17.35a	11.2a	0.24a
Burned soil (BS)	1.84a	0.79b	1.06a	1.34a	66.9b	1.12a	18.37a	16.4b	0.39b

CHS, CFA, CHA, LF and FLF: expressed as % in function of 100 g of TOC Different letters (a–b) in the same column indicate significant differences (p< 0.05) according to Tukey test.

Table 2. Carbon content in each humic substances fraction (CHS, CFA, and CHA), carbon light fraction content (LF), nitrogen and carbon content in the carbon light fraction, and the free lipidic fraction (FLF) content in burned and unburned soil

The variation in the CHS content after the wildfire is not statistically significant; this could be due to the original humic materials transformations into an alkali-insoluble macromolecule material (Gonzalez-Perez et al., 2004; Fernandez et al., 2004), which is in agreement with the amount of hydrophobic fraction (FLF) found in both soil samples (table 2).

The CFA increases around 15% after fire and CHA decrease around 12% in the soil exposed to high temperatures. The increase of the CFA content indicate the newly formed compounds, with more aliphatic chains, in general, with less molecular size, produced by the breakup of the more aggregated structures of the humic acids and thus, the carbon humic fraction decrease. The Humification Index (HI) (Table 2) is reduced about 12% indicating, also, the alteration in the humic substances by wildfire.

The ratio CHA/CFA (Table 2), also known as "degree of polymerization or polymerization index", decrease around 25% in the burned soil, reflecting the breakdown of the complex and more aggregated structures of unheated soil humic fraction, indicating that the wildfire lead to an important change in the structure and the properties of the humic substances fraction (Debano et al., 2000; Shakeesby &Doerr 2006).

3.3 Spectroscopic properties of soil alkaline extracts

The scattering of monochromatic light in a diluted solution of macromolecules or colloidal particles is closely related to weight, size, aggregation and interaction of particles in solution. The UV-Visible absorption of humic substances was used to evaluate the condensation degree of the aromatic compounds (Chen el al., 1977; Stevenson, 1982; Polak et al., 2009).

Sutton & Sposito (2005), suggest that the apparent size of humic materials do not change due to tight coiling (or uncoiling), but instead change due to disaggregation (or aggregation) of clusters of small molecules.

The absorption at 280 nm was also introduced to represent total aromaticity, because the π-π* electron transition occurs in this UV region, for phenolic arenes, benzoic acids, aniline derivatives polyenes and polycyclic aromatic hydrocarbon with two or more rings (Uyguner & Bekbolet, 2005).

The absorption at 470 nm is related with the fragment produced for the depolimerization or disaggregation of the supramolecular structure or material with a low humification degree (Sellami et al., 2008; Zbytniewski & Buszewski, 2004).

The absorbance at 664 nm is characteristics of high oxygen content, aromatic compound, strongly humified material with a high degree of condensed groups (Sellami et al., 2008).

Lipski et al.(1999) defined E2/E4 ratio (the ratio of absorbance at 280 and 400 nm) to characterize the degradation of phenolic/quinoid core of humic acids to simpler carboxylic aromatic compounds. This ratio may represent an alternative parameter for the elucidation of the photocatalytic degradation efficiency.

The value of the quotient E4/E6 (the ratio of absorbance at 400 and 665nm) and E2/E6 (the ratio of absorbance 280 and 665 nm) coefficient are related with aromatic condensation; suggest the aggregation level, phenolic and benzene-carboxylic group content, among other characteristics. A low ratio reflects a high degree of aromaticity, aggregation and high humification level; large values are associated with the presence of smaller size organic molecules, more aliphatic structures, high content of functional groups, high disaggregation level (Chen et al., 1977; Pertusati & Prado, 2007, Zbytniewski & Buszewski, 2004).

The value of the coefficient E2/E4 for CHS and CFA (Table 3) obtained in the alkaline extracts for the burned and unburned soil, don't have a great variation, suggesting that the degradation of core structure of humic substances, depolymerization or the disaggregation of the supramolecular structure was not significant, probably several aggregate disruption was produced by heating the soil (Uyguner & Bekbolet, 2005; Sutton & Sposito, 2005).

The values of the quotient E2/E6 and E4/E6 are around 20-30% greater for both fraction (CHS and CFA) in the burned soil than in the unburned (Table 3). This variation suggest that the temperatures reached during the fire event, probably around 250-300ºC, produced some degree of disaggregation effect and also, the increasing in the quotient value could be due to the newly organic compounds produced by the litter and vegetal residues burned during the wildfire.

Sample	E2/E4	E2/E6	E4/E6	E2/E4	E2/E6	E4/E6
	CHS	CHS	CHS	CFA	CFA	CFA
Unburned soil (UBS)	7.7a	0.67a	5.2a	28.65a	0.63a	18.1a
Burned soil (BS)	7.5a	0.84b	6.33b	29.4a	0.81b	23.9b

Different letters (a–b) in the same column indicate significant differences (p<0.05) according to Tukey test.

Table 3. Alkaline extracts Absorbance ratio of burned and unburned soil samples

The greater content obtained for the CFA (table 2) is in agreement with the disaggregation observed through the E2/E6 and E4/E6 values after the fire event.

3.4 Humic Acids characterization
3.4.1 Elemental composition
Elemental composition (ash and moisture-free basis) O/C, H/C (atomic ratios) and E2/E4, E2/E6 and E4/E6 ratio of the HA extracted from unburned and burned soil are shown in Table 4.

The increase in the carbon content after fire could be produced by the incorporation of the incompletely burned necromass to the original supramolecular structure. The decrease in the oxygen content after fire suggests that the environment could have reducing properties.

The atomic ratio of O/C and H/C are often used to monitor structural changes of humic substances (Gonzalez-Perez et al., 2004; Adani et al., 2006).

Sample	C	H	N	O	S	O/C	H/C	E280	E460	E660	E4/E6
HA-UBS	49.67a	5.46a	4.97a	39.46b	< 0.4a	0.59b	1.32b	1.78a	0.34a	0.09a	3.77a
HA-BS	53.89b	5.28a	4.93a	35.45a	< 0.4a	0.49a	1.18a	2.15b	0.47b	0.12b	3.92a

Different letters (a–b) in the same column indicate significant differences (p<0.05) according to Tukey test.

Table 4. Elemental composition (ash and moisture-free basis) O/C, H/C (atomic ratios) and E2/E4, E2/E6 and E4/E6 ratio of the HA studied

The decrease in the atomic H/C ratio observed for HA-BS, suggest a diminution in the peripheral aliphatic chains with low thermal stability and thus, an increase in the aromaticity because this domains was found resistant to the effects of fire. The decrease in the O/C ratio indicates a substantial loss of oxygen-containing functional groups. The mains change observed in HA heated in laboratory or in natural fire are the dehydration and decarboxyilation which explain the progressive alteration in the colloidal properties of soil affected by fire (Gonzalez-Perez et al., 2004).

3.4.2 Spectroscopic properties

UV-Visible spectra were recorded for both HA analyzed, the specific absorbance decreases steadily with increasing wavelength. The spectra are close to those presented in other studies related to the chemical nature of humic acids (Senesi et al., 1989; Fuentes et al., 2006). The absorption properties are conventional and versatile for the characterization and were used to evaluate the condensation degree of the humic aromatic nuclei. Various absorption wavelengths at 270, 280, 300, 400, 465 nm, among other, and their ratios have been cited for the spectral differentiation of humic substances (Sellami et al., 2008; Uyguner et al., 2005).

By analyzing the absorption spectrum of UV-Visible, three important regions were observed at 280, 460 and 660 nm. The absorbance at 280 nm (E280) is related to lignin, aniline derivatives, polyenes and polycyclic aromatic hydrocarbon with two or more rings (Uyguner & Bekbolet, 2005). The absorbance at 460 nm (E460) is the result of organic macromolecules with a low polymerization degree, and the absorbance at 660 nm (E660) is characteristic of high oxygen content, aromatic compound, high size and molecular weight (Sellami et al., 2008; Uyguner et al., 2005).

The absorbance of the HA extracted from the burned soil is greater than the absorbance of the HA isolated from the unburned soil, similar to those obtained for Vergnoux et al.(2011a). This behavior indicate that the HA isolated from the soil exposed to high temperatures have greater content of different fraction of organic compounds. The increase of the absorption at 280 nm (Table 4) indicate the presence of fraction like lignin derivatives and compounds with aliphatic chains; the absorption at 460 (Table 4) suggest the increment of compounds with a low polymerization degree or less condensed structural domains and the increment of the absorption at 660 nm (Table 4) suggest the increase of aromatic compounds with great microbial and / or chemical resistance, structures that have refractory character (Vergnoux et al. 2011a; Sellami et al., 2008; Santin et al 2008; Gonzalez-Perez et al., 2004).

The growth observed in the content of all these fractions could be due through the incorporation of the compounds produced by an incomplete combustion of the vegetation, and therefore, a considerable amount of newly formed C forms were adding together to the thermal modified C forms previously existing in the ecosystem (Cofer et al., 1997; Gonzalez-Perez et al., 2004). Through the E4/E6 value for both HA, burned and unburned HA, (3.92 and 3.77 respectively), in general, is possible to suppose that the nuclei of the macromolecule of HA, the aromaticity, the size, the weight were not disrupt by the temperature reached in this event fire, instead, the wildfire could have enough energy to produce a disruption onto the linkage which retain together the small fraction of the supramolecular structure and thus a disaggregation could take place; this behavior is shown through the increment of the absorbance values.

3.4.3 Potentiometric titration: Acid base properties and charge evolution

The charges-pH curves (-Q versus pH) of the HA isolated, between pH 3 and 11, obtained from potentiometric titration, corrected for blank solution and fitted with sixth degree polynomial according to Machesky (1993) and Campitelli & Ceppi (2008), are shown in the Figure 1a. This smoothing function was selected for their simplicity.

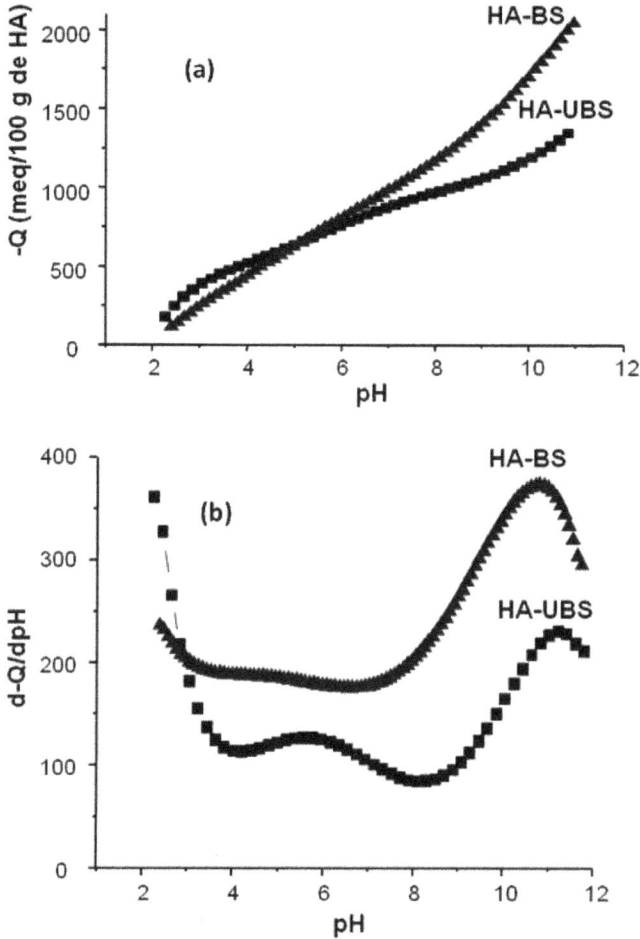

Fig. 1. (a): Charge-pH curves of humic acids extracted from burned (HA-BS) and unburned soil (HA-UBS). The charge developments were calculated on the basis of the sixth polynomial equation (with R^2 values exceeding 0.999 in all cases). (Charge development were calculated taking into account the ash content); (b): Apparent proton-affinity distribution of humic acids extracted from from burned (HA-BS) and unburned soil (HA-UBS) obtained from the first derivatives through charge-pH curves $[d(-Q)/d (pH)]$ smoothing with sixth degree polynomial equation through the experimental data in the range of 3–10

The charge development of HA isolated from the burned soil (HA-BS) is greater than for the humic acids extracted from unburned soil (HA-UBS) in the region of pH 6 to 11 and lower at the more acidic region (3 to 6). Total acidity is about 60% greater in the HA extracted from burned soil than those of the unburned soil. In the acidic pH region (3 to 6) the lower charge development for HA isolated from burned soil could be due to the loss of strong acidic sites produced by the disruption of the supramolecular structure.

The disaggregation produced by temperature could be the reason for the increment of the negative charge development up to pH 6, because the negative charged groups increase as the size of the fractions decrease (Tombacz, 1999). This behavior is in agreement with that observed through the spectroscopic analysis.

Through the first derivative of the –Q versus pH curves (-dQ/dpH) obtained from the titration curves smoothed with the polynomial equation (Figure 1b), is possible: i) to obtain the average of apparent proton-dissociation constant (pKaap) of each set of acidic groups, ii) to analyze the chemical heterogeneity of each class of acidic group present in the HA macromolecule, iii) to estimate the concentration of each set of acidic groups by the calculus of the area under each peak and iv) to estimate the buffer capacity developed by each class of acidic site (Nederlof et al., 1994; Koopal et al., 2005; Campitelli et al., 2006; Campitelli & Ceppi, 2008). In this way, is possible to follow how the acid-base characteristics, i.e, the evolution in quantity and quality for the principal acidic groups (carboxylic and phenolic), were changed for the fire event.

The number of site classes (set of acidic groups) is then equal to the number of peaks and the peak position could be used as an average of the apparent dissociation constant (pKaap)(de Wit et al.,1993)

The samples of HA extracted from unburned soil (HA-UBS) show two main peaks, the first would be assigned to the carboxylic groups (strong acidic sites) and the second to the phenolic groups (weak acidic sites). For the HA isolated from burned soil (HA-BS) the first peak is only a shoulder and the second peak is well defined. In both HA samples (HA-BS and HA-UBS), also is observed, a small or developing peaks at more acidic pH values ($\leq$ 4), indicating, probably, a presence of stronger acidic sites; this behavior is more clear in HA-BS. This is in agreement with previous results obtained studying HA extracted from soil (Campitelli et al., 2006; Campitelli & Ceppi, 2008).

HA isolated from burned soil (HA-BS) presents the first peaks or shoulder, not well defined, with the maxima at around pH 3.5 and the second with a maximum at pH 10.8; the first could be assigned to strong acid sites (carboxylic groups) and the second to weak acidic sites (phenolic groups). The peak at pH=3.5 was wider than the peak at pH=10.8. The minimum was not well defined, and the partial overlapping of peaks indicate that there is no significant differences among the acidic sites in the surface, in terms of proton dissociation strength. This results suggest a large chemical heterogeneity on the HA present or the production of small organic compounds during the fire event.

These small organic compounds could be produced by the incomplete combustion of the vegetation present; Knicker et al. (2007) suggested that around 250 ^{0}C new molecular structures are produced; the principal structures could be aliphatic C; phenol and/or furan C; Sharma et al. (2004) suggested that some decarboxylation could occur at higher temperatures (>250^{0}C) but the aromatic rings still remain essentially intact. This behavior could justify the decrease in the negative charge development at pH values lower than 6 and their increase at higher pH values (pH > 6).

HA isolated from unburned soil (HA-UBS) have two well defined peaks, the first with the maximum at pH 5.6 and the second at pH 11.2, these values are similar to other obtained for soil derived humic acids (Campitelli et al., 2006; Campitelli & Ceppi 2008)

The pKaap for the carboxylic and phenolic groups in the HA derived from the burned soil (HA-BS) are lower than the corresponding for HA extracted from unburned soil (HA-UBS), this could be due to the disruption of the supramolecular structure of the humic acids, and in this way the carboxylic groups that remains in the surface are those with very strong acidic characteristics, probably those in the aromatic structures, like o-COOOH or in greater fractions, and the phenolic groups are those in the small fraction produced by the disaggregation (Table 5) (Knicker et al., 2007; Sharma et al., 2004). For both type of acidic groups (o-COOH and OH-Phenolic), the contribution could be from the partial combustion of vegetation and then extracted with the alkaline media, without discrimination (Adani et al., 2004).

Humic acids	o-COOH	pKaap	-COOH	pKaap	phenolic-OH	pKaap
HA-UBS	320a	2.3	473	5.6	567a	11.2
HA-BS	588b	3.5	---	---	1318b	10.8

Acidic groups: cmol kg^{-1} Different letters (a–b) in the same column indicate significant differences (p<0.05) according to Tukey test.

Table 5. Acidic functional groups content (o-carboxylic, carboxilic and phenolic) content calculated by integration of the area under each maximum of the curves (d-Q/dpH) obtained through the first derivative of smoothed experimental data. The pKaap values correspond to the maximum of each peak

The fire event altered the concentration of acidic sites (Table 5) and therefore the buffer capacity. For the burned soil, the buffer capacity of HA was neglectable at soil pH value around 6 (Table 1) and for pH value ranging between ≈ 3 – 7. This can be attributed to the great heterogeneity of HA in this pH range and to the lost of carboxylic groups with pKaap values around 5.

In the zone up to pH 8 (weak acidic sites) the buffer capacity is greater than that observed for HA from unburned soil, but this groups, in both cases, are not dissociated at soil pH values, thus they have not a significant contribution to the soil buffer capacity. The fire event produced important changes in the acid-base properties, principally in the buffer capacity of the HA.

The loss of carboxylic groups onto this HA structure produced by fire event (Table 5), i.e. the decrease of negative charge development below pH 6, cause a deficiency of charged site to make linkage between the inorganic and organic fraction through cation-bound; and thus, the formation of soil aggregates. In this way, this characteristic could be the key factor promoting soil erosion (Mill & Fey 2004). The fire event could generate important modification in the physicochemical properties of the HA

At the lowest pH measured (sites domains below 4), the HA-UBS shows a developing peak (Fig 1b) indicating that very acidic sites could be present in the macromolecule, in HA-BS it seems that this sites are the only present (Table 5). The minimum around pH 4, which could be considered as a separation of both type of acidic sites (like COOH) from the very acidic sites (like o-COOH), is clearer in the HA extracted from unburned soil (HA-UBS) than in the HA from the burned soil (HA-BS), this indicate, also, the heterogeneity of the acidic groups present in the HA extracted from soil exposed to high temperatures, due to the disaggregation produced by the temperature developed during the wildfire.

3.4.4 Capillary zone electrophoresis
The main characteristics of HA are the occurrence of acidic site with different strength, the principal groups are the strong (carboxilic groups) and weak (phenolic groups) acidic site.

For these HA analyzed the average Pkaap value are around 3.5 – 5.5 for the carboxylic groups and 10 – 11 for the phenolic groups (Table 5).

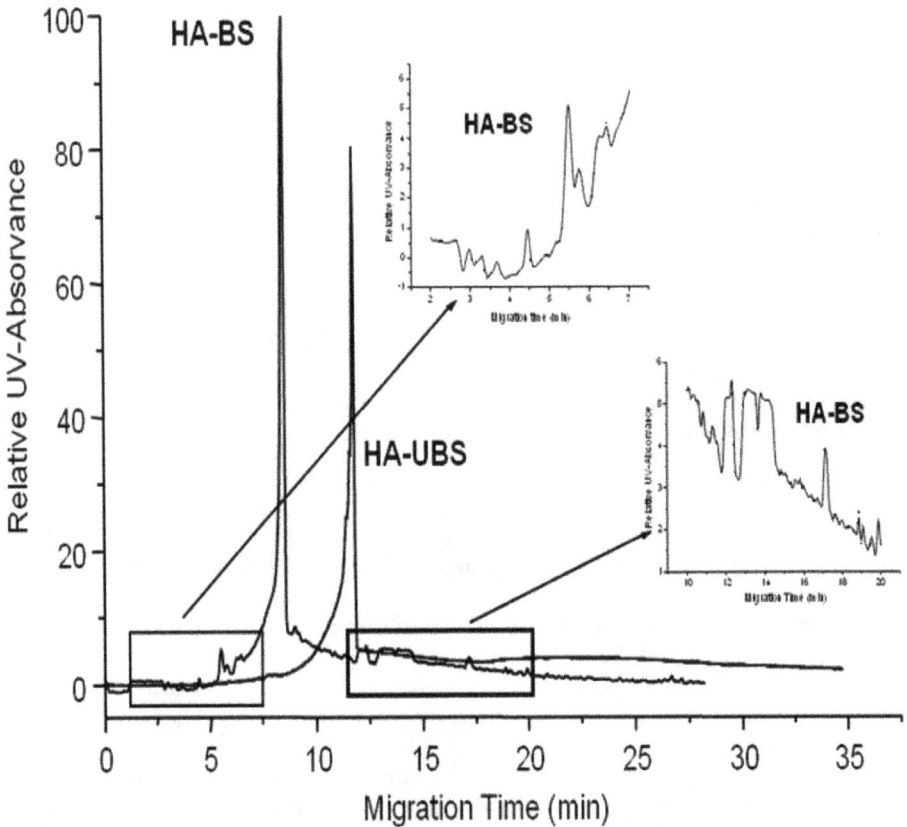

Fig. 2. Electropherograms of acids extracted from burned (HA-BS) and unburned soil (HA-UBS) in buffer borate 20 mmol L^{-1} (pH=9.3), temperature 25^{0}C, the concentration of the AH solutions were 1000 ppm. CZE conditions: voltage of -30kV, injection hydrodynamic 5000 Pa for 20 s, detection at 260 nm, fused-silica capillary, 73 cm total length, 75 μm i. d. (effective length 64.5 cm). Total run time 30 min (for time migration higher than 30 min no significant peak were observed)

At the experimental condition (pH ≈ 9) all of the strong acidic groups and approximately, the half of the weak acidic groups of HA are deprotonated (negatively charged). The presence of negative charges permit to separate HA by electrophoresis in an electrical field (+) to (-) in which the EOF (electro osmotic flow) is responsible for the movement of the analyte (Peuravouri et al., 2004).

The electropherogram of HA extracted from unburned soil (HA-UBS) (fig 2) shows a principal and well defined peak at time migration 11.73 min and the characteristic hump at time migration around 7 – 8 min, just before the main peak, is shown as a tail; at migration time higher than 12 min no peaks were distinguished.

The electropherogram of HA isolated from burned soil (HA-BS) presents the main peak at lower time migration (8.40 min) than in HA of unburned soil (HA-UBS), and several peaks are detected before and after the main peak (fig 2); at migration time higher than 25 min no peaks were distinguished.

The peak at 11.73 min observed in the electropherogram corresponding to the HA-UBS, could indicate that in these experimental conditions the macromolecule migrate as a unbroken entity, the tailing observed at lower time migration, could be assigned to some structure with low mass/charge ratio difficult to be separated; i.e. the macromolecule is not easy to be separated in subfraction with different electrokinetic properties, similar behavior was observed for Fetch & Havel (1998); Pokorna et al.(2000); Peuravouri et al.(2004).

The different time migration for the principal peak of the HA from burned soil (HA-BS) and the peaks detected at both side of the peak at 8.40 min could indicate changes in the macromolecule structure and the presence of subfraction.

The BGE, borate, could react with phenols, phenols carboxylic, polycarboxilic acids, dihidroxy or perihydroxy groups present in the solution and thus the separation of each fraction would be improved (Fetsch & Havel 1998). The phenolic groups present in HA isolated from burned soil (HA-BS) is greater than that quantified in HA extracted from unburned soil (HA-UBS), this characteristic could produce the interaction between the BGE and this acidic groups and enhance the separation.

The electropherogram profile of the HA extracted from burned soil (AH-BS) indicates the presence of distinct subfraction, which could be produced by the disaggregation of the macromolecule of HA and/or the formation of newly small carbon compounds after heating, suggesting that the temperature reached during the fire event, breaks, disaggregates or creates new structure, with lower and higher mass/charge ratio and diverse electrokinetic mobility. This behavior confirms the large heterogeneity, the disaggregation and the new carbon compound produced for the wildfire and are in agreement with those observed through the other different analytical techniques used to study these HA.

4. Conclusions

The temperature reached in the fire event was enough to produce several changes in the organic matter characteristics, i.e. changes in the quantity and/or quality of their fraction: light fraction, humic acids, fulvic acids, free lipidic fraction.

The fire event produced important changes in the structure of the macromolecule of humic acids, like break and/or disaggregation which generate compound with lower size, weight, mass/charge ratio and/or newly formed carbon compounds originated by the incomplete combustion of the vegetal materials.

The fire event could generate important modification in the physicochemical and acid-base properties of the HA.

The amount of acidic functional group was changed: the COOH sites were decreased and the OH phenolic sites were increased by the fire event. The pKaap values were modified, in general, the acidic site are stronger after fire than in the unburned soil. The COOH groups with pKaap value about 5 were lost after fire. The buffer capacity is lower or practically missing at soil pH (≈ 6) after fire.

The negative charge development decrease significantly at field pH (≈ 6) after the fire event, producing a deficiency on sites to make linkage between the organic and inorganic soil fraction, and in this way a reduction of aggregates formation. This characteristic could be the key factor promoting soil erosion.

5. Acknowledgements

SeCyT-UNC are gratefully acknowledged for financial support.

6. References

Abad, M., Noguera, P., Puchades, R., Maqueira, A. & Noguera, V., 2002. Physico-chemical properties of some coconut coir dusts for use as a peat substitute for containerised ornamental plants. *Bioresour Technol.*, 82: 241-245

Adani, F & G. Ricca 2004 The contributionof alkali soluble (humic acid-like) and unhydrolyzed-alkali soluble(core-humic acid-like) fraction extracted from maize plant to the formation of soil humic acid *Chemosphere* 56: 13-22.

Adani, F., Ricca, G., Tambone, F. & Genevini, P., 2006. Isolation of the stable fraction (the core) of humic acids. *Chemosphere* 65:1300-1307.

Alexis, M. A., Rasse, D. P., Rumpel, C., Bardoux, G., Pechot, N., Schmalzer, P., Drake, B., & Mariotti, A. 2007 Fire impact on C and N losses and charcoal production in a scrub oak ecosystem. *Biogeochemistry* 82: 201-219

Almendros, G., Gonzalez-Vila F. J., Martin, F., Frund, R. & Ludemann H. D. 1992 Solid state NMR studies of fire induced changes in the structure of humic substances. *Sci Total Environment* 117-118: 63-74

Almendros, G., Gonzalez-Vila F.J. & Martin, F. 1990 Fire-induced transformation of soil organic matter from an oak forest: an experimental approach to the effects of fire on humic substances. Soil Science 149: 158-168.

Almendros, G., knicker, H. & Gonzalez-Vila F. J. 2003 Rearrangement of carbon and nitrogen forms in peat after progressive isothermal heating as determined by solid-state 13C and 15N-NMR spectroscopy. *Organic geochem* 34: 1559-1568

Almendros, G., Martin, F. & Gonzalez-Vila, F. J. 1988 Effects of fire on humic and lipid fraction in a Dystric Xerochrept in Spain *Geoderma* 42:115-127

Arocena, J. M. & Opio, C. 2003 Prescribed fire-induced changes in properties of sub-boreal forest soil *Geoderma* 113:1-16

Badía, D. & Martí, C. 2003 Plant ash and heat intensity effects on chemical and phisical properties of two contrasting soils. *Arid Land Res Management* 17:23-41

Cade- Menun, B. J., Berch, S. M., Preston, C. M. & Lavkulic, L. M. 2000 Phosphorus forms and related soil chemistry of Podzolic soils on northern Vancouver Island. II. The effects of clear-cutting and burning. *Can J Forest Research* 30:1726-1741

Campitelli, P, A., Velasco, M, I. & Ceppi, S, B., 2006. Chemical and physicochemical characteristics of humic acids extracted from compost, soil and amended soil. *Talanta*, 69:1234–1239.

Campitelli, P. & Ceppi, S., 2008. Effects of composting technologies on the chemical and physicochemical properties of humic acids. *Geoderma*, 144:325–333.

Campitelli, P., Velasco, M. & Ceppi, S., 2003. Charge development and acid-base characteristics of soil and compost humic acids. *J Chil Chem Soc.*, 48:91-96

Certini, G. 2005 Effects of fire on properties of forest soil: a review. *Oecologia* 143: 1-10

Ciavatta, C., Antisari, V. & Sequi, P. 1988 A first approach to the characterization of the presence of humified materials in organic fertilizers. *Agrichimica* 32:510-517

Cofer III W. R., Koutzenogii, K. P., Kokorin & A. Ezcurra, A. 1997 *Biomass burning emission and the atmosphere.* In Clark J. S., Cachier, H., Goldammer J. G., Stocks, B. editors. Sedimient records of biomass burning and global change. NATO ASI Serie, Vol I Berlin Germany: Springer

Czimczik, C. I., Schmidt, M. W. I. & Schulze, E. D. 2005 Effects of increasing fire frecuency on black carbon and organic matter in Podzols of Siberian Scots pine forest. *European J of soil Sci.* 56: 417-428

Chen, Y., Senesi, N. & Schnitzer, M., 1977. Information provide on humic substances by E4/E6 ratio. *Soil Sci Soc Am J.* 41: 352-358.

Chen, Y., Senesi, N. & Schnitzer, M., 1978. Chemical and physical characteristics of humic and fulvic acids extracted from soils of the Mediterranean region. *Geoderma,* 20:87-104.

Dai, X., Boutton, T. W., Glaser, B., Ansley, R. J. & Zech, W. 2005 Black carbon in a temperate mixed-grass savanna. *Soil Biology and biochemistry* 37: 1879-1881

Debano, l. F. 2000 The role of fire and soil heating on water repellence in wildland environment: a review. *J Hydrol* 231: 195-206

Duguy, B. & Rovira, P. 2010 Differential thermogravimetry and differential scanning calorimetry of soil organic matter in mineral horizons: Effect of wildfire and land use. *Organic Geochemestry* 41: 742-752

De Wit, J. C. M., Van Riemsdijk, W. H. & Koopal L. K. 1993 Chemical heterogeneity and adsorption models. Environm. Sci. Technol. 27: 2015-2022

Fernandez, I., Cabaneiro, A. & Carballas, T. 1997 Organic matter changes immediately after a wildfire in an Atlantic forest soil and comparison with laboratory soil heating. *Soil Biology and Biochemestry* 29: 1-11

Fernandez, I., Cabaneiro, A. & Gonzalez-Prieto, S. J. 2004 Use of 13C to monitor soil organic matter transformations caused by a simulated forest fire. *Rapid Commun Mass Spectrom* 18:435-442

Fuentes, M., Gonzalez-Gaitano, G. & García-Mina, J. M. 2006 The usefulness of UV-vis and fluorescencespectroscopies to study the chemical nature of humic substances from soil and compots. *Org. Geochem* 37:1949-1959

Giovannini, G. Lucchesi, S. & Giachetti, M. 1988 Effect of heating on some physical and chemical parameters related to soil aggregation and erodibility. Soil Sci 146: 255-261

Golzalez-Vila, F. J. & Almendros, G. 2009 *Thermal transformation of soil organic matter by natural fires and laboratory controlled heating.* In: Natural and Laboratory Simulated Thermal Geochemical Processes. R. Ikan (ed.) Kluver Academic Publisher. Netherlands.

Gonzales-Perez, J. A., Gonzalez-Vila, F. J., Almendros, G. & Knicker, H. 2004 The effect of fire on soil organic matter: a review *Environmental International* 30: 855-870

Hayes, M. H. B. & Malcom, R. L. 2001 *Considerations of the compositions and of aspects of the structure of humic substances.* In: Clapp, C.E., Hayes, M. H. B., Senesi, N., Bloom, P. R., Jardine, P. M. (Editors) Humic substances and chemical contaminations. Soil Science Society of America Inc. Madison (pp 3-39)

Hernandez, T., Garcia, C. & Reinhardt, I. 1997 Short-term effect of wildfire on the chemical, biochemical and microbiological properties of Mediterranean pine forest soil. *Biol. Fertil. Soil* 25:109-116

Janzen, H. H., Campbell, S. A., Brand, S. A., Laford, G. P. & Townley-Smith, A. 1992 Light fraction organic matter in soil from long-term crop rotation. *Soil Sc. Soc of Am J.* 56:1799-1806

Kincker, H., Gonzalez-Vila, F. J., Polvillo, O., Gomzalez, J. A. & Almendros, G. 2005 Fire induced transformation of C and N forms in different organic soli fractions from a Dystric Cambisol under Mediterranean pine forest (Pinus pinaster). *Soil Biol. Biochem.* 37: 701-718

Knicker, H. 2007 How does fire affect the nature and stability of soil organic nitrogen and carbon? A review. *Biogeochemistry* 85-118

Knicker, H., Gonzalez-Vila, F. J., Plovillo, O. & Gonzalez, J. A., Almendros, G. 2005a Fore-induced transformation of C and N forms in different organic soil fraction from a Dystric Cambisol under a mediterranean pine forest (pinus pinaster). *Soil Biol Biochemestry* 37:701-718

Knicker, H. & Skjemstad, J. O. 2000 carbon and nitrogen functionality in protected organic matter of some Australian soils as revealed by solid-state 13C and 15N NMR . *Australian J of Soil Sc*, 38: 113-127

Kononova, M. M. 1982. *Materia Orgánica del Suelo.* Vilassar de Mar. Barcelona. España

Koopal, L. K., Saito, T., Pinheiro., van Riemsdijk, W. H. 2005 Ion binding to natural organin matter: General considerations and the NICA-Donnan model. *Colloids and Surface A: Physicochem. Eng aspects* 265: 40-54.

Kutiel, P. & Imbar, M. 1993 Fire impact on soil nutrients and soil erosion in a mediterranean pine forest plantation. *Catena* 20:129-139

Lipski, M., Slawinski, D. & Zych, D. 1999 Chnages in the luminescent properties of humic acid induced by UV radiation. *J Fluorescence* 9: 133-138

Machesky, M., 1993. Calorimetric acid-base titrations of aquatic and peat derived fulvic and humic acids. *Environm Sci Technol.*, 27: 1182-1198.

Mill, A. J. & Fey, M. V. 2004 Frequent fire intensity soil crusting: physicochemical feedback in the pedoderm of long-term burn experiments in South Africa. *Geoderma* 121: 45-64

Neary, D. G., Klopatec, C. C., deBano, L. F. & Fgolliott P. F. 1999 Fire effects on belowground sustainability: a review and synthesis. *Forest Ecol manag.* 122:51-71

Nederlof, M. M., van Riemsdijk, W. H. & Koopal, L. K., 1994. Heterogeneity analysis for binding data using adapted smoothing spline techniques. *Environ Sci Technol.*, 28: 1037-1047.

Pertusatti, J. & Prado, A. G. S., 2007. Buffer capacity of humic acid: Thermodynamic approach. *J Colloid and Interface Sci.*, 314:484-489.

Polak, J., Bartoszek, M. & Sulkowski, W. W. 2009 Comparison of some spectroscopic and physico-chemical properties of humic acids extracted from sewage sludge and bottom sediments. *J. of Molecular Structure.* 924-926: 309-312

Richter, M. & Von Wistinghausen, E. 1981 Unterscheidbarkut von humusfraktione in boden be unterschiedlicher Bewirtschaftung, Z. Pflanzenernaehr Bodenk 144:395-406

Roletto, E., Barberis, R., Consignlid, M. & Jodice, R. 1985 Chemical parameters for evaluation of compost maturity. *Biocycle March*: 46-48

Santín, C., Knicker, H., Fernandez, S., Mendez-Duarte, R. & Alvarez, M. A. 2008 Wildfire influence on soil organic matter in an Atlantic mountainous región (NW of Spain). *Catena* 74: 286-295

Sapeck, B. & Sapeck, A. 1999. *Determination of optical propierties in weakly humified samples.* In: Dziadowiec, H., Gonet, S.S. (Eds.), The Study of Soil Organic Matter-the Methodical Guide. Warszawa, Poland.

Schnitzer, M. 2000 A lifetime perspective on the chemical of soil organic matter. *Adv. Agron.* 68: 3-58

Sellami, F., Hachicha, S., Chtourou, M., Medhioub, K. & Ammar, E., 2008. Maturity assessment of compostded olive mill waste using UV spectra and humification parameters. *Bioresour Technol.*, 99:6900-6907.

Senesi, N., 1989. Composted materials as organic fertilizer. *Sci Total Environm.*, 81:521-542.

Shakeesby, R. A. & Doerr, S. H. 2006 Wildfire as a Hydrological and geomorphological agent. *Earth Science Rev* 74:269-307

Sharma, R. K., Wooten, J. B., Baliga, V. L., Lin, X., Chan, W. G. & Hajaligol, M. R. 2004 Characterization of chars from pyrolysis of lignin. *Fuel* 83: 1469-1482

Sims, J. R. & Haby, V. A., 1971. Simplified colorimetric determination of soil organic matter. *Soil Sci.*, 112:137-141.

Spaccini, R., Piccolo, G., Haberhauer, G. & Gerzabek, M. H. 2000 Transformation of organic matter from maize residues into labile and humic fraction of three European soil as reveled by 13C distribution and CPMAS-NMR spectra. *European J soil Sci.* 51: 583-594

Stevenson, F. J., 1982. *Humus chemistry. Genesis, composition, reactants.* John Wiley and Sons N. York.

Sutton, R. & Sposito, G., 2005 Molecular structure in soil humic substances: the new view. *Environm Sci Technol.*, 39: 9009-9015.

Tombacz, E. 1999 Colloidal properties if humic acids and spontaneos changes of their coloidal state under variable solution conditions *Soil Sci* 164:814-824

Uyguner, C. S., Bekbolet, M. 2005 Evaluation of humic acid photocatalytic degradation by UV-vis and fluorescence spectroscopy. *Catalysis Today* 101: 267-274

Vergnoux, A., Di Rocco, R., Domeizel, M., Guiliano, M., Doumenq, P., Theraulaz, F. a) 2011 Effects of fire on water extractable organic matter and humic substances from Mediterranean soil: UV-vis and flourecsence spectroscopy approaches *Geoderma* 160:434-443

Vergnoux, A., Dupuy, N., Guiliano, M., Vennetier, M. Theraulaz, F. & Doumenq, P. 2009 Fire impact on forest soil evaluated using near-infrared spectroscopy and multivariate calibration. *Talanta* 80: 39-47

Vergnoux, A., Guiliano, M. Di Rocco, R., Domizel, M., Theraulaz, F. & Doumenq, P. b) 2011 Quantitative and mid-infra-red changes of humic substances from burned soils. *Environmental Research* 111: 193-198

Vergnoux, A., Malleret, L., Domeizel, M. Theraulaz,F. & Doumenq, P. 2007 Effect of forest fire on water extractabke organic matter and polycyclic aromatic hydrocarbon in soil. *Progerss in Environmental Science abd Technology.* Beiijing, China

Zancada, M. C., Almendros, G., Sanz, J., & Romám, R. 2004 Speciation of lipids and humus-like coloidal compounds in a forest soil reclaimed with municipal solid waste compost. *Waste management and research* 22: 24-34

Zbytniewski, R. & Buszewski, B. 2004 Characterization of natural organic matter (NOM) derived from sewage sludge compost. Part 1: chemical and spectroscopic properties. *Bioresource technology* 96: 471-478

Zbytniewski, R. & Buszewski, B. 2005 Characterization of natural organic matter (NOM) derived from sewage sludge compost. Part 2: multivariate techniques in the study of compost maturation. *Bioresource technology* 96: 479-484

5

Pesticide Contamination in Groundwater and Streams Draining Vegetable Plantations in the Ofinso District, Ghana

Benjamin O. Botwe[1], William J. Ntow[2] and Elvis Nyarko[1]
[1]University of Ghana, Department of Oceanography & Fisheries
[2]University of California, Department of Plant Sciences
[1]Ghana
[2]USA

1. Introduction

1.1 Ghana's geographical location and climate

Ghana, officially the Republic of Ghana and formerly the Gold Coast, is a West African country with a geographical location of 5°36' N, 0°10' E. It shares borders with Cote d'Ivoire to the west, Burkina Faso to the north and Togo to the east. To the south of the country is the Gulf of Guinea of the Atlantic Ocean. The climate is tropical equatorial ranging from the bimodal rainfall equatorial type in the south to the tropical unimodal monsoon type in the north. It is influenced by the hot, dry and dusty-laden air mass that moves from the north-east across the Sahara and by the tropical maritime air mass that moves from the south-west across the southern Atlantic ocean. The annual rainfall ranges from 1015 to 2300 mm with annual mean temperature and relative humidity of 30°C and 80% respectively (Ntow & Botwe, 2011). Ghana has a total land area of about 23,853, 900 ha and a population of about 24.2 million. The arable land covers an area of about 13,628,179 ha (approx. 57% of total land area) of which approximately 44% is under cultivation.

1.2 Economic importance of agriculture in Ghana

Agriculture is Ghana's most important economic sector, employing more than 60% of the labour force. Currently, agriculture contributes about 33% of Ghana's gross domestic product (GDP) and accounts for over 40% of export earnings. Ghana's agriculture is predominantly smallholder, traditional and rain-fed. The major agro-ecological zones in Ghana are Rain Forest, Deciduous Forest, Forest-Savannah Transition, Coastal Savannah and Northern (Interior) Savannah which comprises Guinea and Sudan Savannahs. The type of agricultural activity carried out in each zone is determined largely by rainfall. In the south, there is a major and a minor growing season due to the bimodal rainfall pattern in the Forest, Deciduous Forest, Transitional and Coastal Savannah zones whereas in the Northern Savannah, the unimodal rainfall pattern results in a single growing season. Within the agricultural sector, vegetable production plays an important socio-economic role, having developed from a mainly subsistence activity to a commercial activity. Vegetable production in Ghana typically occurs in intensely managed vegetable plantations characterized by an extensive network of drainage systems through which surplus water may flow out (Ntow *et al.*, 2008). Vegetables cultivated in Ghana include tomato (*Lycopersicon esculentum*), eggplant (*Solanum melongena*), pepper (*Capsicum annum*) and onion (*Allium cepa*), although some regions are more efficient and specialised in the production of only one or two vegetable crops (Ntow, 2001).

1.3 Pesticide use in vegetable cultivation in Ghana

Vegetables generally attract a wide range of pests and diseases, and require intensive pest management (Dinham, 2003), which includes all aspects of the safe, efficient and economic use and handling of pesticides. In Ghana, pest and disease control practices in vegetable production involve the use of chemical pesticides. A total of 43 pesticides, comprising insecticides, fungicides and herbicides, have been found in use in vegetable farming in Ghana. Among these pesticides, the herbicides class of pesticides is the most used (44%), followed by insecticides (33%) and then fungicides (23%) (Ntow *et al.*, 2006). Although it is recognized that better management of pesticides results in high crop productivity while greatly reducing adverse environmental impacts, most of the local farmers lack adequate training in the proper application of pesticides. Pest and disease control therefore involves relatively high inputs of highly toxic chemical pesticides which are most of the time misapplied (Ntow *et al.*, 2006). The average pesticide application rate is estimated to be 0.08 litres active ingredient (a.i.) per hectare (Ntow *et al.*, 2008). Misapplication and intensive use of pesticides in vegetable cultivation can result in pesticide contamination of the environment.

1.4 Impacts of pesticide use in vegetable agro-ecosystems

Water pollution by pesticides has long been recognized as a major environmental impact associated with agriculture due to the potential adverse effects on aquatic life and on humans if contamination extends to drinking waters (Skinner *et al.*, 1997). Most vegetable farms in Ghana are sited a few meters from streams for easier access to water for irrigation purposes. The close proximity of streams to vegetable farms is of particular concern as there is high potential for pesticides to move offsite into surrounding streams via run-off through the extensive system of drainage canals that characterize these farms. Persistent pesticides, particularly the organochlorine group of pesticides, can be transferred to aquatic organisms at all trophic levels within the food chain due to their bioconcentration and bioaccumulation potential. Many organochlorine pesticides are known to mimic hormones and disrupt reproductive cycles of humans and wildlife (Colborn and Smolen, 1996) and therefore they can be detrimental to a wide variety of aquatic wildlife populations (Robinson, 1991). Even non-persistent pesticides, such as the pyrethroids, carbamates and organophosphate group of pesticides, can be highly toxic to aquatic life (Castillo *et al.*, 2006). Pesticides can also enter groundwater via seepage or soil percolation. Pesticides contamination of streams and groundwater also presents health threat to the rural communities as they depend on streams and groundwater for drinking and other domestic purposes.

Concerns over the adverse ecological and human health impacts of pesticides have led to the institution of very strict programs to control and monitor pesticide contamination in water sources in developed countries such as the United States and members states of the European Community (García de Llasera and Bernal-González, 2001). These programs have, however, not been implemented in most developing countries such as Ghana. Few studies conducted in Ghana have focused on the organochlorine pesticides (Osafo & Frimpong, 1998; Ntow, 2001, 2005). However, pesticides from the organophosphate group, which are now commonly used in Ghana following the ban on persistent organochlorine pesticides, have not been determined in water quality studies. In this chapter, pesticide contamination in groundwater and streams draining vegetable plantations in the Ofinso District of Ghana are assessed and the ecotoxicological significance of the pesticides contamination evaluated.

1.5 Study area
The present study was conducted in the Ofinso District of the Ashanti Region of Ghana (Fig. 1). The Ofinso District is located in the extreme North-Western part of the Ashanti Region, with about half of its boundaries bordered by Brong Ahafo Region (in the north and west).

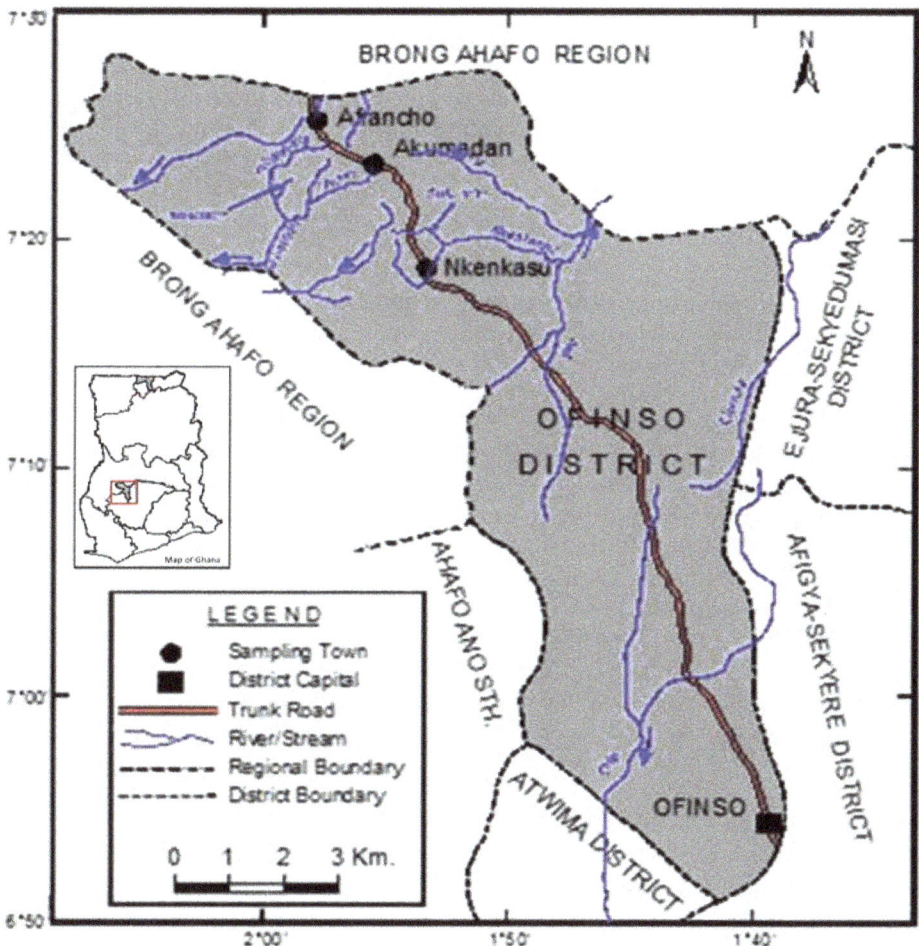

Fig. 1. Map showing the study area

It is bordered to the east by Ejura-Sekyedumasi District, to the south by Afigya Sekyere, Ahafo Ano South and Atwima Districts. The district has 126 settlements and a population of about 35,190 with New Ofinso as its capital. The district has five towns namely Abofour, Nkenkasu, Afrancho, Akumadan and New Ofinso. The study area is within the Ofin, Pru and Afram river basins. In the present study, vegetable plantations were selected from Akumadan, Nkenkasu, and Afrancho. Agriculture is the main economic activity in these areas with over 70% of the active population being farmers. The district is well known for the cultivation of vegetable crops. Other major crops cultivated include cassava, maize, plantain and cocoa. More than 23 different active ingredients formulated as insecticides, herbicides and fungicides have been used in the cultivation of vegetables in the district.

The five most frequently used insecticides include two organophosphates (chlorpyrifos and dimethoate), two pyrethroids (lambda-cyhalothrin and cypermethrin), and one organochlorine (endosulfan). Farmers use these highly toxic pesticides under primitive field conditions with insufficient protective equipment and training. Pesticide applications occur frequently, all year round, and are relatively intensive (500-1000 ml/ha). Pesticides are also sprayed in combinations, with many farmers (60%) spraying their crops on calendar basis, at 7-day intervals (Ntow *et al.* 2006). Streams within the catchments of vegetable farmlands are vulnerable to pesticide contamination as a result of spray drift and surface runoff (Maule *et al.*, 2007). The quality of these water resources is of critical interest as they serve as aquatic habitats and drinking water sources.

2. Methods

2.1 Sampling of stream water, sediment and groundwater

Twenty-one streams flowing in and/or around vegetable plantations, stream-bed sediments and 9 drinking water wells in the Ofinso District of the Ashanti Region of Ghana were sampled between February 2008 and January 2009 as part of a pesticide monitoring programme in vegetable agro-ecosystems in Ghana. Streams sampled included Akumadan, Nkenkasu and Afrancho (Fig. 1) which flow in and around vegetable farmlands. The other streams sampled in the district were ephemeral and these included Srani, Bosompong, Sukubrim, Siasu, Ankonom, and Naasu (not shown). For each stream, 1 L water samples were collected into 1-L glass amber coloured bottles with Teflon-lined caps from upstream, mid-stream and downstream. During the same period, stream-bed sediment samples of about 200 g were collected into wide-neck glass jars. Groundwater samples were extracted from drinking water wells located within farming communities at Akumadan, Nkenkasu, and Afrancho into 1-L glass amber coloured bottles with Teflon-lined caps. None of the wells sampled was in a farmed section of the study area. All the wells are shallow wells (< 15 m) and represent unconfined aquifers. The wells receive water from the soil and upper porous rock zones that characterise the Ofinso District. Prior to sampling, pumps were run for about 5 min to clear the casing of standing water and to bring in fresh water from the aquifer. During this period, field measurement parameters (e.g. temperature) were stabilised. The sampling bottles were rinsed with well water before taking the water samples. Three replicates were collected from each well. The samples were transported to the laboratory within 24 to 48 h on ice in clean ice chests and stored in the laboratory refrigerator at 4°C until analysed. The samples were extracted within 24 h of arrival at the laboratory. Field blanks were prepared with distilled water and were analysed only when pesticide residues were detected in primary samples. Sampling was conducted throughout rainy and dry seasons and was not timed to applications of different pesticides or to rain events.

2.2 Chemical analysis
2.2.1 Sample extraction
The extraction and analyses of water samples were performed following the Association of Official Analytical Chemists 990.06 and 970.52 methods as described by Ntow et al. (2008). Briefly, water samples were extracted sequentially three times with 25 mL n-hexane each time. The extract was dried with anhydrous sodium sulphate and concentrated down to 10 mL by means of ultrasonic bath type concentrators (Turbo Vap II). Extract clean up was done, using a chromatographic column, packed with florisil, previously activated for 3 h in an oven at 130°C, and anhydrous sulphate (all rinsed with petroleum ether). The extract was transferred to the column. Three fractions were obtained after elution with 6, 15, 50% ethyl ether in petroleum ether. Maximal flux rate of elution was 5 mL/min. Each eluate was evaporated and the extracts (re-dissolved in 1.5 ml n-hexane and made up to 2 ml with more n-hexane) were injected into a gas chromatographic system for identification and quantification of the pesticides.

Extraction and analysis of sediment samples followed the method described by Ntow et al. (2008). Briefly sediment samples were well mixed to obtain a homogeneous sample and then transferred into a pan to air-dry at ambient temperature until a constant weight was obtained. The samples were later ground in a mortar into fine powder such that they could pass through 2 mm sieve. Five grams (dry weight) of the sediment samples were soxhlet extracted in methanol, and cleaned up in florisil in the same way as described above for water. Sampling protocol and analytical procedures were subjected to quality control through field and laboratory blanks and spiking of samples with pesticide standards.

Laboratory glassware used in the sampling and analyses were cleaned as described by Ntow (2001). Pesticide grade solvents used for the analyses were n-hexane (>99%) and acetone (>99.9%) (Sigma, Munich, Germany); methanol (99.8%) and petroleum ether (BDH; VWR International, UK); dichloromethane and ethyl ether (Fluka; Munich, Germany). Deionised water was used from a Milli-Q water purification system (Millipore, Bedford, MA, USA) for blanks, sediment extraction, and spiked samples. Pesticide standards (> 98% purity) were obtained from Dr. Ehrenstorfer (Augsburg, Germany). Standard mixtures were prepared from individual pesticide stock solutions (50-100 mg in 100 ml acetone) and then diluted to working calibration standards at three concentration levels with acetone/cyclohexane (pesticide grade) mixture (1:9).

2.2.2 Pesticide residue analysis
Measurement of pesticide compounds in water and sediment samples was performed on a GC-MS (Agilent 6890 Series GC System) coupled with an Agilent 5973N mass selective detector-electron impact ionization; and fused capillary column (HP-5MS) packed with 5% Phenyl Methyl Siloxane (30 m * 0.25 mm I.D and film thickness 0.25 μm), which was operated in the selected ion-monitoring mode at the following conditions: injection port 250°C (splitless, pressure 22.62 psi; purge flow 50 mL/min; purge time 2.0 min; total flow 55.4 mL/min). Column oven: initial 70°C, held 2 min, programming rate 25°C/min (70 to 150°C); 10°C/min (150 to 200°C); 8°C/min (200 to 280°C) and held 10 min at 280°C. The

carrier gas was nitrogen at 15 psi; detector make-up, 30 mL/min. The injection volume was 1 µL (Agilent 7683 Series injector). Selection of analysed pesticides was done on the basis of pesticide use information provided by Ntow *et al.* (2006). The pesticides analysed included α- and β-endosulfan, endosulfan sulphate, dieldrin, dichlorodiphenyltrichloroethane (p,p'-DDT), dichlorodiphenyldichloroethylene (p,p'-DDE) and chlorpyrifos. For quality control of gas chromatographic conditions, a checkout procedure was performed before sample analysis in which a standard mixture with α-endosulfan content of 400 ng/L was used. Calibration was carried out when the concentration of α-endosulfan in standard mixture deviated significantly from 400 ng/L. Also the linearity of detector response was checked with five standard solutions of concentration 200 - 1000 ng/L. The correlation coefficient, r, obtained was ≥ 0.94. Recovery of the different pesticides ranged between 79% and 104% and their detection limits varied between 0.001 and 0.01 µg/L. The residues are expressed as µg/L (ppb) for surface water and µg/kg dry weight (ppb) for sediment. Because most of the pesticides analysed by GC/MS had a method detection limit at or below 0.01 µg/L, the reporting limit was chosen as 0.01 µg/L for these compounds. This reporting limit was used in calculating incidences of occurrence. A pesticide that has been identified but not quantified is indicated as below the detection limit.

2.2.3 Physicochemical analysis
The pH and temperature of samples were determined in situ using a pH meter. The pH meter was first calibrated with standard pH buffers before immersing the probe into the water or sediment. Temperature was measured concurrently. Total suspended solids and turbidity in water were measured using a turbidity meter (2100P Turbidimeter, Hach Company, Loveland, CO, USA). Calibration of the turbidity meter was done by filtering some water samples through pre-weighed Whatman GF/F (0.45 µm pore-size) glass microfiber filters which were then dried at 60°C for 48 h and re-weighed to determine TSS. Water content (expressed as weight fraction of water) was determined by first weighing wet sediment samples, then oven-drying the sediment samples at 105°C until constant weight, and obtaining the weight difference. Total suspended solids and turbidity were measured concurrently (Ntow *et al.*, 2008). Total organic carbon was obtained from the percentage organic matter in the sediments as percentage loss-on-ignition after drying 1.0 g of the sediment samples (previously acidified for the removal of carbonates) at 550°C in a furnace (Mwamburi, 2003).

2.3 Statistical analysis
A paired Student's *t* test was performed to analyze significant differences between pesticide residue levels in stream water and stream-bed sediment. One-way analysis of variance (ANOVA) was performed to analyze significant differences in pesticide residue levels in water and sediment from different sites. Pearson correlation analysis was performed to determine the relationship between the levels of pesticide residues and sediment characteristics (total suspended solids and total organic carbon) at the 95% confidence level ($p < 0.05$)

2.4 Ecotoxicological significance of measured pesticides in stream water
The effects of pesticides on water quality are commonly assessed by comparing the observed concentrations of individual pesticide compounds in the aquatic system with criteria that have been established to protect the health of aquatic organisms (Castillo *et al.* 2000; Hoffman *et al.* 2000). By comparing the pesticide concentrations in this study with the

toxicity values listed in Table 1, the acute and chronic risk ratios were calculated for the water samples. A ratio of 1 means the individual pesticide has reached its criteria concentration in the streams. Risk for acute toxicity is based on the highest pesticide concentration found compared to the LC_{50} (Table 1). Risk for chronic toxicity is calculated based on the average concentration of all positive observations and the water quality criteria (Table 1).

Pesticides	Main use	Lowest LC_{50} for crustaceans or fish (μgL^{-1}) (EXTOXNET 1996)	Water-quality criterion (μgL^{-1}) (USEPA 1999)
α-Endosulfan	Insecticide	1.20	0.056
β-endosulfan	Insecticide	1.20	0.056
Endosulfan sulphate	Insecticide	1.20	0.056
Dieldrin	Insecticide	-	0.056
Chlorpyrifos	Insecticide	0.01	0.041
p,p'-DDE	Insecticide	0.18	0.001
p,p'-DDT	Insecticide	0.18	0.001

Table 1. Reference toxicity values for pesticides analyzed

3. Results and discussion

3.1 Physicochemical characteristics of water and sediment samples analyzed

The physicohemical characteristics of groundwater, stream water and sediment samples analyzed are presented in Table 2.

Physicochemical parameter	Groundwater	Stream water	Sediment
pH	5.8 - 6.4	6.6 - 8.3	5.6 - 6.8
Temperature (°C)	21.4 - 22.7	23.2 - 27.4	-
Total suspended solids (mg/l)	0	5.8 - 20.6	-
Turbidity (NTU)	0.74 - 2.93	2.2-32.5	-
Moisture content (%)	-	-	18.0 - 26.3
Total organic carbon (%)	-	-	2.1 - 13.6

Table 2. Physicohemical characteristics of groundwater, stream water and sediment samples analyzed

The pH and turbidity of groundwater were within acceptable levels for human consumption. The pH, temperatures, levels of total suspended solids and turbidity of stream water were also suitable for aquatic life.

3.2 Pesticide residue levels in groundwater

Pesticide residues were not detected in all the groundwater samples analyzed (Table 3). The non-detection of pesticide residues in groundwater could be due to their high adsorption to soil particles which does not facilitate their infiltration into groundwater. This is an indication that groundwater consumption may not contribute to community exposure to these pesticides.

Pesticide components	Groundwater (n =81)		Stream water (n = 192)		Sediment (n = 180)	
	Mean ± SD (μgL^{-1})	I.R (%)	Mean ± SD (μgL^{-1})	I.R (%)	Mean ± SD (μgKg^{-1} dw)	I.R (%)
α- Endosulfan	ND	-	0.027 ± 0.015	27.8	0.38 ± 0.24	90.0
β-Endosulfan	ND	-	0.021 ± 0.010	13.9	0.18 ± 0.09	97.5
Endosulfan-sulphate	ND	-	0.022 ± 0.010	21.5	0.53 ± 0.24	98.3
Dieldrin	ND	-	ND	-	0.16 ± 0.04	35.0
p,p'-DDE	ND	-	ND	-	3.77 ± 1.90	25.0
p,p'-DDT	ND	-	ND	-	ND	-
Chlorpyrifos	ND	-	ND	-	1.23 ± 0.40	68.3

Table 3. Concentrations (Mean ± SD) and incidence ratios of pesticide residues in groundwater, stream water and stream-bed sediment samples analyzed. [n = number of samples analyzed; SD = standard deviation; ND = below detection limit (0.01 μgL^{-1} or μgKg^{-1}); I.R. = incidence ratio; dw = dry weight]

3.3 Pesticide residue levels in stream water

The mean concentrations and incidence of occurrence of pesticides detected in stream water are summarized in Table 3. Only 67 (35%) of the 192 stream water samples analyzed had pesticide residue detections. α-endosulfan, β-endosulfan and endosulfan sulphate were the only pesticide residues detected with mean concentrations of 0.027 ± 0.015, 0.021 ± 0.010, and 0.022 ± 0.010 μgL^{-1} (or ppb), respectively. The incidence of occurrence of these organochlorine pesticide residues were α-endosulfan (27.8%), β-endosulfan (13.9%) and endosulfan (21.5%). Technical endosulfan is a mixture of α- and β-endosulfan in a ratio of 7:3. Endosulfan sulfate is the principal metabolite of endosulfan and it is highly toxic. Endosulfan sulfate levels in stream water samples were nearly equal to those of the parent compounds (α- and β-endosulfan), suggesting current use of the pesticide. The occurrence of endosulfan residues in streams may be the result of direct overspray, spray drift, atmospheric transport of volatilized pesticides, agricultural runoff, pesticide misuse, and improper disposal of pesticide containers (Maule *et al.*, 2007; Wan *et al.*, 2005; Ntow *et al.*, 2008). Inflows from shallow groundwater originating in the agricultural areas are however unlikely sources of pesticide contamination in the streams studied since pesticide residues were not detected in groundwater samples analyzed in the present study.

Pearson correlation analysis revealed an association ($r^2 > 0.6$) between endosulfan residue concentration and total suspended solids in stream water for most sites. Thus, increase in the level of suspended solids (sediment) resulted in a corresponding increase in concentration of endosulfan. This partitioning behavior of endosulfan in the streams studied may be influenced by physicochemical properties. Endosulfan has low water solubility (0.32 mgL at 22ºC) and high affinity for sediment as indicated by the high soil adsorption coefficient of 2,400 mLg^{-1} which can be attributed to its high octanol-water partitioning coefficient (log$K$ow = 3.6) (see Table 4). Therefore, with these properties, there is a high tendency for endosulfan to adsorb onto suspended sediments in the water column than to remain in solution as the study has shown.

Pesticide name	Water solubility at given temperature (mgL⁻¹)	LogKow	Soil adsorption coefficient, K_{oc} (mLg⁻¹)	Soil half-life
Endosulfan (α, β and sulfate)	0.32 (22°C)	3.6	2,400	50 d
Dieldrin	-	5.5	-	-
DDE	< 1 (20°C)	5.8	100,000	2-15 years
Chlorpyrifos	2 (25°C)	4.7	6,070	35-78 d

Table 4. Properties of pesticides detected in streams. Source: EXTOXNET (1996)

Endosulfan is banned or restricted in many countries because of its human health and environmental impacts. In the United States, for example, endosulfan is applied to grains, tea, fruits, vegetables, tobacco, and cotton (DeLorenzo *et al.* 2001). In Ghana, endosulfan has a restricted use that does not include vegetables (it has only been registered for use on cotton), yet it is used on vegetables. According to Ntow *et al.* (2006), endosulfan is one of the most commonly used pesticides in the study area. Different formulations of the active ingredient are sold in the study area under different trade names such as Thionex 35 EC/ULV, Thiodan 50 EC, Endosulfan, Endocoton, Caiman 350 EC, Phaser and Novasulfan 35 EC. Vegetable farmers in Ghana spray endosulfan on tomato, pepper, okra, egg-plant (garden eggs), cabbage and lettuce. Although there are numerous pests and diseases prominent on vegetables (for instance, there are 13 fungal pathogens on tomato alone), the use of endosulfan was not necessarily to control diseases. Application of endosulfan to control diseases was done on a trial-and-error basis because the local farmers were not able to identify the pests causing damage (Ntow *et al.*, 2006). The use of endosulfan on vegetables by Ghanaian farmers is of great concern due to the persistence and extreme toxicity of endosulfan to fish and aquatic invertebrates (Pérez-Ruzafa *et al.*, 2000). The presence of endosulfan in stream water also has implications for public health as rural communities depend on stream water for drinking. The levels of endosulfan residues obtained in the present study are comparable to those obtained in a previous study by Ntow (2001) (see Table 5).

Pesticide components	Stream water (n = 50)		Sediment (n = 42)	
	Mean ± SD (μgL⁻¹)	I.R (%)	Mean ± SD (μgKg⁻¹ dw)	I.R (%)
α- Endosulfan	0.062 ± 0.007	64	0.19 ± 0.02	90.0
β-Endosulfan	0.031 ± 0.011	60	0.13 ± 0.01	97.5
Endosulfan-sulphate	0.031 ± 0.012	78	0.23 ± 0.01	98.3
Dieldrin	-	-	-	35.0
p,p'-DDE	ND	-	0.46 ± 0.24	25.0
p,p'-DDT	-	-	-	-
Chlorpyrifos	-	-	-	68.3

Table 5. Concentrations (Mean ± SD) and incidence ratios of pesticide residues in stream water and stream-bed sediment. Source: Ntow (2001)

3.4 Pesticide residue levels in stream-bed sediment

The mean concentrations and incidence of occurrence of pesticides detected in stream-bed sediments are summarized in Table 3. Several pesticide residues were detected in the stream-bed sediment samples analyzed compared with stream water samples analyzed. α-endosulfan, β-endosulfan, endosulfan sulfate occurred in at least 90% of all the sediment samples analyzed while dieldrin, p,p'-DDE and chlorpyrifos occurred in 25%, 35% and 68% of all the sediment samples analyzed, respectively. Chlorpyrifos is an organophosphate pesticide while all the other pesticide residues detected belong to the organochlorine group of pesticides.

DDT is well-known to persist in the environment, even in tropical environments (Kidd *et al.* 2001). Although p,p'-DDT was not detected in stream water and sediment, its metabolite p,p'-DDE was detected in sediment at an average concentration of 3.77 ± 1.90 μgKg^{-1} dry wt. DDE is more persistent in the environment than DDT. Thus, when the use of DDT in a country ceases, its levels are expected to decrease more rapidly while the levels of DDE increases, thereby producing an increasing DDE/DDT ratio. The DDE/DDT ratio is often used as an indicator of recent DDT inputs into the environment; a ratio < 1 indicates recent input (Ballschmiter & Wittlinger, 1991). The absence of DDT and presence of DDE in sediment could imply the disuse of the parent compound, DDT in Ghana. The relatively high levels of p,p'-DDE detected in the present study is a justification of the ban of DDT from agricultural use in Ghana. The non-detection of DDT could also confirm the efficacy of the ban on the agricultural use of DDT in Ghana.

Dieldrin and chlorpyrifos were also detected in sediment with mean concentrations of 0.16 ± 0.04 and 1.23 ± 0.40 μgKg^{-1} dw, respectively, although they were not detected in stream water samples. Apart from its usage, dieldrin can occur in the environment as a result of the degradation of a related pesticide, aldrin. Aldrin and dieldrin are persistent in the environment and they have been banned from agricultural use in Ghana (Ntow & Botwe, 2011). The occurrence of dieldrin in sediment could therefore be due to previous use of dieldrin and/or aldrin. Chlorpyrifos recorded the highest incidence of occurrence (68%) in sediment samples. Chlorpyrifos is a broad-spectrum organophosphorus pesticide. Chlorpyrifos, under the trade name Dursban 4E, is a registered insecticide in Ghana for the control of scale borers in cereals, vegetables and ornamentals, and for public health purposes. The occurrence of chlorpyrifos in sediment could be as a result of their current use in vegetable plantations. Residues of chlorpyrifos have also been measured in vegetables from the Ashanti Region (Amoah *et al.*, 2006; Darko & Akoto, 2008).

Generally, the detected pesticides accumulated in sediment to several times their ambient water concentrations (Fig. 2). Thus, sediment is a better indicator of pesticide pollution than the overlying water. For example, endosulfan (α-endosulfan + β-endosulfan + endosulfan sulfate) accumulated to over 15 times its ambient water concentration. p,p'-DDE was also not detected in stream water although it occurred in relatively high concentrations in sediment (3.77 ± 1.90 μgKg^{-1}). There was also a significant correlation ($r^2 > 0.6$) between levels of pesticide residues and organic carbon content of sediment. This agrees well with the finding that sediment organic matter is the preferential site for the sorption of hydrophobic pollutants (Pignatello, 1998), which includes organochlorine pesticides.

The relatively higher levels of pesticides in sediment than the overlying water can be explained by the fact that pesticides are sequestered by sediments in aquatic systems

(Reinert *et al.*, 2002), which leads to the accumulation of pesticide residues in sediment over a period of time. The distribution of pesticide residues in water and sediment could be related to their physicochemical properties such as water solubility, soil adsorption coefficient and persistence as shown in Table 4. For example, endosulfan and DDE (organochlorine pesticides) have low water-solubility, high soil adsorption coefficients (Koc) and high persistence in soil, with half-lives between 50 days and 15 years (EXTOXNET, 1996). They are therefore expected to exhibit low degradation in sediment and so were frequently detected in sediments than water. These characteristics imply there could be a direct contribution to the streams from erosion of soil contaminated with these compounds (Munn & Gruber, 1997). The accumulation of chlorpyrifos in stream-bed sediment is in accordance with its high soil adsorption coefficient, K_{oc}, of 6,070 mLg^{-1} and its half-life of 35 to 78 d in the water-sediment system.

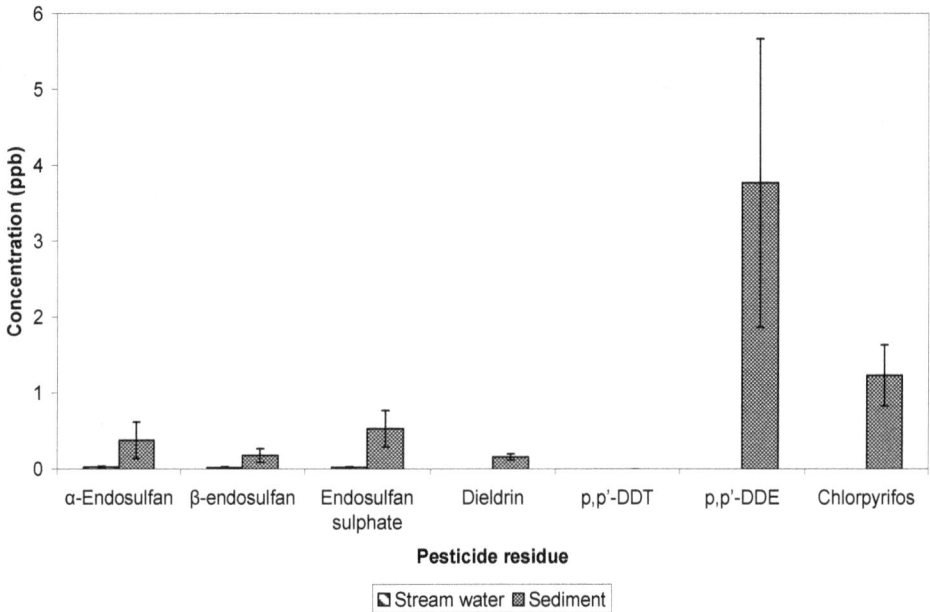

Fig. 2. Pesticide residue concentrations in stream water and underlying sediment

There were also differences in the distribution patterns of endosulfan, dieldrin, DDE and chlorpyrifos in sediment which could be related to the differences in their physicochemical properties. DDE, with the highest Koc (100,000 mLg^{-1}), recorded the highest levels in sediment, followed by chlorpyrifos (Koc = 6,070 mLg^{-1}) while endosulfan with the lowest Koc (2,400 mLg^{-1}) recorded the least concentration. The reverse trend was observed for stream water. The mean level of total endosulfan (α-endosulfan + β-endosulfan + endosulfan sulfate) in sediment from the present study (1.09 ± 0.57 µgKg^{-1} dw) was not significantly higher ($p > 0.05$) than that obtained from the previous study (0.54 ± 0.04 µgKg^{-1}

dw) by Ntow (2001). However, the mean sediment DDE level obtained from the present study (3.77 ± 1.90 µgKg^{-1} dw) was significantly higher ($p < 0.05$) than that obtained from the previous study by Ntow (2001), possibly due to the accumulation of the residue in the environment over time.

3.5 Ecotoxicological significance of measured pesticides in stream water

To evaluate the ecotoxicological significance of pesticides contamination in streams, acute (ARR) and chronic (CRR) risk ratios were calculated for the water samples by comparing the pesticide concentrations in the samples with their toxicity values (Table 1).

The calculated risk ratios for acute toxicity are shown in Fig. 3. It was found that none of the detected pesticides had an acute risk ratio greater than 1. Using the quantification limit of 0.01 µgL^{-1}, chlorpyrifos had a value of 1 in the streams. This means that when chlorpyrifos is detected in water, its concentration is already equal to its acute risk criteria. Thus, any occurrence of chlorpyrifos in water could pose a risk of acute toxicity to fish and crustaceans, and especially, species such as cladocerans, which have been observed to be highly sensitive to chlorpyrifos (Brock *et al.*, 1992; van Wijngaarden *et al.*, 2005). According to the fringing communities, fish is scarce in the streams within the catchments although fingerlings and other aquatic organisms such as frogs and crabs are present. Considering that the maximum concentrations found in this study are not the highest possible concentrations that can occur, compounds with a factor > 0.1 could pose a moderate risk of acute toxicity. Also, for many compounds, there is not a large data set of toxicity values for aquatic organisms of different trophic levels. Furthermore, the great majority of compounds have not been tested with tropical organisms.

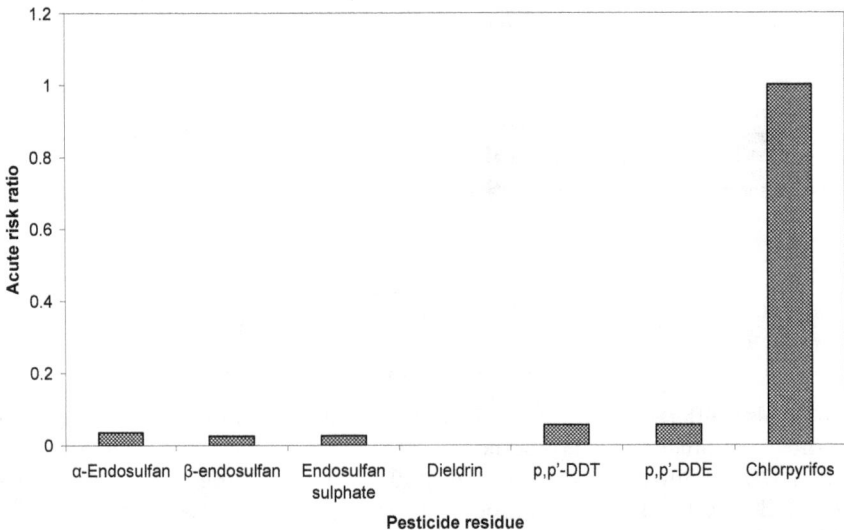

Fig. 3. Acute risk ratios for detected pesticides in stream water

The calculated risk ratios for chronic toxicity are shown in Fig. 4. It was found that DDT and DDE exceeded their chronic risk criteria in the streams. Although DDT and DDE were not detected in water at the quantification limit of 0.01 µgL^{-1} in this study, the toxicity factor was considered relevant to estimate since the quantification limits for these pesticides were generally above their respective water quality criteria. For example, the water quality criterion for DDT and DDE is 0.001 µgL^{-1} (Table 5) and the quantification limit was 0.01 µgL^{-1}. This means that when DDT and DDE are detected, they have already exceeded their water quality criteria many times (see Figs. 3 & 4). The quantification limit was therefore used to calculate the toxicity factors. Thus, any occurrence of DDT and/or DDE in the streams is significant.

Endosulfan, dieldrin, DDE and chlorpyrifos are among the pesticides that are very toxic to fish and many aquatic invertebrate species. There were no records of fish or amphibian kills in the streams at the time of the study. However, simultaneous exposure to multiple contaminants is known to produce an additive, and sometimes even synergistic and complex effects in organisms which can affect the abundance and diversity of non-target species and alter trophic interactions (Rovedatti et al., 2001). Sediment is an important reservoir of contaminants, acting as both an ultimate sink and potential source via a series of biogeochemical processes (Guo et al., 2009). Pesticide contamination of sediments may thus lead to exposure of sediment-dwelling organisms to repeated pulses or fluctuating concentrations of pesticides (Reinert et al., 2002). There is therefore the need to assess the impact of water and sediment contamination on species abundance and diversity in these aquatic systems.

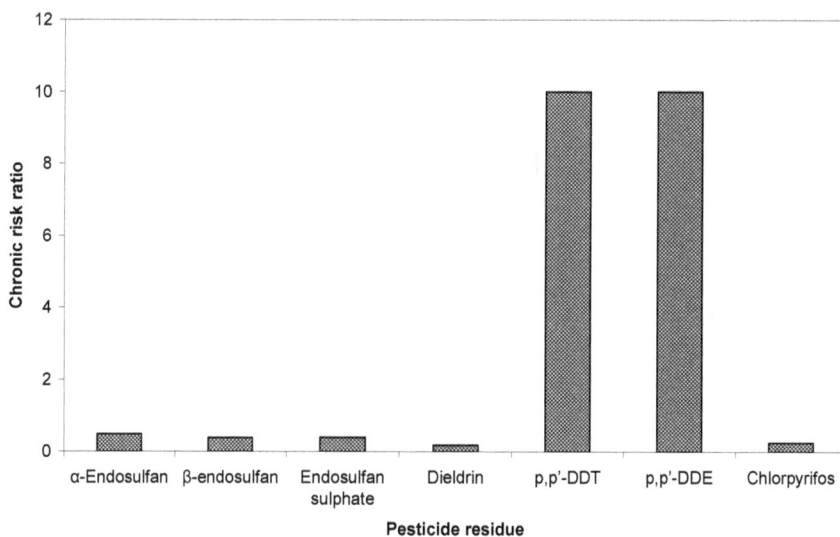

Fig. 4. Chronic risk ratios for detected pesticides in stream water

4. Conclusion

The results of this study have provided an insight into the levels of pesticide residue contamination in streams flowing in and around vegetable plantations in the Ofinso District of Ghana. Among the pesticides detected, endosulfan was the compound with the highest incidence of occurrence in both water and sediment, which is also the most frequently used pesticide in the study area. Sediment samples exhibited greater number and higher concentrations of pesticides residues than stream water samples. Although acute and chronic risk ratios indicated that the concentrations of the detected pesticide residues in streams did not surpass aquatic quality criteria, the presence of endosulfan in stream water has implications for public health. The use of endosulfan in agriculture should continue to be carefully monitored given its persistence, bioaccumulation, and continued release into streams. An extension of both the study areas and range of pesticides residues analyzed should be considered in future work.

5. Acknowledgement

The authors wish to express their gratitude to the Royal Netherlands Academy of Arts and Sciences (KNAW) for financial support. The Kinneret Limnological Laboratory, Migdal, Israel, is acknowledged for technical assistance in the use of GC/MS.

6. References

Amoah, P.; Drechsel, P.; Abaidoo, R. C. & Ntow, W. J. (2006). Pesticide and pathogen contamination of vegetables in Ghana's urban markets. *Archives of Environmental Contamination and Toxicology*, Vol. 50, pp. 1–6.

Ballschmiter, B. & Wittlinger, R. (1991). Interhemispheric exchange of hexachlorocyclo-hexanes, hexachlorobenzene, polychlorobiphenyls, and 1,1,1-trichloro-2,2-bis(*p*-chlorophenyl) ethane in the lower troposphere. *Environment, Science &Technology*, Vol. 25, pp. 1103-1111.

Brock, T. C. M.; Crum, S. J. H.; van Wijngaarden, R.; Budde, B. J.; Tijink, J.; Zuppelli, A. & Leeuwangh, P. (1992). Fate and effects of the insecticide Dursbans 4E in indoor Elodea-dominated and macrophyte-free freshwater model ecosystems: I. Fate and primary effects of the active ingredient chlorpyrifos. *Archives of Environmental Contamination and Toxicology*, Vol. 23, pp. 69–84.

Castillo, L. E.; Ruepert, C. & Solis, E. (2000). Pesticide residues in the aquatic environment of banana plantation areas in the North Atlantic Zone of Costa Rica. *Environmental Toxicology and Chemistry*, Vol. 19, No. 8, pp. 1942-1950.

Colborn, T. & Smolen, M. J. (1996). Epidemiological analysis of persistent oranochlorine contaminants in cetaceans. A review. *Environmental Contamination and Toxicology*, Vol. 146, pp. 92–172.

Darko, G. & Akoto, O. (2008). Dietary intake of organophosphorus pesticide residues through vegetables from Kumasi, Ghana. *Food and Chemical Toxicology*, Vol. 46, pp. 3703–3706

DeLorenzo, M. E.; Scott, G. I. & Ross, P. E. (2001). Toxicity of pesticides to aquatic microorganisms: A review. *Environmental Toxicology and Chemistry*, Vol. 20, No. 1, pp. 84-98.

Dinham, B. (2003). Growing vegetables in developing countries for local urban populations and export markets: Problems confronting small-scale producers. *Pest Management Science*, Vol. 59, pp. 575–582.

EXTOXNET (Extension Toxicology Network) (1996). Pesticides Information Profiles. 17.11.2008, Available from http://extoxnet.orst.edu/pips/ghindex.html.

García de Llasera, M. P. & Bernal-González, M. (2001). Presence of carbamate pesticides in environmental waters from the Northwest of Mexico: Determination by Liquid Chromatography. *Water Research*, Vol. 35, pp. 1933-1940.

Guo, Y.; Yu, H-Y. & Zeng, E. Y. (2009) Occurrence, source diagnosis, and biological effect assessment of DDT and its metabolites in various environmental compartments of the Pearl River Delta, South China: A review. *Environmental Pollution*, Vol. 157,pp. 1753–1763

Hoffman, R. S.; Capel, P. D. & Larson, S. J. (2000). Comparison of pesticides in eight U.S. Urban streams. *Environmental Toxicology and Chemistry*, Vol. 19, No. 9, pp. 2249-2258.

Kidd, K. A.; Bootsma, H. A. & Hesslein, R. H. (2001). Biomagnification of DDT through the benthic and pelagic food webs of Lake Malawi, East Africa: Importance of trophic level and carbon source. *Environmental Science and Technology*, Vol. 35, pp. 14-20.

Maule, A. G.; Gannam, A. L. & Davis, J. W. (2007). Chemical contaminants in fish feeds used in federal salmonid hatcheries in the USA. *Chemosphere*, Vol. 67, pp. 1308–1315.

Mwamburi, J. (2003). Variations in trace elements in bottom sediments of major rivers in Lake Victoria's basin, Kenya. *Lakes & Reservoirs: Research and Management*, Vol. 8, pp. 5–13.

Munn, M. D. & Gruber, S. J. (1997). The relationship between land use and organochlorine compounds in streambed sediment and fish in the Central Columbia Plateau, Washington and Idaho, USA. *Environmental Toxicology and Chemistry*, Vol. 16, pp. 1877-1887.

Ntow, W. J. & Botwe, B. O. (2011). Contamination status of organochlorine pesticides in Ghana. In: *Global Contamination Trends of Persistent Organic Chemicals*, Eds. Loganathan, B. G. & Lam, P. K. S., pp. 393-411, CRC Press, Boca Raton, FL., USA.

Ntow, W. J.; Drechsel, P.; Botwe, B. O.; Kelderman, P & Gijzen, H. J. (2008). The impact of agricultural runoff on the quality of two streams in vegetable farm areas in Ghana. *Journal of Environmental Quality*, Vol. 37, pp. 696–703.

Ntow, W. J.; Gijzen, H. J.; Kelderman, P. & Drechsel, P. (2006). Farmer perceptions and pesticide use practices in vegetable production in Ghana. *Pest Management Science*, Vol. 64, No. 4, pp. 356-365.

Ntow, W. J. (2005). Pesticide residues in Volta Lake, Ghana. *Lakes & Reservoirs: Research and Management*, Vol. 10, No. 4, pp. 243-248.

Ntow, W. J. (2001). Organochlorine pesticides in water, sediment, crops and human fluids in a farming community in Ghana. *Archives of Environmental Contamination and Toxicology*, Vol. 40, pp. 557-563.

Osafo, S. & Frimpong, E. (1998). Lindane and endosulfan residues in water and fish in the Ashanti region of Ghana. *Journal of the Ghana Science Association*, Vol. 1, pp. 135-140.

Pérez-Ruzafa, A.; Navarro, S.; Barba, A.; Marcos, C.; Cámara, A.; Salas, F. & Gutiérrez, J. M. (2000). Presence of pesticides throughout trophic compartments of the food web in the Mar Menor lagoon (SE Spain). *Marine Pollution Bulletin*, Vol. 40, pp. 140-151.

Pignatello, J. J. (1998). Soil organic matter as a nanoporous sorbent of organic pollutants. Advances in Colloid Interface Science, Vol. 76-77, pp. 445-467.

Reinert, K. H.; Giddings, J. M. & Judd, L. (2002). Effects of analysis of time varying or repeated exposures in aquatic ecological risk assessment of organochemicals. *Environmental Toxicology & Chemistry*, Vol. 21, pp. 1977-1992.

Robinson, A. Y. (1991). Sustainable Agriculture: The Wildlife Connection. *American Journal of Alternative Agriculture*, Vol. 6, No. 4, pp. 161-167.

Rovedatti, M. G.; Castañé, P. M.; Topalián, M. L. & Salibián, A. (2001). Monitoring of organochlorine and organophosphorus pesticides in the water of the Reconquista River (Buenos Aires, Argentina). *Water Research*, Vol. 35, pp. 3457-3461.

Skinner, J. A.; Lewis, K. A.; Bardon, K. S.; Tucker, P.; Catt, J. A. & Chambers, B. J. (1997). An overview of the environmental impact of agriculture in the U.K. *Journal of Environmental Management*, Vol. 50, pp. 111–128

United States Environmental Protection Agency (USEPA). (1999). National recommended water quality criteria. USEPA 822-Z-99-001. Washington, DC.

van Wijngaarden, R. P. A.; Brock, T. C. M. & Douglas, M. T. (2005). Effects of chlorpyrifos in freshwater model ecosystems: the influence of experimental conditions on ecotoxicological thresholds. *Pest Management Science*, Vol. 61, pp. 923–935.

Wan, M. T.; Kuo, J.; Buday, C.; Schroeder, G.; Van Aggelen, G. & Pasternak, J. (2005). Toxicity of α-, β-, (α + β)-endosulfan and their formulated and degradation products to *Daphnia magna*, *Hyalella azteca*, *Oncorhynchus mykiss*, *Oncorhynchus kisutch*, and biological implications in streams. *Environmental Toxicology & Chemistry*, Vol. 24, pp. 1146–1154.

Part 3

Soil Fertility and Irrigation

6

Forest Preservation, Flooding and Soil Fertility: Evidence from Madagascar

Bart Minten[1] and Claude Randrianarisoa[2]
[1] *International Food Policy Research Institute, Addis Ababa,*
[2]*United States Agency for International Development (USAID),*
[1]*Ethiopia*
[2]*Madagascar*

1. Introduction

In several developing countries, forest preservation programs have been put in place with an economic justification based on the local ecological services that they provide (Pagiola et al., 2002). It is argued that the presence of forests preserve the hydrological balance; reduce soil erosion due to increased soil stability; reduce flooding and regulate flows (Perrot-Maitre and Davis, 2001; Johnson et al., 2002; Pattanayak and Kramer, 2001a,b). However, other authors dispute the domestic benefits of forests and state that natural scientists often overvalue forests (Chomitz and Kumari, 1998; Aylward and Echeverria, 2001; Calder, 1999). Assuming that an externality costs of deforestation exists, policy makers have started to look at how to correct for this and how a workable system can be put in place to pay for ecological services locally. Increasing attention is going towards the direct payment for environmental services (Ferraro and Simpson, 2002; Durbin, 2002; Pagiola et al., 2002).

We look at this issue in a case study in Madagascar. Multiple studies have shown the high and accelerating deforestation rate in Madagascar (McConnell, 2002). Causes of deforestation are multiple and have been linked to poverty (Zeller et al., 2000), conversion of forest land to pastures (McConnel, 2002), use of wood for charcoal (Casse et al., 2004), wood exports or household fuel consumption (Minten and Moser, 2003), slash-and burn agriculture (Barrett, 1999; Keck et al., 1994; FOFIFA, 2001; Casse et al., 2004; Terretany, 1997), rural insecurity (Minten and Moser, 2003), and land tenure problems (Freudenberger, 1999). While deforestation threatens the unique eco-system of Madagascar, it has also been linked to higher incidences of flooding and greater soil erosion and damages therefore the agricultural resource base domestically (Freudenberger, 1999, Kramer et al., 1997). Overall, it is estimated that the damage of soil erosion in Madagascar is high (Kramer et al., 1997; World Bank, 2005) although the numbers that have been suggested might have been exaggerated (see f.ex., Larson, 1994).

In this analysis, we study the potential domestic benefits of forests on lowland agriculture. While we do not try to establish explicit linkages between deforestation and sedimentation off-site, we do look at the effects of flooding and sedimentation downstream as perceived by rice farmers. The analysis is based on a small-scale survey in Northern Madagascar where we try to monetize the cost to farmers of flooding and sedimentation on their rice fields downstream.[1] If the link between forest cover and flooding would exist in this area and if the link is strong, a positive willingness to pay might then justify investments in conservation measures upstream.

We contribute to the literature in two ways. First, we show that an important percentage of rice farmers benefit from flooding and sedimentation (as shown in higher land values after sedimentation and refusal for contribution towards conservation) and that current economic returns to investment in forest preservation, largely beneficial because of averted rice productivity declines, might thus be overestimated.[2] Second, in the rural scarcely monetized settings of developing countries where land transactions are rare, we develop an alternative to the hedonic price analysis of land values using willingness-to-accept scenario's explicitly allowing for uncertainty.

The structure of the paper is as follows. First, we discuss the conceptual framework. Second, the methodology, data sources and the structure of the survey are presented. Third, we look at descriptive statistics describing households as well as sedimentation and flooding incidence. Fourth, the determinants of land values, incorporating the impact of sedimentation, and the results of a willingness to pay question to avoid flooding and sedimentation are discussed. We finish with the conclusions.

2. Conceptual framework

Assume an expenditure minimization problem where expenditures are minimized subject to the constraint that utility equal or exceed some stated level, U^0. The solution to this minimization problem is the restricted expenditure function

$$e = e(p^0, T^0, U^0, \varepsilon^0)$$

where p^0 can be thought of as a vector of prices, T^0 is land availability to the household and ε^0 represents uncertain factors not reflected in p^0, T^0 and U^0.

In a first offer, the household is asked to sell land for a total payment of P^1. In a second offer, the household is asked to pay for conservation for a total payment of P^2. The change from T^0 to T^i in either of the two scenarios will result in a new expenditure function with a new set of prices and environmental and resource flows, i.e. $e = e(p^1, T^1, U^0, \varepsilon^1)$ in the first scenario and $e = e(p^2, T^2, U^0, \varepsilon^2)$ in the second scenario. It seems reasonable if you take away land or income, and given imperfect markets, that the shadow prices and wages are likely to change, i.e. we do not assume the price vector to be independent in the two scenarios.

In such a set-up, the welfare change - the Hicksian compensating surplus - is defined as the difference between the two expenditure functions,

$$e(p^i, T^i, U^0, \varepsilon^i) - e(p^0, T^0, U^0, \varepsilon^0)$$

where i is 1 (scenario 1) or 2 (scenario 2). The value of the welfare change is established by using contingent valuation measures and the Willingness to Accept/Pay (W) at the farm household level might be represented by W_j for household j

$$W_j = e(p^i, T_j^i, U_j^0, \varepsilon_j^i, X_j) - e(p^0, T_j^0, U_j^0, \varepsilon_j^0, X_j) + \eta_j$$

---

[1] While there is some rice cultivation on upland, the majority happens in the lowlands.

[2] For example, the World Bank (2005) estimates in its economic calculations that most of the benefits of the national environmental program (EP3) are obtained from avoiding productivity losses on rice fields.

where X_j is a vector of socio-economic characteristics for household j and η_j is an error term. Such a model can be further refined to allow for dynamic behavior (Holden and Shiferaw, 2002). If we let W_2 represent the subjective present value of future land productivity gains by switching from no interventions to conservation efforts in the uplands, the following equation holds in the case of the maximization of an expected intertemporal utility function:

$$U_j^0 (C_j^0) - U_j^0 (C_j^0 - W_j) = \sum_{t=1}^{\infty} (1+\delta_j)^{-t} EU_j^t(C_{1j}^t - C_{0j}^t)$$

Where δ_j is individual j time preference, EU_j^t is the expected utility for individual j in time t, and $U_j^t(C_{1j}^t - C_{0j}^t)$ is the utility gain in time t when switching from no interventions to conservation efforts in the uplands. Nonseparability in a dynamic context implies that intertemporal markets do not work well and that W would then vary over time with household discount rates that can be very high for poor liquidity constrained households.[3]
W can then be specified as a random variable which is a continuous function of observational variables that appear in the expenditure function such as farm, technology and socio-economic characteristics. W can thus be written as

$$W_j = Z_j\beta + \mu_j$$

where $\mu_j \sim (0, \sigma^2)$

where Z is a vector of explanatory variables, μ_i is the error term and σ is the standard deviation.

3. Methodology and data

An agricultural household survey was organized in November 2001 in an area northwest of Maroantsetra, in the northeast of Madagascar. The area was selected on the basis of the high diversity in watershed forms and areas and the perceived clear link between upstream activities and lowland impacts. First, a census of all the watersheds was done. In total, 65 watersheds were identified. Due to logistical reasons, only 52 watersheds were sampled. In each watershed, a stratified sample of rice plots was done. Rice plots were stratified based on the distance to the main river. In each watershed, around six fields were sampled, depending on the size of the watershed. In total, data on 268 rice farmers were obtained. The questionnaire that was implemented consisted of four parts. The first part dealt with plot characteristics (including a land valuation question), the second with questions on the rice harvest of last year on that plot, and the third on the overall structure of the agricultural firm. The final part described a willingness to pay scenario where households were asked to value their desire to avoid flooding and sedimentation in their rice fields.
Instead of the widely used and recommended dichotomous choice valuation question (Arrow et al., 1993), a stochastic payment card method (Wang and Whittington, 2005) was implemented for different reasons: (1) Given logistical constraints, a relatively small sample had to be relied upon. The payment card format gives the benefit of having extra information beyond the yes/no question (For papers that discuss the benefits of information

---

[3] Dasgupta (1993) has demonstrated this theoretically and Pender (1996) and Holden, Shifraw and Wik (1998) provide good empirical evidence on this.

beyond dichomotous choices, see Blamey et al. (1999) and Ready et al. (2001)). (2) Whittington (1998) and Wang and Whittington (2005) show that a main problem in contingent valuation studies is that the range that is offered is often not large enough to allow for a robust estimation of the valuation function. Moreover, as we had little a priori knowledge about the valuation function, we had to make sure that extreme levels were included in the bids on the payment card. Given the small sample, this could not have been achieved in the dichotomous choice variable format. (3) Uncertainty (for example on the future price evolution of agricultural products) and imperfect information (household chief had to answer immediately during the interview and could not consult with family members and/or village leaders) is allowed for in this format. Wang (1997), Wang and Whittington (2005) and Alberini et al. (2003) show the benefits of the explicit modeling of uncertain responses in contingent valuation data.

Two valuation questions were asked. The valuation questions were set up in such a way to reduce as much as possible the problem with starting point bias and with yea-saying: therefore, it started with an open-ended question (no starting point bais) followed by a payment card (additional information). In the case of the land valuation, a willingness to accept scenario was described where a certain monetary payment was given in exchange for the plot studied. As previous surveys in Madagascar had shown the reluctance of farmers to give a sales price for land - they would often report they would be unwilling to sell the plot whatever happened - it was made clear from the beginning that this was a hypothetical situation where we like to know their approximate financial value of the plot in their farming enterprise. The respondent was presented with a payment card in local currency but with references to values of local rice units, bikes, and value of livestock. On this payment card, the enumerator proceeded to fill in for every amount that was mentioned a code corresponding to 1. Accept to pay for sure; 2. A little bit in doubt but would say yes; 3. Not yes or no, do not know; 4. A little bit in doubt but would say no; 5. Will not pay for sure.

In the case of the question on willingness to pay for reduction in flooding and sedimentation, the valuation scenario was constructed as follows. Respondents were first asked if they thought if flooding and sediments had a negative, neutral or positive influence on rice productivity, in general and on the specific plot that was studied. A scenario was then described in the following way:

" *Suppose that we leave the situation as it is and we leave damage as it is without any intervention to limit deposits on this rice field or to reduce the frequency of flooding on this field. In a second situation, actions will be undertaken in the watershed upstream of your fields. In this case, you will not suffer anymore from problems of flooding and sediments. However, you know that these actions will cost money. We would like to know how much you would be willing to pay for these actions, taking into account your possibilities. If you do not pay as much than what you would be really able to pay, actions will not be sufficient to reduce flooding and sedimentation. On the other hand, if you give a level that is higher than you can afford, functional interventions can not be agreed upon. How much would you be willing to pay? x sobika of rice?"*

The question was formulated in local units of rice as this measure was easily recognizable by farmers. To finish the valuation section, a question was asked to the farmers on where they would get the rice from for the amount that they were willing to contribute. It was hoped that this would remind them of their budget constraint. Corrections on the payment card were allowed for afterwards.

While non-responses were not a problem in the plot valuation question, about one third of the respondents did not answer the willingness to pay question to avoid flooding or sedimentation. The characteristics of the respondents that refused to answer are not randomly distributed and might therefore cause inconsistency and inefficiency in the estimation of the coefficients in the regression of the willingness to pay question. A common method to control for non-responses to the willingness to pay question is to estimate a sample selection model (Messonier et al., 2000; Mekkonen, 2001), usually referred to as the Heckman two stage approach (Heckman, 1997). In this case, we estimate:

$$Y^* = \beta'X + \varepsilon$$

$$Y = 0 \text{ if } Y^* \leq 0$$

$$\text{and } Y = Y^* \text{ otherwise}$$

$$Z = \alpha'V + \mu$$

$$Z = 1 \text{ if } Z^* > 0$$

$$\text{and } Z = 0 \text{ if } Z^* \leq 0$$

where Y is willingness to pay (censored at 0); X is a vector of explanatory exogenous variables that explain Y; Z is 1 when there is a valid response and 0 otherwise; V is vector of explanatory exogenous that influence the probability of giving a valid response; α and β are parameters to be estimated; ε and μ are disturbances; Y^* and Z^* are latent variables.

4. Descriptive statistics

The Maroantsetra area in the Northeast of Madagascar is a humid area characterized by two types of agriculture: slash-and-burn cultivation ("tavy") on the hillsides and lowland rice cultivation. The area is isolated from the rest of Madagascar and is highly dependent on agriculture for income. The region is also still highly forested and is one of the largely untouched areas in Madagascar. Table 1 shows the basic descriptive statistics of the households in the survey. The head of households have a low average level of education, i.e. only three years. 10% of the households are female headed and these are mostly poorer households (Razafindravonona et al., 2000). The average size of the household is six members. Almost all the households are natives from the region and all the households report to depend on agriculture for their livelihood.

An average household in the sample possesses 62 ares[4] of lowland and 73 ares of upland. As in most of Madagascar, the main staple is rice. The average production is just below 1 ton which is estimated to be sufficient for subsistence by almost 70% of the population. However, most households - even some that declare to be self-sufficient in rice - reduce overall consumption during the lean period. The average length of this lean period is estimated to be three months. A household possesses on average 2 zebus. Total annual monetary household income is estimated at 2.7 M Fmg[5], i.e. around 415$US, i.e. low but

---

[4] 1 are = 0.01 ha

[5] Malagasy Franc; 1 USD$ = 6500 Fmg at the time of the survey

94

consistent with the high poverty levels and the low GNP of Madagascar (Razafindravonona et al., 2001).

Tables 2 presents the descriptive statistics of the rice plots that will be analyzed in more detail later on. The average plot size is small, 2.1 ares, with a range between 1 and 25 ares. Most of the plots are reported to be irrigated through a dam (96%). When asked about production problems in the last agricultural year, 28% of the farmers complained of droughts, 21% of sedimentation problems, and 14% of floods. Average yields during the previous agricultural year were estimated at 3.3 ton per hectare, high compared to the rest of the country but consistent with the excellent country-wide production conditions in 2001.[6]

variable	Unit	N	mean	median	min	max
size of household	number of people	268	5,65	5	1	14
education level head of hh	years	268	3,13	3	0	12
age	years	268	45,55	44	15	81
gender	man=1	268	0,90	1	0	1
native of region	yes=1	268	0,99	1	0	1
lowland	ares	268	61,87	50	0	340
upland	ares	268	73,40	50	0	1000
forest savoka	ares	268	33,09	0	0	500
primary forest	ares	268	30,06	0	0	600
zebus	number	268	1,75	0	0	18
total production of rice	kg	268	913,46	720	60	4500
total income	1000 Fmg	268	2695,57	1635	0	30100
rice production is enough	yes=1	268	0,27	0	0	1
length of lean period	number of months	268	2,81	3	0	10
potential access to credit	1000 Fmg	268	706,03	100	0	25000

Table 1. Descriptive statistics of household variables

Two major cyclones hit the area in the last five years: Huddah in 2000 and Gloria in 1997. The majority of the farmers state that production of plots was not affected by these events. Even when plots were affected, the perceived impact was reported to be small. Only 12% and 3% of the farmers declare that these cyclones had an impact on their rice yields in 2000 and 1997 respectively. Of these farmers, only 3% and 1% state that the impact on rice yield had been very high. Hence, it seems that the direct overall impact of these cyclones has been very small. This might be because the cyclones normally hit outside the regular growing period in Maroantsetra.[7]

---

[6] However, few farmers use modern inputs yet.
[7] The reported median harvest month is around November – in contrast to the rest of the country where main harvest are in April/May - while cyclones often hit in the beginning of the year.

variable	Unit	N	mean	median	min	max
Parcel characteristics						
area	ares	268	2,16	1,2	0,1	25
distance from house	minutes	268	15,40	10	1	90
isolated parcel	yes=1	268	0,04	0	0	1
parcel along river	yes=1	268	0,14	0	0	1
traditional perimetre	yes=1	268	0,82	1	0	1
parcel far from river	yes=1	268	0,57	1	0	1
parcel in terras	yes=1	268	0,17	0	0	1
parcel close to river (<100m)	yes=1	268	0,15	0	0	1
parcel between 100 and 200m of river	yes=1	268	0,10	0	0	1
interior of bend of river	yes=1	268	0,04	0	0	1
exterior of bend of river	yes=1	268	0,19	0	0	1
parallel to river	yes=1	268	0,56	1	0	1
irrigated by rainfall	yes=1	268	0,04	0	0	1
irrigated by dam	yes=1	268	0,96	1	0	1
distance river parcel	meters	268	103,45	40	0,2	1200
height difference parcel river	meters	266	2,51	2	0,2	20
order in irrigation (rank)	number	268	9,51	5	1	99
soil depth	cm	267	26,34	20	3	120
Sedimentation and flooding						
no deposits	yes=1	268	0,44	0	0	1
deposits of clay	yes=1	268	0,26	0	0	1
deposits of sand	yes=1	268	0,30	0	0	1
Cyclone Huddah 2000						
length flooding	days	218	1,66	1	0	30
maximal depth of water	cm	202	116,46	100	0	600
no impact on yields	yes=1	268	0,56	1	0	1
little impact on yields	yes=1	268	0,05	0	0	1
medium impact on yields	yes=1	268	0,04	0	0	1
strong impact on yields	yes=1	268	0,03	0	0	1
Cyclone Gloria 1997						
length flooding	days	144	1,31	1	0	17
maximal depth of water	cm	122	117,37	100	0	500
no impact on yields	yes=1	268	0,41	0	0	1

variable	Unit	N	mean	median	min	max
little impact on yields	yes=1	268	0,01	0	0	1
medium impact on yields	yes=1	268	0,01	0	0	1
strong impact on yields	yes=1	268	0,01	0	0	1
This harvest						
problems with flooding	yes=1	268	0,14	0	0	1
problems with drought	yes=1	268	0,28	0	0	1
problems with deposit sand	yes=1	268	0,21	0	0	1

Table 2. Descriptive statistics parcel, flooding, and sedimentation

Runoff and erosion happen often during rare events such as cyclones and heavy, intense rainfall (Kaimowitz, 2000; Brand et al., 2002). While direct impact on productivity might be small, long-term impacts through increased sedimentation might be large. In the next section, we will evaluate the values these rice farmers attach to sedimentation and flooding. We will estimate these through well-established methods in environmental economics: (1) an indirect valuation method using the hedonic pricing methodology and (2) a direct valuation method using the contingent valuation technique.

5. Regression results

5.1. Land valuation

To evaluate to what extent farmers incorporate physical and environmental amenities in land valuation, a modified hedonic pricing analysis was done. Given that land sales are rare in the region and good land valuations are therefore more difficult to get at, a stochastic payment card method was implemented to arrive at approximate land valuations of the rice plot in the sample. The stated price at which households are willing to sell their plot for sure is used as dependent variable in the regression analysis. The results of this regression are shown in Table 3.

The results illustrate that farmers are well aware of the effect of the physical characteristics on the value of their plots. As expected, area is shown to be a significant determinant of value (see Figure 1). A doubling in area increases the value of the plot by only 0.54, i.e. significantly different from one. This result indicates that larger plots are relatively less valuable than smaller plots, controlling for physical characteristics. On first sight, this implies that there are potential profits to be made by repacking plots in smaller units.[8] While returns to scale would result in relatively higher values for larger plots, a potential explanation might be that farmers prefer different smaller plots compared to one big plot as in this way, farmers are able to diversify their risk.[9] The likelihood that small plots, that are spatially segregated, are all hit by calamities at the same time - such as flooding, drought, sedimentation problems or plant diseases - is less than for one big plot. This risk averseness, typical for poor small farmers, might be an important explanation of the concave land price relationship.

---

[8] Similar results have been found in other countries as mentioned by Lin and Evans (2000).

[9] Blarel et al. (1992) study this phenomena in depth in Ghana and Rwanda.

variables	Unit	Coefficient	t-value	P>|t|
plot characteristics				
area	log(ares)	0,503	**7,590**	**0,000**
parcel in terras	yes=1	-0,343	**-1,840**	**0,067**
parcel along river	yes=1	-0,617	-1,470	0,142
tradional perimeter	yes=1	-0,215	-0,530	0,596
interior bend of river	yes=1	-0,371	-1,600	0,111
exterior bend of river	yes=1	-0,045	-0,300	0,765
distance river parcel	log(meters)	0,027	0,700	0,488
height difference parcel river	log(meters)	0,040	0,300	0,762
soil depth	log(cm)	0,260	**2,140**	**0,034**
irrgation directly from river	yes=1	0,216	**1,700**	**0,091**
irrigated by dam	yes=1	-0,305	-1,000	0,320
clay deposit after cyclones	yes=1	0,299	**1,990**	**0,048**
sandy deposits after cyclones	yes=1	0,429	**2,980**	**0,003**
household characteristics				
education head of household	years	0,016	0,740	0,463
age of head of household	years	0,001	0,130	0,893
gender head of household	man=1	-0,237	-1,110	0,267
annual monetary income	log(Fmg)	0,041	1,330	0,185
length of lean period	months	0,017	0,690	0,492
potential access to credit	log(Fmg)	-0,010	-0,870	0,384
owned number of zebus	log(number)	0,300	**3,460**	**0,001**
owned agricultural land	log(ares)	-0,070	-0,950	0,341
intercept		12,248	**14,700**	**0,000**
Number of observations	256			
F(21, 234)	9,62			
Prob > F	0			
R-squared	0,3929			
Root MSE	0,9256			

Table 3. Hedonic price regression
(dep. var. = log (value of land); robust standard errors)

Most of the physical variables turn out not significant at the conventional statistical levels, indicating that these are not major determinants of sales prices. However, there are a few exceptions. Plots in terraces, at the top of the river, are less valuable. This might be because these plots are more likely to be affected by drought. The impact is shown to reduce the value of the plot by around 34%. The perceived cultivable soil depth is a highly important determinant of land prices. A doubling of soil depth increases the value of rice land by 26%. Agronomic evidence suggests that soil depth is crucial for root development which has been shown to be an important constraint on rice production in Madagascar.

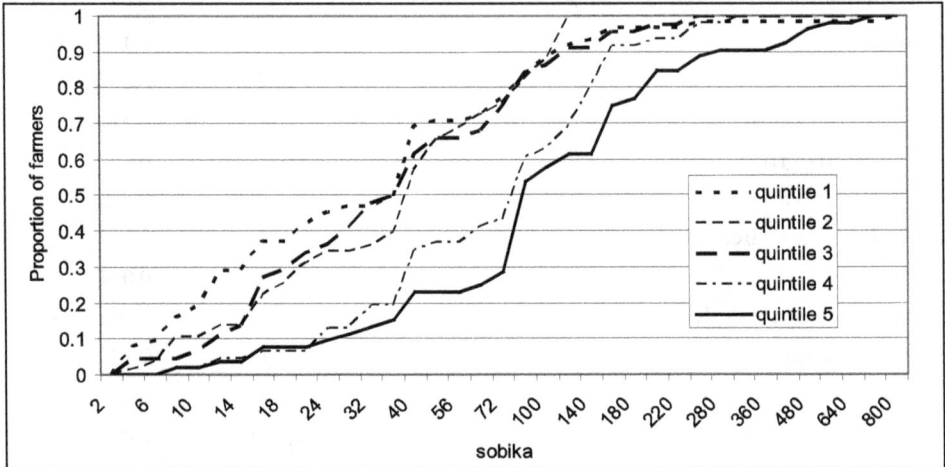

Fig. 1. Willingness-to-accept the sales price 'for sure' (by plot size quintile)

In line with de Janvry et al. (1991), we assume imperfect or missing markets where farm households are the decision makers and production and consumption decision are not separable. This implies that land prices would also depend on household characteristics and they were thus included in the regression. Few of these variables come out significant. Only the ownership of cattle leads to significant higher land values. This seems linked to the importance of ownership of cattle to access to manure, an important lasting fertility and land quality enhancing input in these environments (Minten et al., 2007; Barrett et al., 2002).

To measure the effect of sedimentation, we created dummies for clay and sand deposits during recent floods. Compared to soils without deposits during floods, these plots are estimated to be significantly more valuable. The plots affected by soil and sand deposits are estimated to be respectively 30% and 43% more valuable. The latter results might seem surprising at first sight. However, sand deposits come usually together with organic material that might significantly improve the fertility of soils. Farmers also often remove the more damaging sand from the plot. These results indicate overall that sedimentation does not reduce the value of the plot per se, ceteris paribus. We discuss this in more detail below.

5.2 Willingness-to-pay to avoid sedimentation and flooding

All sedimentation is not perceived to be bad for rice productivity. In fact, erosion and heavy rainfall might induce runoff of the good topsoil of the uplands that ends up in the lowland ricefields (Chomitz and Kumari, 1998). This seems also to be the case in the lowlands of the Maroantsetra region. When asked about the perceived effect of flooding and sedimentation on rice yields overall, 53% of the farmers reported that they thought this relation was negative (Table 4). However, 38% of the farmers thought that it was actually good for rice yields (while 9% thought its effect was neutral). In a follow-up question, it was asked what the rice farmers expected of the effect of sedimentation and flooding on the rice plot in the sample. Farmers were evenly divided on the question: 37% thought that the effect would be negative, 38% expected a positive effect and 26% reported to expect a neutral effect.

variable	Unit	N	mean
Overall effect sediment/flooding on rice yield...			
positive	yes=1	268	0,38
neutral	yes=1	268	0,09
negative	yes=1	268	0,53
Effect on studied parcel of sediments/flooding on rice yield...			
positive	yes=1	268	0,37
neutral	yes=1	268	0,26
negative	yes=1	268	0,38

Table 4. Perceived effect of sedimentation/flooding

Finally, farmers were asked what they were willing to pay to avoid flooding and sedimentation. Figure 2 illustrates, for the respondents that were willing to pay, how the willingness to pay varies for the different levels that were offered to the respondent. We see that the median willingness to pay (at 95% for sure) to avoid flooding is just over 2 sobika, the local unit for a rice basket containing 12 kgs on average per household per year. This amounts to around 4$. This implies that if a vote would be held in the region, more than 4$ would not be accepted by a majority of the population. 50% of the farmers would refuse to pay more than 4.5 sobika for sure. On average, this corresponds to 7% of their total rice production of last year.

The number of farmers that were undecided about accepting or refusing the offer is largest in the middle of the graph, as could be expected (see Wang (1997; p. 223)). For some bids, the indecision domain contains up to 15% of the farmers. This high number indicates the importance of allowing farmers to convey information beyond the simple yes/no format in contingent valuation studies as has been shown by other authors (Blamey et al., 1999; Ready et al., 2001; Alberini et al., 2003).

Regressions were run to look at the determinants of the willingness to pay to avoid flooding and sedimentation on the plot in the sample. These results serve to validate the WTP answers. A two-step approach was used. In a first step, a selection equation was run to explain the characteristics of the households that are willing to contribute to avoid flooding and sedimentation. In this step, variables are included that are potential determinants of the likelihood of the plot to be subject to flooding and sedimentation. In a second step - controlling for the characteristics of the plot and the household which explain if it is willing to contribute - economic variables are included in the regression to measure to what extent they are able to contribute, taking into account their socio-economic background. A selectivity coefficient was then included in the second-stage willingness to pay regression.

This set-up would allow us to obtain efficient and unbiased estimates in the second stage regression.

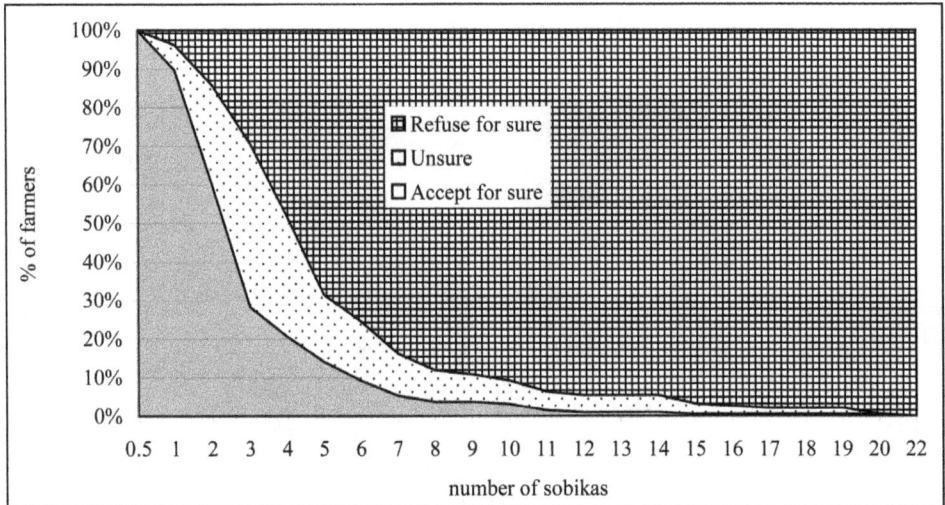

Fig. 2. Willingness to pay to avoid flooding/sedimentation

The results are largely conforming to expectations. The coefficient that measures the expected effect of sedimentation and flooding on the plot and the household perceived effect of sedimentation are significant determinants of the probability that the household is willing to contribute.[10] Households with plots on the exterior bend of the river, bigger soil depth and irrigated by dams are significantly more willing to contribute. These plots might be more exposed to risk or are more valuable. It is interesting to note that negative experiences with the last two cyclones make the household less likely to contribute. These households might believe that there is not much that can be done or, alternatively, that this type of adversity might easily be overcome.

The results on the amount that households are willing to contribute – the second stage regression - suggest that wealthier households are willing to pay more. Different measures of wealth were included. A doubling of the area of lowland in possession would increase the willingness to pay significantly by 11%. A lean period that lasts one month longer as measured by the period that they do not have sufficient rice, an indicator of poverty of the household (Barrett and Dorosh, 1998; Minten and Zeller, 2000), reduces the willingness to pay of the household by 6%. Potential access to credit increases the willingness to pay significantly. However, its coefficient is small. Overall income and the number of zebus owned by the household show the expected positive sign but are not significant at the 10% level. Household characteristics, such as level of education, gender and age of the head of household, do not influence the willingness to pay significantly.

---

[10] However, the match is not perfect. 72% of the households that expected a positive impact were not willing to pay to participate. This compares to 91% of the households that expected a neutral impact and 8% of the households that expected a negative impact.

variables	Unit	Coef.	z	P>z
dep. var.: willingness to pay in log(Fmg)				
area of the plot	log(ares)	-0,014	-0,230	0,816
education head of household	years	-0,016	-0,820	0,410
age of head of household	years	-0,006	-1,320	0,187
gender head of household	man=1	-0,108	-0,610	0,542
owned lowland	log(ares)	0,112	**1,950**	**0,051**
annual monetary income	log(Fmg)	0,036	1,090	0,276
length of lean period	months	-0,058	**-2,460**	**0,014**
potential access to credit	log(Fmg)	0,021	**1,910**	**0,057**
owned number of zebus	log(number)	0,030	0,400	0,691
intercept		0,116	0,220	0,822
selection equation				
expected effect of sedimentation on plot	1=pos; 2=neutral; 3=neg.	0,683	**4,990**	**0,000**
area of the plot	log(ares)	-0,161	-1,440	0,150
parcel along river	yes=1	-1,074	-1,500	0,134
traditional perimetre	yes=1	-0,704	-1,060	0,287
parcel in terras	yes=1	-0,442	-1,460	0,144
parcel close to river (<100m)	yes=1	-0,448	-1,500	0,134
parcel between 100 and 200m of river	yes=1	-0,030	-0,070	0,943
interior of bend of river	yes=1	-0,216	-0,400	0,689
exterior of bend of river	yes=1	0,836	**3,200**	**0,001**
irrigated by dam	yes=1	0,970	**1,960**	**0,050**
distance river parcel	meters	0,005	0,080	0,939
height difference parcel river	meters	0,113	0,810	0,420
order in irrigation (rank)	number	-0,087	-0,910	0,362
slope (distance one sees w/o obstacle)	meters	0,056	0,640	0,525
soil depth	cm	0,464	**2,740**	**0,006**
estimated age of plot	years	-0,003	-0,410	0,680
little impact on yields of cyclone 1	yes=1	-0,087	-0,180	0,859
medium impact on yields of cyclone 1	yes=1	-1,182	**-1,780**	**0,075**
strong impact on yields of cyclone 1	yes=1	-1,642	**-2,510**	**0,012**
little impact on yields of cyclone 2	yes=1	-0,333	-0,390	0,695
medium impact on yields of cyclone 2	yes=1	-0,087	-0,060	0,949
strong impact on yields of cyclone 2	yes=1	0,434	0,290	0,770

variables	Unit	Coef.	z	P>z
overall perceived impact of sedimentation	1=pos; 2=neutral; 3=neg.	0,385	**2,580**	**0,010**
intercept		-3,186	1,018	-3,130
rho		-0,744	0,119	
sigma		0,778	0,050	
lambda		-0,578	0,118	

LR test of indep. eqns. (rho = 0): chi2(1) = 13.88 Prob > chi2 = 0.0002	
Number of obs	265
Censored obs	82
Uncensored obs	183
Wald chi2(9)	21,33
Prob > chi2	0,011
Log likelihood	-296,3996

Table 5. Willingness to pay to avoid flooding on rice plot (Heckman selection model)

The coefficients on the explanatory variables show that the stated amount is consistent with economic logic. Households with access to liquidity and who perceive to suffer from flooding are willing to pay more. To further test for robustness, regressions were run without the selectivity coefficient and with the refusal to pay for sure as dependent variable. The coefficients obtained - but not reported - confirm the results discussed earlier.

We end this section with a final note on the significance of these results at the national level. There are two main differences of the surveyed farmers with the rest of the country. First, the watersheds in this area are small and sedimentation downstream can easy be linked to upstream activities. This is however not the case in the rest of Madagascar and the link between sedimentation downstream and corrective measures upstream are more difficult to make as watersheds are larger.[11] Second, the rice harvest in the Maroantsetra region is at the end of the year, i.e. before major cyclones hit the country. This might reduce the willingness to pay for a reduction of floods. In the rest of the country, the main harvest is in the beginning of the year and it might thus more directly be affected by rice losses due to floods and submersion. The run-off of good soils might then only affect the subsequent harvests.

Based on the data of the national household survey of 2001, it is found that 15% of agricultural households cultivate lowland, 21% upland and 64% both. This compares to respectively 25% and 75% of the households that we interviewed in the Maroantsetra area (because of the study subject, only rice farmers were selected). The majority of the households in Madagascar and in our dataset cultivate thus both uplands and lowlands and

---

[11] Brand et al. (2002) show how the size and the shape of watersheds are important determinants for run-off.

it seems that farmers that might cause erosion and those that suffer from it are often the same households.[12] This makes a compensation mechanism cumbersome.

Secondary sources of information further seem to indicate that siltation and erosion might be a relatively minor problem in Malagasy agriculture overall and this despite the high recent deforestation rates in Madagascar.[13] The 2001 national household survey asked farmers about the biggest constraints they faced to improved agricultural productivity. The same question was asked in the 2004 national household survey, based on a different sampling frame and with a bigger sample. Respondents had to rank options from 'not important' to 'very important'. The results are presented in Table 6 ordered in decreasing percentage of households that identified the constraint as 'quite' or 'very' important. Answers were strikingly consistent between the two surveys, three years apart and with a different sample. The most and least frequently cited constraints were common to both surveys. Access to agricultural equipment, access to cattle for traction and transport and access to labor are ranked among the top four constraints in both surveys. The clear pattern in these answers is that inputs that complement labor and boost its productivity are most limiting in farmers' opinion (Minten et al., 2007). By contrast, less than 40 percent of households identify the siltation of land as an important constraint and it is more commonly identified as not a constraint on agricultural productivity. Farmers were further asked for each plot in the national household survey of 2001 about the production problems in the year preceding the survey. Siltation was mentioned as a problem on less than 1% of the rice plots.

6. Conclusions

Flooding and sedimentation downstream are often linked to deforestation upstream. While the debate is on-going and results seem to be variable and site specific (Chomitz and Kumari, 1998; Calder, 1999), policy makers are looking for ways to solve this externality problem to ensure sustainable financing for ecological services of conservation efforts such as reforestation and soil conservation measures. Based on interviews with almost 300 rice farmers - users of land downstream - in the Northeast of Madagascar, this paper tries to shed light on the willingness to pay for ecological services for forests, in this case to avoid flooding and sedimentation.

The results of our analysis show that the rice farmers are clearly aware of the effect of sedimentation on production. Sedimentation is not perceived to be unambiguously bad for lowland productivity. Policy interventions that focus on only correcting the perceived negative relationship are therefore misguided. A hedonic pricing analysis on riceland

---

[12] Lowlands are further divided depending on the type of irrigation scheme. The World Bank (2005) estimates the total lowland area at 1.1 million hectares, representing 40% of the cultivated area. The bulk of these lowland areas, 800,000 hectares or 70 percent of the total irrigated lands, are very small in terms of average superficies (a few hectares), and are not equipped with improved irrigation infrastructure. 300,000 hectares is equipped with infrastructure meant to improve water management. Uplands can further be divided in uplands that are cultivated on a permanent basis and land that is cultivated for three to four years and is then followed by long fallow periods.

[13] It is estimated that Madagascar lost about 12 million ha of forest between 1960 and 2000, effectively reducing forest cover by 50% in just 40 years (World Bank, 2003).

values shows that farmers take sedimentation into consideration in the valuation of their rice plots but that rice plots with sedimentation are valued significantly higher, ceteris paribus.

Variables	Percentage of households that state this constraint is ... important			
	not	a bit	quite	very
Constraints to overall agricultural productivity				
EPM 2001, 2470 agricultural households				
Access to agricultural equipment	19	18	27	35
Access to land	27	19	29	25
Access to cattle for traction and transport	24	23	29	24
Access to labor	22	28	30	20
Access to credit	36	19	23	22
Degradation of irrigation infrastructure due to environmental problems	29	31	22	18
Access to agricultural inputs (e.g. fertilizer)	34	26	19	21
Access to cattle for fertilizer	42	23	19	16
Land tenure insecurity	44	26	22	8
Silting of land	46	29	18	7
EPM 2004, 3543 agricultural households				
Access to agricultural equipment	11	14	32	43
Access to irrigation	13	21	29	37
Access to cattle for traction and transport	16	20	35	29
Access to labor	17	22	37	24
Avoid droughts	20	19	27	34
Access to agricultural inputs (e.g. fertilizer)	24	20	26	30
Phyto-sanitary diseases	19	25	30	26
Avoid flooding	25	20	26	29
Access to cattle for fertilizer	28	22	25	25
Access to credit	31	23	22	24
Silting of land	33	29	23	15
Land tenure insecurity	38	24	23	15

Table 6. Farm households' reported constraints on improved agricultural productivity

The results of the survey further show that, while 10% of the farmers believe that flooding and sedimentation has no effect, a significant part of the farmers (almost 40% of the rice farmers in the sample) feels that their plots actually benefit from flooding and sedimentation. This seems related to the fact that flooding occurs outside the main harvest period and thus therefore not seem to cause any large immediate production damage. Damage depends then on the type of deposits as flooding can actually cause valuable soils and organic material to be transported to the ricefield and to be beneficial for rice productivity. The negative or positive effect of flooding seems to depend on spatial determinants, i.e. location with respect to the main river that irrigates the rice fields matters.

However, a significant part of the farmers also realize the bad effects that sedimentation can have on their rice production. Therefore, they are willing to contribute to avoid flooding and sedimentation on their fields. These farmers are willing to contribute 4$ per household per year. The magnitude of the amount that they are willing to pay corresponds to spatial as well as economic rationales. Households that are richer, not credit constrained, and that suffer less from seasonality problems are willing to pay significantly more to avoid this flooding and sedimentation damage. Given beneficial effects of sedimentation for some farmers and small willingness to pay by other farmers, our results overall thus suggest that current economic rates of return on forest preservation projects in Madagascar, largely beneficial because of across-the-board domestic agricultural benefits on lowlands, might be overestimated.[14]

7. Acknowledgement

We would like to thank Jurg Brand, Tim Healy, and Andy Keck for help with the set-up of survey, helpful discussions and comments on preliminary results. The field work for this research was financed by ONE (Office National de l'Environnement), PAGE (Projet d'Appuie à la Gestion de l'Environnement) and by the Ilo program. The two last projects were financed by USAID-Madagascar. Any remaining errors are solely the authors' responsibility.

8. References

Alberini, A., Boyle, K., Welsh, M., Analysis of Contingent Valuation Data with Multiple Bids and Response Options allowing Respondents to express Uncertainty, Journal of Environmental Economics and Management, 2003, 45: 40-62.

Arrow, K., Solow, P., Portney, P. Leamer, E.E., Radner, R., Schuman, K., Report of the NOAA Panel on contingent valuation, NOAA, 1993

Aylward, B., Echeverria, J., Synergies between livestock production and hydrological function in Arenal, Costa Rica, Environment and development economics, 6, 2001, pp. 359-381

---

[14] One caveat of the analysis is whether poor farmers' own high discount rates should be the basis for assessing the severity of this type of environmental problem or whether the perceived damage from society's perspective should be based on the much lower social discount rates. While this will change the overall end result, it can be expected, given positive and negative effects of sedimentation, that this is partly canceled out, no matter low or high discount rates. In any case, the objective of the analysis is not to do a full-blown cost-benefit analysis (in which case we would also have to value the loss of production on the uplands) but to show that some of the assumptions in the calculation of current economic rates of return might be questionable.

Barrett, C.B., Stochastic Food Prices and Slash-and-Burn Agriculture, Environment and development economics, vol. 4, no. 2, 1999, pp. 161-176

Barrett, C.B., Place, F., Aboud, A., Natural Resource Management in African Agriculture: Understanding and Improving Current Practices, CAB International, 2002

Barrett, C.B., Dorosh, P., Farmers' welfare and changing food prices: Nonparametric evidence from rice in Madagascar, American Journal of Agricultural Economics, 78, 1996, pp. 656-669

Blamey, R.K., Bennett, J.W., Morrison, M.D., Yea-Saying in Contingent Valuation Studies, Land Economics, February 1999, 75(1), pp. 126-141

Blarel, B., Hazell, P., Place, F., Quiggin, J., The economics of farm fragmentation: Evidence from Ghana and Rwanda, World Bank Economic Review, vol. 6, no. 2, may 1992, pp. 233-254

Brand, J., Minten, B., Randrianarisoa, C., Etude d'impact de la deforestation sur la riziculture irriguée, Cahier d'études et de recherches en économie et sciences sociales, No. 6, December 2002, FOFIFA, Antananarivo

Calder, I.R., The Blue Revolution: Land Use and Integrated Water Resources Management, 1999, Earthscan publications Ltd, London

Casse, T., Milhoj, A., Ranaivoson, S., Randriamanarivo, J.R., Causes of deforestation in southwestern Madagascar: What do we know?, Forest Economics and Policy, 2004, 6(1): 33-48

Chomitz, K.M., Kumari, K., The domestic benefits of tropical forests: A critical review, World Bank Research Observer, Vol. 13, No. 1, February 1998, pp. 13-35

Dasgupta, P, An Inquiry into Well-Being and Destitution, Clarendon Press, 1993, Oxford, England.de Janvry, A., Fafchamps, M., Sadoulet, E. (1991), Peasant household behavior with missing markets : Some paradoxes explained, The Economic Journal, 101: 1400-1417

Durbin, J., 'The potential of conservation contracts to contribute to biodiversity conservation in Madagascar', 2001, mimeo.

Ferraro, P.J., Simpson R.D., The cost-effectiveness of conservation payments, Land Economics, 2002, 78(3): 339-353

FOFIFA, Culture sur brûlis: vers l'application des résultats de recherche, actes de l'atelier de EPB-BEMA, 2001

Freudenberger, K., 1999, Flight to the forest: a study of community and household resource management in the commune of Ikongo, Madagascar

Heckman, J., Sample selection bias as a specification error, Econometrica, 47, 1979, pp. 153-161

Holden, S.T., Shiferaw, B., Peasants' willingness to pay for sustaining land productivity, in Barrett, C., Place, F., Aboud, A., Natural Resources Management in African Agriculture: Understanding and improving current practices, CAB international, 2002

Holden, S.T., Shiferaw, B., Wik, M., Poverty, market imperfections and time preferences: of relevance for environmental policy?, Environment and Development Economics, 1998, 3:105-130

Johnson, N., White, A., Perrot-Maître, Developing Markets for Water Services from Forests: Issues and Lessons for Innovators, Forest Trends, 2002

Kaimowitz, D., Useful Myths and Intractable Truths: The politics of the link between forests and water in Central America, CIFOR, 2000, mimeo

Keck, A., Sharma, N.P., Feder, G., Population growth, shifting cultivation, and unsustainable agricultural development: a case study from Madagascar, World Bank Discussion Paper, No. 234, Africa Technical Department Series, The World Bank, Washington DC, 1994

Kramer, R.A., Richter, D., Pattanayak, S., Sharma, N., Ecological and economic analysis of watershed protection in Madagascar, Journal of Environmental Management, 49, 1997, pp. 277-295

Larson, B.A., Changing the economics of environmental degradation in Madagascar: Lessons from the national environmental action plan process, World Development, Vol. 22, No. 5, May 1994, pp. 671-689

Lin, T., Evans, A.W., The relationship between the price of land and size of plot when plots are small, Land economics, August 2000, pp. 386-394

McConnell, W.J., Madagascar: Emerald Isle or Paradise Lost?, Environment, October 2002, vol. 44, No. 8, pp. 10-14

Mekkonen, A., Valuation of community forest in Ethiopia: A contingent valuation study of rural households, Environment and Development Economics, 2001, 5:289-308

Messonier, M.L., Bergstrom, J.C., Cornwell, C.M., Teasley, R.J., Cordell, H.K., Survey response-related biases in contingent valuation: concepts, remedies, and empirical application to valuing aquatic plant management, American Journal of Agricultural Economics, 83, May 2000, pp. 438-450

Minten, B., Moser, C., Forêts: Usages et menaces sur une ressource, in (eds) Minten, B., Randrianarisoa, J., Randrianarison, L., Agriculture, pauvreté rurale et politiques économiques à Madagascar, Programme Ilo/Cornell University, 2003, pp. 86-89

Minten, B., Zeller, M., Beyond market liberalization: welfare, income generation and environmental sustainability in rural Madagascar, Ashgate, 2000

Minten, B., Randrianarisoa, J., Barrett, C., Productivity in Malagasy rice systems: Wealth-differentiated constraints and priorities, Proceedings of the IAAE conference, 2007.

Pagiola, S., J. Bishop and N. Landell-Mills, Setting Forest Environmental Services: Market-based Mechanisms for Conservation and Development, London: Earthscan, 2002.

Pattanayak, S.H., Kramer, R.A., Worth of watersheds: a producer surplus approach for valuing drought mitigation in Eastern Indonesia, Environment and Development Economics, 6, 2001, pp. 123-146

Pattanayak, S.H., Kramer, R.A., Pricing ecological services: Willingness to pay for drought mitigation from watershed protection in eastern Indonesia, Water resources research, Vol. 37, No. 3, 2001, pp. 771-778

Pender, J.L., Discount rates and credit markets: Theory and Evidence from rural India, Journal of Development Economics, 1996, 50:257-296

Perrot-Maître, D., Davis, P., Case studies of markets and innovative financial mechanisms for water services from forests, 2001, Forest Trends, mimeo

Razafindravonona, J., Stifel, D., Paternostro, S., Changes in poverty in Madagascar: 1993-1999, Instat, Antananarivo, 2000.

Ready, R.C., Navrud, S., Dubourg, W.R., How do respondents with uncertain willingness to pay answer contingent valuation questions?, Land economics, August 2001, vol. 77, no. 3, pg 315-326

Terretany, Cahier Terretany no. 6, 1997, Un système agro-écologique dominé par le tavy: la region de Beforona, Falaise Est de Madagascar

Wang, H., 'Treatment of don't know responses in contingent valuation surveys: a random valuation model', Journal of Environmental and Economic Management, 1997, 32: 219-232.

Wang, H., Whittington, D., 2005, 'Individuals' valuation distributions using a stochastic payment care approach, Ecological Economics, 55, pp. 143-155

Whittington, D., Chapter 16: Environmental Issues, in Eds. Grosh, M., Glewwe, P., Designing Household Survey Questionnaires for Developing Countries: Lessons from Ten Years of LSMS Experience, World Bank, 1998

World Bank, Review of agricultural and environmental sector, 2003, Washington.

World Bank, Environmental Program 3, 2005

World Bank, Madagascar: The impact of public spending on irrigated perimeters productivity (1985-2004), Economic and Sector Work, 2005, Washington DC.

Zeller, M., Lapenu, C., Minten, B., Randrianaivo, D., Ralison, E., Randrianarisoa, C., Rural development in Madagascar : Quo vadis ? Towards a better understanding of the critical triangle between economic growth, poverty alleviation and environmental sustainaibility, Quarterly Journal of International Agriculture, 1999, Vol.2

7

Nutrient Mobility and Availability with Selected Irrigation and Drainage Systems for Vegetable Crops on Sandy Soils

Shinjiro Sato[1] and Kelly T. Morgan[2]
[1]Department of Environmental Engineering for Symbiosis, Soka University, Tokyo,
[2]Southwest Florida Research and Education Center, University of Florida, Florida,
[1]Japan
[2]USA

1. Introduction

A wide variety of vegetable crops is produced on varying types of soils including sandy soils where the production can be maximized as long as proper fertilization, irrigation and drainage systems are implemented. However, most sandy soils have low water- and nutrient-holding capacities, hence appropriate irrigation scheduling is critical for proper plant health as well as for minimizing water requirement. Healthy crops are better able to withstand pest and disease pressures, as well as produce a high quality commercial product. Irrigation management should be geared towards maintaining optimum moisture and nutrient concentrations within the plant root zone. If this goal is achieved, crops will take up their maximum amounts of water and nutrients with minimum wastage. Equally important, excessive irrigation will reduce water use efficiency, as well as require more water and contribute to potentially negative environmental impacts.

It is crucial to recognize how nutrients move and transform in soils after the application for improved application efficiencies and reduced environmental losses. However, different irrigation and drainage systems practiced on sandy soils for vegetable production can complicate the dynamics of mobility and availability of nutrients and water. Yet, the number of researches on this matter has not been as many as needed. Therefore, this review attempts to summarize characteristics of sandy soils for vegetable production (Section 2), clarify pros and cons of different irrigation and drainage systems practiced on sandy soils (Section 3), and elucidate the nutrient mobility and availability for vegetable production under different irrigation systems specifically on sandy soils (Section 4), in which the soil environment can greatly differ from other soil types in terms of nutrient dynamics in soil.

2. Characteristics of sandy soils for vegetable production

2.1 Types and physiochemical properties of sandy soils

Soils on which crops are grown greatly influence how irrigation water, nutrients, and other agrichemicals should be managed to maximize the production while minimizing resource use and effects on the environment. Soil properties that influence soil water management include soil texture, hydraulic conductivity, water-holding capacity, and natural drainage, which also affect soil nutrient management that differs depending on soil organic matter (OM) content, pH, cation exchange capacity (CEC), and coatings on sand grains.

Soil texture is the relative proportion of sand, silt, and clay in a mineral soil. Texture influences how much water a soil can hold against drainage by gravity and how quickly water drains away if it has an outlet. Sandy soils contain 80% or more sand in the root zone (Shirazi and Boersma, 1982). The high sand contents make irrigation water management extremely difficult because sands are dominated by large pores that have little capacity to hold water through capillarity (Kern, 1992). Therefore, if too much water is applied to the sandy soil, the excess is lost below the root zone and can induce nutrient leaching.

Soil OM includes anything that was once alive, from freshly deposited plant residues to highly decomposed humus. In their native state, sandy soils may contain as much as 5% OM under grass vegetation, and somewhat less under forest cover (Six et al., 1998). Cultivated soils usually contain less OM than native soils, typically less than 3%, due to decreased plant diversity and the use of herbicides or plastic mulches that reduce weed growth. Under well-drained conditions, soil OM is rapidly lost as carbon dioxide by oxidation in warm and humid climates, and is not replaced in large quantities by crop production because relatively low area is covered by plant materials at any given time. In sandy soils, OM is an extremely valuable component because it provides both water and nutrient-holding capacities, and its decomposition provides recycled nutrients to plants (Khaleel et al., 1981).

Soil water-holding capacity is provided by the smaller pores that exist between and within the smallest fraction of soil and OM particles (Khaleel et al., 1981). Therefore, the water-holding capacity is directly related to amounts of silt, clay, and OM present. Since sandy soils contain only minimal amounts of these components, their water-holding capacity is rarely greater than 2.5 cm per 30 cm of soil depth, and are often less than 1.9 cm per 30 cm.

2.2 Characteristics of subsurface layers

Argillic and spodic layers can be found underneath many surface sandy soils, and have considerably different physicochemical properties from the surface soils. The argillic layer is created by the deposition of clay particles and is usually mottled gray in color and sandy or sandy loam in texture. This horizon can be either acidic or alkaline with high clay content. The spodic layer is composed of OM that is leached down the profile by both physical and chemical means and deposited in the lower part of the soil profile. This distinct brown or black layer is often high in OM, aluminum, and iron, usually with a low pH, and almost always sandy in texture. Both argillic and spodic layers impede vertical water percolation and causes water to accumulate above these horizons because their permeability is low. This water accumulation is referred to as a perched water table, and is beneficial for maintaining a constant water table for subsurface irrigation for vegetable production (Muchovej et al., 2005). In addition, the water-holding capacity and CEC are typically higher in these subsurface layers than in the surface soils (Obreza & Collins, 2002).

The nutrient and irrigation managements can be different and may be complicated when these layers are excavated and mixed in as a result of the bedding process. The subsurface layers can be found relatively deep in some Alfisols and Spodosols in USA such as Holopaw (70 to 162 cm depth), Pineda (95 to 130 cm), Immokalee (90 to 137 cm), and Oldsmar series (95 to 125 cm), hence remain undisturbed following the bedding process. Other sandy soils such as Riviera (57 to 135 cm), Winder (30 to 122 cm), Pomona (52 to 65 cm), and Wabasso series (62 to 85 cm) have relatively shallow argillic or spodic layers that can be excavated during the bedding process (Obreza & Collins, 2002; Gilbert et al., 2008). As a result, these subsurface materials are sometimes mixed into the root zone affecting physicochemical properties of the surface sandy soils (Obreza & Collins, 2002).

3. Irrigation and drainage systems on sandy soils

Irrigation can be defined as the artificial application of water to the soil for assisting in growing crops and is considered one of the most important cultivation practices in dry or limited rainfall areas and during periods with no or little rainfall. An approach to conserving water is to maximize the irrigation efficiency and to minimize water loss. Irrigation efficiency is a measure of the effectiveness of an irrigation system in delivering water to a crop and/or the effectiveness of irrigation in increasing crop yields. Good irrigation practices imply good irrigation efficiency and can be achieved by maintaining a good irrigation water application uniformity and improve water uptake efficiency of the irrigation water. Uniformity can be defined as the ratio of the volume of water used or available for use in crop production to the volume pumped or delivered for use. Crop uptake efficiency may be expressed as the ratio of increase in yield over non-irrigated production to the volume of irrigation water used. Irrigation efficiencies thus provide a basis for the comparison of irrigation systems from the standpoint of water beneficially used and from the standpoint of yield per unit of water used (Haman et al., 2005). Irrigation system efficiency depends primarily on design, installation and maintenance, and management. Thus, a properly designed and maintained system can be inefficient if mismanaged just as a well-designed system can be inefficient if managed effectively with poor maintenance. Irrigation management of vegetable crops includes: 1) combination of target irrigation volume, 2) measure of soil moisture to adjust this volume based on crop age and weather conditions, 3) knowledge of how much the root zone can hold, and 4) assessment of how rainfall contributes to replenishing soil moisture. (Hochmuth, 2007).

Concerns about the environmental impact of water and fertilizer uses by agriculture have dramatically increased in the past few decades. Crop production is linked to leaf photosynthesis and canopy size, and water stress drastically reduces both components (Kramer & Boyer, 1995). Adequate water supply is, therefore, critical in maximizing crop production, nutrient use efficiency (NUE), and quality of most horticultural crops. Efficient water use may promote an increase in fertilizer retention in the effective root zone, maximizing crop production and minimizing the potential of groundwater degradation (e.g., nitrate-nitrogen (NO_3^--N) leaching) (Scholberg et al., 2002). A simple goal of the ideal irrigation scheduling would be to increase crop production with the least amount of water, therefore minimizing water loss by deep percolation, runoff or evaporation. However, no irrigation system has the capability of completely avoiding water losses, although several irrigation methods and techniques can be adopted to minimize losses and increase the water use efficiency by crops.

One of the most important irrigation management factors is irrigation uniformity, which is how evenly water is distributed across the field. Non-uniform distribution of irrigation water may create over- and/or under-irrigated areas which can lead to yield reduction due to excessive nutrient leaching or plant water stress. For a sprinkler irrigation system, the uniformity of application can be evaluated by placing containers in a geometric configuration and measuring the amount of water caught in each container. Dukes et al. (2006) utilized this type of testing to show the effect of pressure and wind speed on operating performance of two types of center pivot sprinkler system nozzle packages. Furthermore, Dukes and Perry (2006) showed that uniformity of a variable rate control system was not different from a traditional control system on two typical center pivot/linear move irrigation systems used in the southeast USA. However, the problem with sprinkler systems is that the water application pattern is susceptible to distortion by the wind. While wind speed and direction are not controlled variables, their effect on irrigation uniformity is significant, so that sprinkler system design must be done with anticipated wind conditions. Drip irrigation systems are very efficient in terms of water distribution and reduction of water losses. The uniformity is directly related to the pressure variation within the entire system and the variability of the emissions of each individual emitter. Several factors contribute to reduce the uniformity of water application such as excessive length of laterals, excessive pressure losses due to changes in elevation along the laterals, emitter clogging, and soil characteristics. Limited lateral water mobility in sandy soils under drip irrigation drastically affects root distribution (Zotarelli et al., 2009), and nutrient interception in the sides of the raised bed. This could be a problem for double row crops like peppers and squash when a single drip tape is placed in center of the bed.

Non-uniform distribution of water in the bed may also compromise the acquisition of nutrients by the root system. Since NO_3^--N is a highly mobile, non-adsorbing ion, low rooting densities may not be sufficient for NO_3^--N acquisition, and a larger fraction of the N applied through fertigation can escape below the root zone. The basis for this lies in previous field observations which demonstrated that the displacement of irrigation water and nutrients is primarily vertical and confined to a 30–38 cm wide zone, due to the extremely high hydraulic conductivity of sandy soils (Zotarelli et al., unpublished data). The use of appropriate irrigation scheduling facilitates more frequent applications of small volumes of water and improves matching of water supply and crop water demand which is critical to reduce potential crop water stress and leaching losses in sandy soils (Zotarelli et al., 2008a, 2008b, 2009). Since applying frequent small volume irrigation with conventional systems tends to be labor-intensive and/or technically difficult to employ, sensor-based irrigation systems may facilitate the successful employment of low volume-high frequency irrigation systems in commercial vegetable systems. In addition, reduction in emitter spacing and also the use of double drip tapes placed closer to the crop rows may improve the uniformity of water and nutrient distribution along the beds, while reducing the amount of water required. However, there is a lack of information about the effectiveness of this system for double row crops.

3.1 Irrigation types and performance characteristics
Irrigated acreage world-wide spans a range of irrigation delivery systems depending on the type of crop and cultural conditions. Irrigation can be grouped into the following general categories: low volume (also known as microirrigation, trickle irrigation, or drip irrigation), sprinkler, surface (also known as gravity or flood irrigation), and seepage (also known as subsurface irrigation or water table control). These irrigation systems vary by application efficiency with surface and seepage being less efficient than microirrigation (Table 1).

Irrigation system	Application efficiency
Microirrigation	80-95%
Sprinkler	60-80%
Surface/Seepage	20-70%

Table 1. Application efficiency for water delivery system (Simonne & Dukes, 2009)

Microirrigation systems: Application efficiencies of microirrigation systems are typically high because these systems distribute water near or directly into the crop root zone, and water losses due to wind drift and evaporation are typically small (Boman & Parsons, 2002; Locascio, 2005). This highly efficient water system (90% to 95%; Table 1) is widely used on high value vegetables and tree fruit crops. The advantages of microirrigation over sprinkler include reduced water use, ability to apply fertilizer with the irrigation, precise water distribution, reduced foliar diseases, and the ability to electronically scheduled irrigation on large areas with relatively smaller pumps. If micro-sprinkler systems are operated under windy conditions on hot, dry days, wind drift and evaporation losses can be high. Thus, management to avoid these losses is important to achieving high application efficiencies with these systems. The most common application of microirrigation in Florida, USA is that of under-tree micro-sprinkler systems for citrus. Less efficiency has been found for micro-sprinkler system compared to drip irrigation system. Application efficiencies of drip and line source systems are primarily dependent on hydraulics of design of these systems and on their maintenance and management (Boman & Parsons, 2002).

Sprinkler system: Sprinkler systems are designed to use overlapping patterns to provide uniform coverage over an irrigated area. Sprinklers are normally spaced 50-60% of their diameter of coverage to provide uniform application in low wind conditions. Studies have shown that 1.5% to 7.6% of irrigated water can be lost due to wind drift and evaporation during application (Dukes et al., 2010). Application efficiencies of sprinkler systems are relatively low at less than 80% (Table 1). Because networks of pressurized pipelines are used to distribute water in these systems, the uniformity of water application and the irrigation efficiency is more strongly dependent on the hydraulic properties of the pipe network. Thus, application efficiencies of well-designed and well-managed pressurized sprinkler systems are much less variable than those of gravity flow irrigation systems, which depend heavily on soil hydraulic characteristics. Therefore, during water applications, sprinkler irrigation systems lose water due to evaporation and wind drift (Haman et al., 2005). More water is lost during windy conditions than calm conditions. More is also lost during high evaporative demand periods (hot and dry days) than during low demand periods (cool, cloudy, and humid days). Thus, sprinkler irrigation systems usually apply water more efficiently at night (and early mornings and late evenings) than during the day. It is not possible to apply water with perfect uniformity because of friction losses, elevation changes, manufacturing variation in components, and other factors. Traveling guns typically have greater application efficiencies than portable guns because of the greater uniformity that occurs in the direction of travel (Smajstrla et al., 2002). Periodic move lateral systems are designed to apply water uniformly along the laterals. No uniformity and low application efficiencies occur when the laterals are not properly positioned between settings. Non-uniformity also occurs at the ends of the laterals where sprinkler overlap is not adequate (Smajstrla et al., 2002).

Surface and Seepage systems: Water is distributed by flow through the soil profile or over the soil surface. The uniformity and efficiency of the irrigation water applied by the surface irrigation system depends strongly on the soil topography and hydraulic properties (Boman & Parsons, 2002). Florida's humid climate requires drainage on high water table soils, and field slope is necessary for surface drainage. But surface runoff also occurs because of field slope. Runoff reduces irrigation application efficiencies unless this water is collected in detention ponds and used for irritation at a later time (Smajstrla et al., 2002). Water distribution from seepage irrigation system occurs below the soil surface. Therefore, wind and other climatic factors do not affect the uniformity of water application. Use of a well-designed and well-maintained irrigation systems can reduce the loss of water and thereby increase application efficiency as well as uniformity (Boman & Parsons, 2002).

3.2 Development and characteristics of "gradient-mulch" system

In the 1960's, a vegetable production system on sandy soils was developed in south Florida using a "gradient-mulch" concept to supply nutrients to plants under seepage irrigation (Geraldson, 1962; Geraldson et al., 1965). This system dominates contemporary vegetable production on Florida's sandy soils. The gradient-mulch system involves soil fumigation and banded application of soluble fertilizers beneath full bed plastic mulch. The system has been proven to provide a controlled environment within the bedded soil for sufficient nutrient supply, optimum soil moisture content, stable root growth, and managements for weed, disease, and insect.

Basic components of the gradient-mulch system include 70- to 90 cm-wide (depending on vegetables) flat topped soil beds raised to 25- to 30-cm above from ground, covered by full plastic mulch (Fig. 1). Soluble fertilizers such as N and potassium (K) are applied as band on or near (top 0 to 4 cm) the soil bed surface with the more insoluble nutrients such as phosphorus (P) and micronutrients mixed in the bed. Seepage irrigation is provided to maintain a constant water table levels that are typically 40 to 45 cm deep in Florida sandy soils. Intermittent ditches are also provided for irrigation and drainage purposes from a precisely leveled field with a slope of about 2.5 cm in 30 m (Fig. 2).

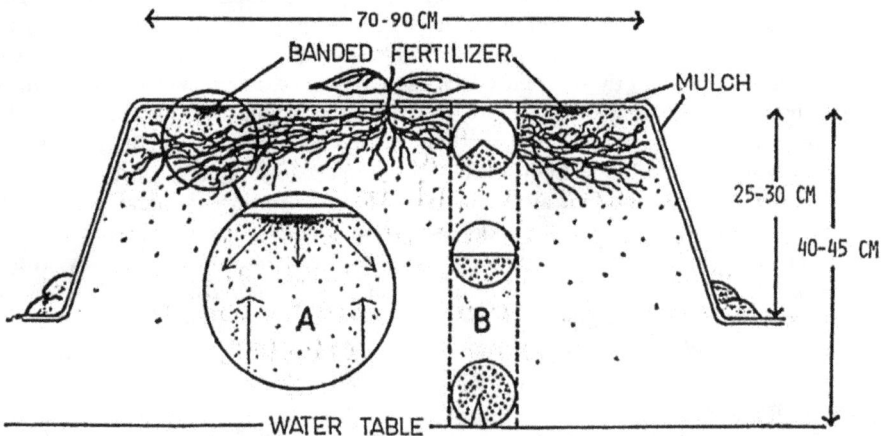

Fig. 1. Diagram of the gradient-mulch system. A. Three-dimensional nutrient gradient where salts diffuse outward from level of highest concentration, and move upward with moisture. B. Two-dimensional moisture-air gradient where moisture moves upward (modified from Geraldson, 1980)

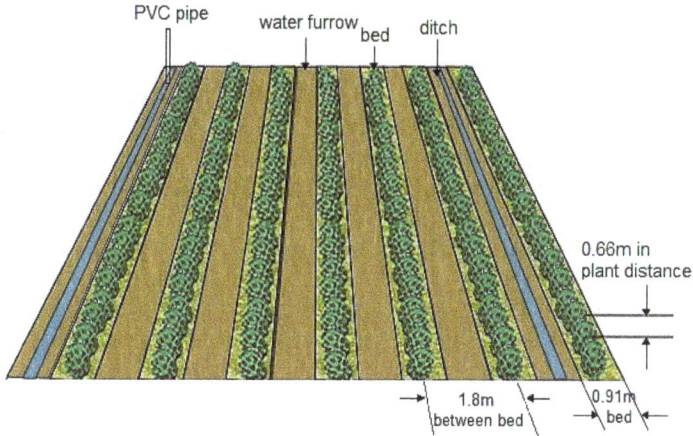

Fig. 2. Diagram of typical gradient-mulch system in a field. A ditch runs between every six raised beds of 91-cm width with 1.8-m distance between beds. For example, tomato plants are 66-cm apart from each other on the bed

Use of a full-bed synthetic mulches on soil beds can serve for minimum nutrient loss by leaching, minimum evaporation loss, optimum soil temperature and moisture/air ratio, and weed and ground rots control (Geraldson, 1981). A reciprocal moisture-air gradient is provided by maintaining the constant water table a given distance below the flat topped soil bed. Thus, a two-dimensional range of decreasing moisture/increasing air is established from a level of saturation to the bed surface. A three-dimensional concentration gradient decreasing with distance from the surface applied fertilizers is superimposed on the moisture-air gradient. Thus, the root from a germinating seed or transplanted seedling can develop in that portion of the bed where the most favorable levels of nutrients, moisture, and air coincide. Once the root system becomes established in a favorable portion of the soil bed, then nutrients and moisture must continue to be supplied to the root as removed by the root; soluble nutrients move by gradient diffusion from the surface to the root. The less soluble nutrients mixed in soil bed continue to become available by equilibrium action, also as removed by the root. Thus, a minimal stress root environment is established and maintained regardless of an increasing crop requirement.

Moisture is similarly supplied from the water table as required. It is important to recognize that a fluctuating water table can alter the stability of both the moisture and nutrient gradients. The depth of the water table can be a function of the design and management of both the drainage and irrigation system (Geraldson, 1981). Many sandy soils in Florida such as Spodosols favor the use of a constant water table which is basic to the functional efficiency of the gradient-mulch system.

The required quantities of fertilizers used under the mulch for intensive production are no problem if used as recommended. However, when finished and the mulch is removed, it would be preferable to have a minimal residue of salts, thus minimal leaching of salts out of the field (minimal pollution) and minimal salts that might accumulate (minimal stress). Residual salts, irrigation water salts, and misplaced fertilizer salts contribute to a salt buildup in the root environment. Accumulation beyond a given concentration progressively reduces production efficiency (Geraldson, 1981).

For tomato production, for example, under the gradient-mulch system with seepage irrigation, all P_2O_5, micronutrients, and 20 to 25% of N and K_2O are broadcast and incorporated into the bed (i.e., "bottom" or "cold" mix). The remaining N and K_2O are placed in narrow grooves 5 to 8 cm deep and 30 to 35 cm offset from the plant bed center (i.e., "top" or "hot" mix). Supplemental N and K_2O at 13.6 and 9.1 kg, respectively, can be applied by liquid fertilizer injection wheel to replace leached N and K_2O (Olson et al., 2009). Therefore, nutrient concentrations can differ considerably with location in the bed and with time throughout the growing season. For example, soil solution NO_3^--N concentrations at the fertilizer band and crop row of a tomato bed at the beginning of the growing season were 4200 and 263 mg L^{-1} at the 0–5 cm depth and 900 and 25 mg L^{-1} at the 10–20 cm depth, respectively. By the end of the growing season, NO_3^--N concentrations at the band and row had significantly decreased to 250 and 129 mg L^{-1} at the shallow depth and 115 and 10 mg L^{-1} at the deeper depth, respectively (Geraldson, 1999).

The gradient-mulch system was an important factor in improving production efficiency in the Florida tomato industry during the 1970's. The productivity improved 2.5 times with the system compared with that without the system, and the value of the 1979-80 Florida tomato crop increased to $228 million compared with $92 million in 1972-73. This system started to de adopted for other crops such as pepper, sweet corn, cauliflower, eggplant, and squash (Geraldson, 1981). Today, as of 2009, the state of Florida has grown to be the second state following California in acreage for fresh tomato production (14,800 and 14,000 ha in California and Florida, respectively), and the leading state in fresh tomato production value in the USA exceeding $520 million which accounted for 26% of the state's total crop production value (USDA/NASS, 2011).

4. Nutrient mobility and availability for vegetable production on sandy soils

4.1 Nutrient availability under drip irrigation systems

The use of plastic mulch and drip irrigation has become more common in high intensity vegetable production on sandy soils than sprinkler and seepage irrigation systems because water application efficiency, defined as the fraction of the water applied and that is available to plant for use, is greater with drip system than with sprinkler and seepage systems (Simonne & Dukes, 2009; Table 1). However, excess irrigation practices can cause reduced water application efficiency and leaching of soluble nutrients out of the root zone. Although irrigation and fertigation practices vary widely among growers, irrigation typically occurs once or twice each day with regularly scheduled time normally and prolonged time during peak growth stages, while fertigation only takes place 1 to 2 times each week.

Soluble nutrients such as NO_3^--N are transported mainly by convection with water mobility, therefore can move through soil profile with water applied using drip irrigation system. It is evident that the reduction in water moving through the root zone corresponds to a reduction in the amount of NO_3^--N lost below the root zone. When drip irrigation management was improved using more controlled irrigation scheduling than traditional fixed time scheduling, the amounts of water thus NO_3^--N leached out of the root zone were reduced (Dukes et al., 2006). In experiments for tomato and green bell pepper grown on Candler and Tavares sands (both 97% sand), USA, N was applied at 192 and 208 kg ha^{-1} to tomato and bell pepper, respectively. Electric probes installed in soil beds measured soil water content in the beds and functioned as a bypass controller to skip a scheduled timed irrigation event if the soil volumetric water content (VWC) was above a preset threshold.

When a probe installed in the tomato bed was set the threshold for soil VWC of 13%, the amount of excess irrigation water leached out of the root zone (below 60 cm) was 84% less compared to that with the fixed time scheduling irrigation system (6.8 vs. 42.8 mm). Similarly, the amount of NO_3^--N leached was reduced to 82% between the controlled and fixed time scheduling irrigation systems (7 vs. 37 kg NO_3^--N ha^{-1}). On bell pepper using the threshold of 10% and 13% VWC, the controlled irrigation system reduced water leaching by 81% and 51%, respectively, as well as NO_3^--N leaching by 84% and 20%, respectively, compared to the fixed irrigation system. While tomato showed an increase in crop yield, bell pepper exhibited a significant reduction in crop yield especially when the VWC was maintained to 10% (Dukes et al., 2006).

Otherwise, the less mobile nutrients such as P are transported mainly by diffusion, hence its mobility in soils is less strongly governed by water mobility than the mobile nutrients. Triple superphosphate (0-45-0) was applied as P fertilizer with four different rates (0, 30, 60, and 90 kg ha^{-1}) and two different water management regimes (drip irrigation and non-irrigation) to tomatoes grown on Granby loamy sands (77 to 82% sand), Canada for two consecutive years (2007-2008) (Liu et al., 2011). For both growing seasons, in the 0-40 cm soil profile, water extractable P (WEP) content was lower in the drip irrigation treatment than in the non-irrigation treatment (Fig. 3a). However, irrigation management did not have significant effects on WEP below 40-cm depth. Similarly, soil WEP content significantly increased with increasing fertilizer P rate applied only in the top 0-40 cm profile, but not below the depth of 40 cm (Fig. 3b). It appeared that the reduced WEP in the top 0-40 cm with the drip irrigation treatment may have caused by increased crop uptake of P with drip irrigation rather than the vertical mobility of P with water. The drip irrigation of P can provide precise amounts of water and nutrients in an efficient manner, optimizing plant uptake and minimizing environmental losses (Hartz & Hochmuth, 1996). However, environmental losses of P through vertical leaching can occur in sandy soils with high hydraulic conductivity, low P adsorption capacity, and shallow water table levels (Leinweber et al., 1999; Djodjic et al., 2004).

It appears that fertigation with 100% water-soluble fertilizers applied through drip irrigation (i.e., drip fertigation) can reduce the amount of leachable fertilizers such as NO_3^--N and K to deeper soil layers, compared to soluble fertilizers applied to soil with water applied by drip irrigation. However, less mobile nutrients such as P tend to be fixed at the point of application. Yet, subsurface drip fertigation can cause higher available P in deeper layers. Tomatoes were grown on sandy loam soil in India with fertilizer rates of 180-66-99.6 kg N-P-K ha^{-1} using urea, single superphosphate, and muriate of potash as normal fertilizer for drip irrigation, and urea, mono-ammonium phosphate (12-26.84-0), and potassium nitrate (13-0-38.18) as 100% water-soluble fertilizer for fertigation, both applied daily through in-line drippers (Hebbar et al., 2004). After 2 years of growing seasons of 116 and 119 days, respectively, lower residual NO_3^--N was observed at 30-45 cm soil layer in the fertigation treatment (55 kg ha^{-1}) than in the drip irrigation treatment (66 kg ha^{-1}). The reduction in the residual NO_3^--N was further enhanced at 45-60 cm soil layer. Similarly, the residual exchangeable K (by 1 N ammonium acetate) accumulation was higher at deeper layers (30-45 and 45-60 cm) in drip irrigation treatment, compared with the fertigation treatment (93 vs. 83 kg ha^{-1} and 95 vs. 72 kg ha^{-1}, respectively). However, the level of available P (by Bray 1) was significantly higher in 15-30 cm (62 kg ha^{-1}), 30-45 cm (36 kg ha^{-1}), and 45-60 cm (22 kg ha^{-1}) depths in the subsurface drip irrigation compared to the drip irrigation treatment.

Because P fertilizer was delivered at 20 cm below the surface by means buried laterals, more concentration of P was observed at deeper depths (Hebbar et al., 2004).

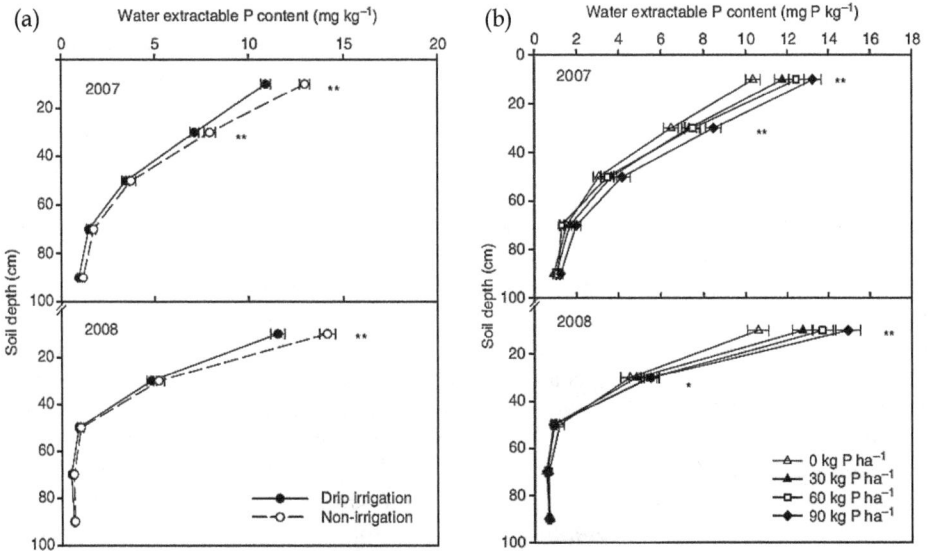

Fig. 3. Post-harvest water-extractable P in the 0- to 100-cm soil profile as affected by (a) different water management regimes and (b) fertilizer P application rates under processing tomato at Harrow, Ontario, 2007-2008 (adapted from Liu et al., 2011)

4.2 Nutrient mobility in soil under seepage irrigation

The spatial and temporal distribution of nutrients and transformation between chemical forms within the soil bed throughout the growing season under seepage-irrigated sandy soils have rarely documented. It is critical, however, to understand dynamics of nutrient mobility in the soil bed for proper fertilization and irrigation managements or best management practices for vegetable production. Tomatoes were grown on Holopaw fine sand (98% sand), USA using the gradient-mulch system with plastic mulch under seepage irrigation (Sato et al., 2009a, 2009b). Total fertilizer rates applied were 224–61–553 kg N-P-K ha[-1]. While N and K were applied with the bottom (17 kg N ha[-1] and 27 kg K ha[-1]) and top mix (207 kg N ha[-1] and 526 kg K ha[-1]), all P was applied only with the bottom mix. Soil samples were collected using an auger weekly or biweekly for 18 wk after planting (WAP) at two locations (the fertilizer band and the bed centerline) at three different depths with 10-cm increment (Fig. 4). Each sampling location was denoted as B1: band and top, B2: band and middle, B3: band and bottom, C1: centerline and top, C2: centerline and middle, and C3: centerline and bottom. The soil samples were analyzed for ammonium-N (NH_4^+-N) and NO_3^--N by 2 M KCl, and available P and K by Mehlich-1 extractions.

The NH_4^+-N concentration at B1 location was highest throughout the season compared with other locations, and in general steadily decreased with time (Fig. 5a). Initially low NH_4^+-N at C1 location peaked during 3 to 5 WAP, and elevated NH_4^+-N concentrations were found at C2 location as well until 5 WAP. Most NH_4^+-N resided at B1, B2, C1, and C2 locations because N was applied in the top mix placed directly above B1 and B2, and also in the bottom mix that was incorporated into the soil (C1 and C2) when the raised bed was formed. However, the N rate applied as the bottom mix (17 kg N ha^{-1}, equivalent to 8.5 mg N kg^{-1} when broadcast in the top 15 cm of soil) was too small to account for the NH$_4^+$-N concentrations found at C1 and C2 locations (ranging from 50 and 100 mg N kg^{-1} until 5 WAP), indicating translocation (lateral mobility) of NH$_4^+$-N most likely from B1 location. The NH$_4^+$-N concentrations below 20-cm depth were consistently low or non-existent throughout the season, indicating that NH$_4^+$-N virtually did not move vertically below 20-cm depth under seepage irrigation (Sato et al., 2009a).

The NO$_3^-$-N concentration at B1 location peaked during 3 WAP, which was 2 to 3 wk later than the NH$_4^+$-N peak, then slowly decreased with time (Fig. 5b). This may indicate that it took about 3 wk for NH$_4^+$-N at B1 to process nitrification that commenced about 2 wk into the season. The most notable behavior of NO$_3^-$-N was an elevated concentration peak during 6 to 8 WAP at every location in the bed except for B1 (which remained much higher than the rest of the locations throughout the season), and subsequent decrease to almost zero at the end of the season. This peak corresponded with a raised water table level and accordingly increased soil water content, especially in the middle and bottom layers during 5 to 7 WAP. Since water is supplied to the root zone by capillarity under seepage irrigation, it is critical to maintain the water table at a depth that supplies sufficient upward flux. The water table depth below the bed surface was relatively stable at recommended levels between 45 and 60 cm for seepage-irrigated tomato bed (Stanley and Clark, 2003) except for two elevations that occurred during 2 WAP (43 cm) and 5 WAP (26 cm). On the other hand, the water table fluctuation did not appear to influence NH$_4^+$-N in any part of the bed since the NH$_4^+$-N steadily decreased after 5 WAP at every location in the bed. This difference could be due mainly to different diffusivity of the two ions, given the same soil properties

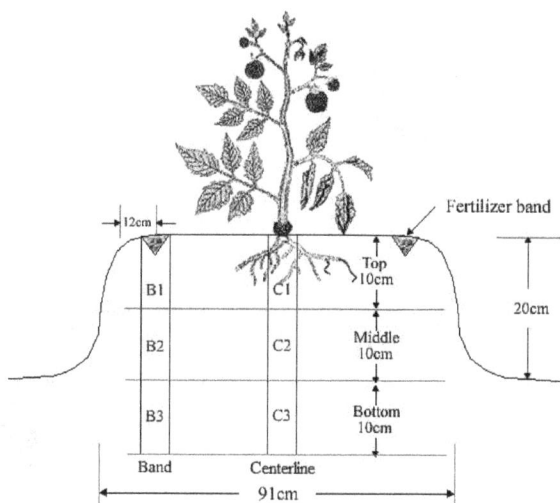

Fig. 4. Cross-sectional diagram of tomato bed and sampling locations in the bed. B1: band and top, B2: band and middle, B3: band and bottom, C1: centerline and top, C2: centerline and middle, and C3: centerline and bottom (modified from Sato et al., 2009a)

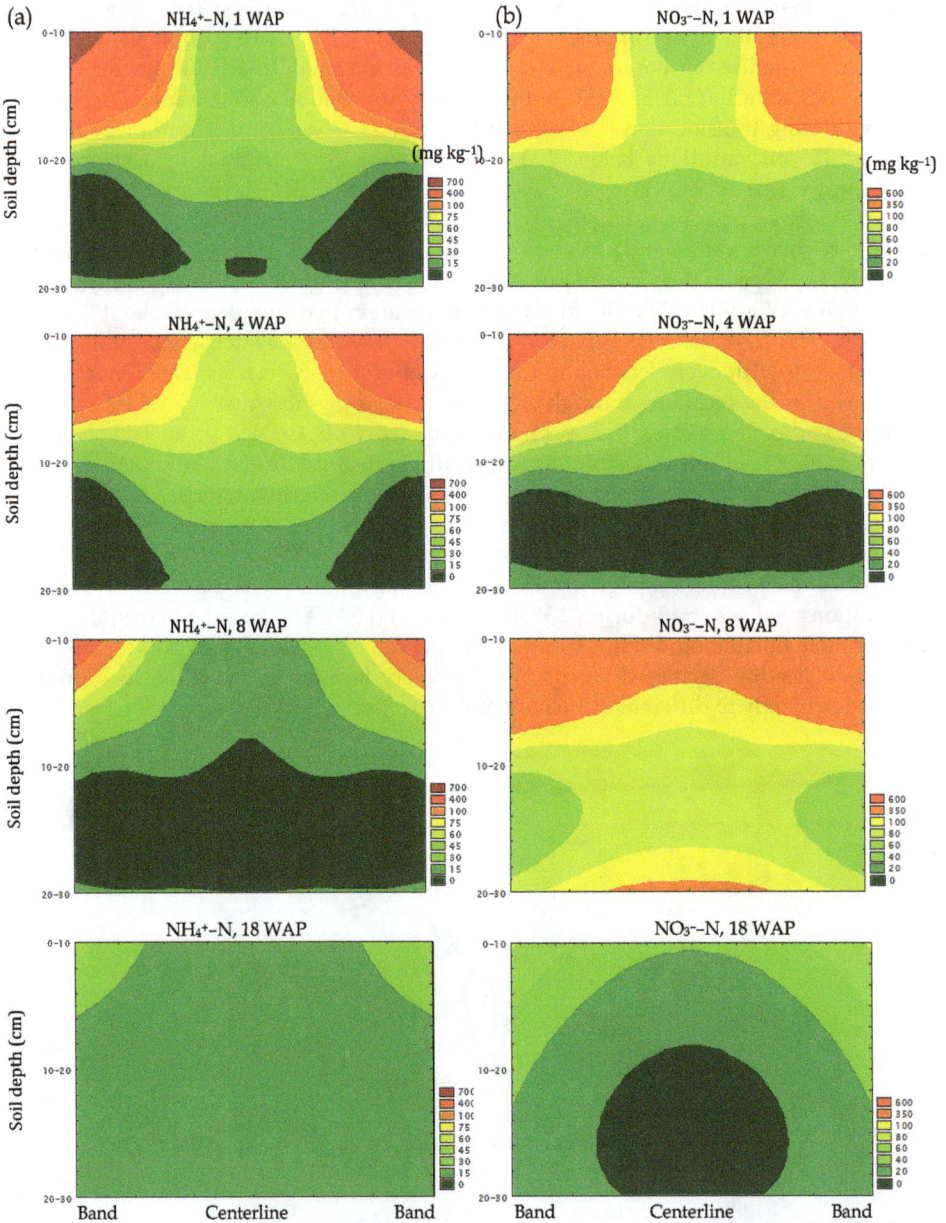

Fig. 5. Cross-sectional diagrams of spatial distribution of (a) NH_4^+–N and (b) NO_3^-–N in soil bed during 1, 4, 8, and 18 wk after planting (WAP). The concentrations in the soil bed are assumed to be symmetrical about the centerline on both sides of fertilizer band

affecting the ion mobility in soil such as soil hydraulic conductivity, water-holding capacity, texture, porosity, and density. Increased diffusivity of NO_3^--N with increased water content in the middle and bottom layers during 5 to 7 WAP would have greatly facilitated both lateral and vertical mobility of NO_3^--N (Sato et al., 2009a). The key to the stability of moisture and nutrient gradients and the root environment in the gradient-mulch system under seepage irrigation is the constant water table that is often difficult to maintain due to periodic rains and complex management of drainage and irrigation (Geraldson, 1980, 1981).

Most of P remained at C1 and C2 locations in the bed ranging between 150 and 400 mg kg^{-1} with gradual decrease at the end of the season, while P in the rest of the bed did not change to a great extent maintaining less than 100 mg kg^{-1} throughout the season. Since it is a common practice under the seepage irrigation system to apply P fertilizer only in the bottom mix to target the root zone, the bottom mix was broadcast on the surface soil before bedding, and the soil was then pushed by discs at the outer edges of the bedding apparatus toward center to form the raised bed. Therefore, most of P was initially placed in C1 and C2 and remained in the root zone. This implies that P did not move outside the root zone regardless of water table fluctuations and soil water content (Sato et al., 2009b).

Most of K in the bed remained in B1 location at an order of magnitude higher concentration (from 4000 up to 12000 mg kg^{-1}) than other bed locations because most of K fertilizer was applied in the top mix. The K concentrations in C1 and C2 locations maintained relatively high concentrations (100 to 180 mg kg^{-1}) until 7 to 8 WAP, then decreased to low values, possibly because of the bottom mix of some of K fertilizer. The K concentrations in the bottom layer were consistently low and did not greatly change throughout the season, except a raised concentration during 7 WAP. Although the substantial amounts of K were present in B1 location at the end of the season, remarkably low K concentrations were found in the rest of the bed after production. This indicates that under gradient-mulch system minimum K leaching occurred during the growing season, except for possible K loss through leaching to some extent during 5 to 7 WAP when water table fluctuation was observed, as similarly seen with NO_3^--N (Sato et al., 2009b).

4.3 Nutrient availability from controlled- and slow-release fertilizers

The use of controlled-release fertilizers (CRFs) and slow-release fertilizers (SRFs) for vegetable productions on sandy soils has been investigated with varying results, and nutrient availability on sandy soils from CRFs and SRFs has been increasingly clarified (Simonne and Hutchinson, 2005). Nitrogen (NH_4^+-N + NO_3^--N) availability from the total of 18 different CRFs (plus 4 different soluble fertilizers as comparison) mixed in Ellzey fine sand (94% sand), USA was evaluated in a plastic pot through which 400 mL of water was leached every 7 d for 12 times. The N rates applied were 6.18 and 4.80 g pot^{-1} for 2001 and 2002 trials, respectively (the 2001 rate corresponded to 224 kg N ha^{-1}). While the soluble fertilizers leached 82–98% of applied N after 12 leaching events, leached N from CRFs ranged between 13–38% in 2001 and 22–49% in 2002. Some CRFs may not release 100% of the coated N as the thickness of some prills may permanently prevent N release. The fraction of N never released is termed "locked-up N" (Simonne & Hutchinson, 2005). Nevertheless, all CRFs tested did not release N rapidly enough to supply adequate N to vegetable crops. However, since all CRFs tested except for 2 types had urea as the only N source, most of the N recovered was in the NH_4^+-N form. The NH_4^+-N release pattern from some CRFs was similar to those of the soluble fertilizers and closely desirable for vegetable production.

Many researches have demonstrated that CRFs and SRFs can reduce N, particularly NO_3^--N leaching on sandy soils compared with soluble fertilizers (Alva, 1992; Wang & Alva, 1996; Paramasivam & Alva, 1997; Fan & Li, 2009). Moreover, not only NO_3^--N but also other nutrients such as P, K, calcium (Ca), magnesium (Mg), and copper (Cu) can CRFs reduce leaching compared with uncoated fertilizers. Four different coated CRFs were compared with an uncoated (soluble) fertilizer on a sandy clay loam of Bungor soil (Typic Paleudult), Malaysia for nutrient leaching in soil columns for 30 d (Hanafi et al., 2002). The percentages of the amount of nutrients leached on the nutrients initially applied ranged 23–33%, 2–4%, 10–19%, 2–9%, 4–9%, and 1–3%, whereas those of the uncoated fertilizer were 80%, 28%, 90%, 29%, 20%, and 6% for N, P, K, Ca, Mg, and Cu, respectively. On the other hand, the distribution of the amount of N, P, and K left in the soil profile after 30 d of leaching differed depending on the type of fertilizers (Table 2). Nitrogen and K left in the soil from the uncoated fertilizer were almost evenly distributed among 0–6, 6–12, and 12–18 cm depths, with almost all P was accumulated only in the top 0–6 cm depth. Almost a half of N and K left in the soil from the coated fertilizers was found only in the top 0–6 cm depth, while up to 90% of P left in the soil was in the top depth with up to 15% found in the 6–12 cm depth. Accumulation of nutrients released from CRFs in upper soil layers appears to enhance the reduction of nutrient leaching from CRFs compared with those from soluble fertilizers.

Soil depth	Uncoated fertilizer			Coated fertilizers		
	N	P	K	N	P	K
cm		%			%	
0–6	35	97	26	47–49	84–90	43–51
6–12	29	2	24	13–15	8–15	19–29
12–18	29	1	21	13–14	1–2	16–19
18–24	4	0	16	13–16	0–1	5–14
24–30	3	0	13	9–12	0	1–8
		mg kg^{-1}			mg kg^{-1}	
0–30	732	172	497	2341–2694	235–267	3410–4043

Table 2. The percentage of the amount of nutrients left at different soil depths after leaching for 30 d on the total amount of nutrients left in the soil profile (0–30 cm). Coated fertilizers show ranges of the percentage of 4 different CRFs tested (made from Hanafi et al., 2002)

Nutrient use efficiency, particularly for N using CRFs or SRFs may be improved compared to that using soluble fertilizers (Shaviv & Mikkelsen, 1993). The NUE for N ranged between 10% and 32% for uncoated fertilizer and between 79% and 94% for coated fertilizers when peanut was grown on a sandy soil (Typic Udipsamment), Japan under drip irrigation with N application rates of 30 to 120 kg ha^{-1} (Wen et al., 2001). Seepage-irrigated Irish potato produced on Elley fine sand (90–95% sand), USA had a significantly higher NUE for N with CRFs treatment compared with soluble fertilizer treatment only when 112 kg N ha^{-1} was applied, but not when 168 and 224 kg N ha^{-1} were applied (Hutchinson et al., 2003). Other studies showed similar results of higher NUE for N with CRFs compared with soluble fertilizers only with lower N application rate applied, but not with higher rate (140 vs. 280 kg N ha^{-1}; Zvomuya et al., 2003; and 146 vs. 225 kg N ha^{-1}; Pack et al., 2006; both for potato production on sandy soils). Nutrient availability caused by the use of CRFs and SRFs over the soluble fertilizer can be improved by the interaction and competition between plant roots, soil microorganisms, chemical reactions, and pathways for loss, and matching nutrient release with plant demand (Shaviv & Mikkelsen, 1993).

4.4 Availability of nutrients applied with organic waste materials

Nutrient availability from organic waste materials such as composts and biosolids when applied in sandy soils is different from, and more difficult to be clarified as organic materials involve more factors and processes in determining nutrient availability than chemical fertilizers (Mylavarapu & Zinati, 2009). Particularly, N is more complicated than other nutrients because the transformation of compost N varies widely among different sources and is affected by soil properties, compost characteristics, and environmental factors (Sims, 1995; Amlinger et al., 2003). The mineralization of one biosolid and two composts with C/N ratio ranging between 5.8 and 38.0 and organic N between 2.9 and 49.0 g kg^{-1} was evaluated for one-year period in field-conditioned columns packed with Oldsmar sand, USA (He et al., 2000). Both rate and the total amount of mineralized N (NH_4^+-N + NO_3^--N) from the biosolid (lower C/N ratio and higher organic N content) were greater than those from the composts (higher C/N ratio and lower organic N content) during the incubation. While the biosolid reached a peak of the mineralized N within the first 90 d of the incubation, the composts had two distinct peaks at about 90 d and about 280 d of the incubation. The first and second peaks of N mineralization from these materials might be the results of its relatively uniform components made of sewage sludge in both materials and grass clippings and wood chips mixed only in the composts, respectively (He et al., 2000). The mobility of the mineralized N in soil column also differed among these materials. While only small portion of the mineralized NH_4^+-N (9% of the total NH_4^+-N for the biosolid) leached out of the column of 20-cm depth, 56% of the mineralized NO_3^--N in the total NO_3^--N for the biosolid leached out almost constantly throughout the incubation. On the other hand, the composts had leaching of the mineralized NH_4^+-N (57–65% of the total NH_4^+-N for the composts) constantly throughout the incubation, on average 75–85% of the mineralized NO_3^--N of the total NO_3^--N for the composts leached out during the second half of the incubation. Application and management of the organic materials when utilized as N fertilization need to be carefully considered to minimize the risk of NO_3^--N leaching on sandy soils, especially when the materials contain high amounts of materials with low C/N ratio such as sewage sludge-derived biosolids.

The biosolids produced by different treatment processes contain varying amounts of mineral and organic N and the organic components may be stabilized to varying degrees. Unstabilized biosolids usually contains high available C, thus high C/N ratio, and may cause a net reduction of mineralized N in amended soils due to immobilization during incubation (Epstein et al., 1978; Parker & Sommers, 1983). The N immobilization in the amended soil may occur when the C/N ratio of the biosolids exceeds 15 (Epstein et al., 1978) or 20 (Parker & Sommers, 1983). Twelve different biosolids produced at 8 different sewage treatment facilities in the UK were incubated for 73 d in a loamy sand soil (86% sand) (Smith et al., 1998). All biosolids showed a concomitant reduction in NH_4^+-N concentration with the formation of NO_3^--N in the amended soil with increasing time of incubation. Indeed, all except dewatered undigested biosolids exhibited an initial rapid NO_3^--N accumulation followed by a slower release reaching maximum amounts of NO_3^--N production. In contrast, the dewatered undigested biosolids, which contained higher amounts of OM among the biosolids tested, showed a significant immobilization of mineral N in the amended soil during the initial stages of incubation. Based on the N mineralization patterns in incubated soil, 4 categories of the biosolids were proposed (Smith et al., 1998). They are: category 1 – liquid digested and lagooned liquid undigested biosolids that have the greatest NO_3^--N accumulation potential due to large content of NH_4^+-N; category 2 – liquid undigested, lagooned liquid digested, and dewatered digested biosolids have low to intermediate NO_3^--N accumulation potential; category 3 – dewatered undigested biosolids are high in available C and may produce an initial net N immobilization followed by NO_3^--N accumulation after soil microbes has metabolized the added substrate C; and category 4 – air-dried digested biosolids are relatively resistant to mineralization and NO_3^--N formation in soil. The C/N ratio of the biosolids of category 1 to 4 is generally ordered from the lowest to the highest.

When the biosolids and manure are applied to soil based on crop N requirements, P in excess of crop needs is usually supplied due to high P content in the organic wastes. Environmental loss of P by surface runoff or leaching from organic wastes application can be significant in areas with shallow groundwater and coarse-textured soils of low P-holding capacities (Eghball et al., 1996; Lu & O'Connor, 2001). Leachability of eight different biosolids was compared with that of a chemical fertilizer (triple superphosphate, TSP) in a column study for 4 months on Candler and Immokalee sandy soils, USA with the application rates of 56 and 224 kg ha^{-1}, corresponding to typical application rates based on P-based and N-based fertility, respectively (Elliott et al., 2002). Candler and Immokalee soils had moderate and very low P-sorbing capacities, respectively, as indicated by the sum of oxalate-extractable Fe and Al. On Candler sand, the percentage of applied P leached ranged between 1.7% and 21.7% in the TSP treatment, and 0.05% and 0.45% in the biosolids treatment among two application rates. The percentage on Immokalee sand increased ranging between 13.6% and 20.7% from TSP, and 0.05% and 11.1% from the biosolids regardless of the application rates. It appears that the leachability of P from the biosolids is lower than that from the chemical fertilizer and considered as minor or negligible in many soils (Peterson et al., 1994; Sui et al., 1999).

However, the P leachability of the biosolids depends on the P-sorbing capacity of the soil; soils with lower sorbing capacity are more susceptible to P leaching. The extent of the biosolids-P leachability also appears to be explained by the P saturation index (PSI) of the biosolids, calculated as the ratio of oxalate-extractable P to the sum of oxalate-extractable Fe and Al (Jaber et al., 2006). The PSI is a measure of the degree to which biosolids P is potentially bound with Fe and Al. Therefore, PSI values < 1 suggest excess Fe and Al for binding P, and values > 1 suggest available P beyond that associated with Fe and Al precipitates. For the biosolids tested in Elliotte et al. (2002), no appreciable P leaching occurred from soils amended with biosolids of PSI < 1.1, and Immokalee soil amended with biosolids of PSI > 1.3 exhibited substantial P leaching. The microbiological processes in the soil also play an important role in reducing P leaching from the biosolids in sandy soils (Yang et al., 2008). Application of the biosolids can result in the mineralization of OM from the biosolids releasing OM-bound P as surplus to leachable P from the biosolids. However, the surplus P can be adsorbed on surfaces of Fe and Al oxides/hydroxides or immobilized to microbial biomass by increased microorganisms due to freshly added organic C from the biosolids, eventually reducing P leaching. More water-soluble P was incorporated into microbial biomass and organic fractions as evidenced by the increased microbial biomass P and microbial biomass carbon in the biosolids-amended soils (Yang et al., 2008). Nevertheless, mineralization of OM in sandy soils is generally fast particularly in humid climate conditions (Kang et al., 2011), therefore organic P including microbial biomass P can be considered to be available or potentially available to crop uptake.

5. Conclusion

Sandy-textured soils generally have low water- and nutrient-holding capacities, which, coupled with different irrigation systems used on sandy soils, make nutrient and irrigation managements difficult for suitable vegetable crop production. The nutrient and irrigation managements can be different and may be further complicated when impermeable layers such as argillic and spodic layers are excavated and mixed in as a result of the bedding process. Sandy soils, however, can be utilized for maximized crop production if proper managements for nutrients, irrigation, and drainage systems are implemented. The "gradient-mulch" system under seepage irrigation developed in 1960's in Florida, USA has become the dominant system to provide a controlled environment within the bedded soil for sufficient nutrient supply, optimum soil moisture content, stable root growth, and managements for weed, disease, and insect. More importantly, nowadays, the gradient-mulch system has been proven to minimize environmental losses of nutrients, particularly NO_3^--N and P below the root zone. Maintaining constant water table levels under seepage irrigation is, however, the most crucial factor for the gradient-mulch system for providing the maximized crop yields and minimized environmental losses of nutrients. In the light of the environmental concerns, CRFs and SFRs have been spotlighted for improved NUE, particularly for N for crop production under sandy soils. However, the effect of application of CRFs on vegetable crop production still need to be clearly understood in order for growers to receive full benefits from the use of these materials. Nutrient availability from organic waste materials such as composts and biosolids when applied in sandy soils is complex as the organic materials involve many factors and processes in determining nutrient availability. Particularly, the understanding on mineralization patterns of the organic materials with different properties under different soil and water managements is critical in determining the nutrient availability in soil and environmental fate of nutrients. Sandy soils can provide proper environmental conditions for appropriate vegetable crop production with suitable nutrient and irrigation management systems, therefore more studies are needed to elucidate the effect of different aspects of the production system in order for the producers to continue vegetable production without environmental damages, particularly from NO_3^--N which can be the most prone nutrient for leaching in sandy soils.

6. Acknowledgement

Parts of the studies in this article was funded by Florida Department of Agriculture and Consumer Services (No. 12280) and institutionally made possible by Southwest Florida

Research and Education Center, University of Florida, USA. The authors thank a farmer in southwest Florida for tomato fields for the experiment, and Ms. Jin Wu for a diagram drawing of tomato bed.

7. References

Alva, A.K. (1992). Differential leaching of nutrients from soluble vs. controlled-release fertilizers, *Environmental Management*, Vol.16, pp. 769-776.

Amlinger, F.; Götz, B.; Dreher, P.; Geszti, J. & Weissteiner, C. (2003). Nitrogen in biowaste and yard waste compost: dynamics of mobilisation and availability – a review, *European Journal of Soil Biology*, Vol.39, pp. 107-116.

Boman, B. & Parsons, L. (2002). Soil Water Measuring Devices, In: *Water and Florida Citrus: use, regulation, irrigation, systems, and management*, B. J. Boman, (Ed.), 148-162, Chapter 5, Institute of Food and Agricultural Sciences, University of Florida, Gainesville, Florida, USA

Djodjic, F.; Borling, K. & Bergstrom, L. (2004). Phosphorus leaching in relation to soil type and soil phosphorus content, *Journal of Environmental Quality*, Vol.33, pp. 678-684.

Dukes, M.D. & Perry, C.D. (2006). Uniformity testing of variable rate center pivot irrigation control systems, *Precision Agriculture*, Vol.7, pp.205-218.

Dukes, M.D.; Zotarelli, L. & Morgan, K.T. (2010). Use of irrigation technologies for production of horticultural crops in Florida - I. Vegetable industry, *HortTechnology*, Vol.20, pp. 133-142.

Dukes, M.D.; Zotarelli, L.; Scholberg, M.S. & Muños-Carpena, R. (2006). Irrigation and nitrogen best management practices under drip irrigated vegetable production, *Proceedings ASCE EMRI World Water and Environmental Resource Congress*, Omaha, Nebraska, USA, May 21-25, 2006

Eghball, B.; Binford, G.D. & Baltensperger, D.D. (1996). Phosphorus movement and adsorption in a soil receiving long-term manure and fertilizer application, *Journal of Environmental Quality*, Vol.25, pp. 1339-1343.

Elliott, H.A.; O'Connor, G.A. & Brinton, S. (2002). Phosphorus leaching from biosolids-amended sandy soils, *Journal of Environmental Quality*, Vol.31, pp. 681-698.

Epstein, E.; Keane, D.B.; Meisinger, J.J. & Legg, J.O. (1978). Mineralization of nitrogen from sewage sludge and sludge compost, *Journal of Environmental Quality*, Vol.7, pp. 217-221.

Fan, X.H. & Li, Y.C. (2009). Effects of slow-release fertilizers on tomato growth and nitrogen leaching, *Communications in Soil Science and Plant Analysis*, Vol.40, pp. 3452-3468.

Geraldson, C.M. (1962). Growing tomatoes and cucumbers with high analysis fertilizer and plastic mulch, *Proceedings of the Florida State Horticultural Society*, Vol.75, pp. 253-260.

Geraldson, C.M. (1980). Importance of water control for tomato production using the gradient mulch system, *Proceedings of the Florida State Horticultural Society*, Vol.93, pp. 278-279.

Geraldson, C.M. (1981). Vegetable production from the Spodosols of Peninsular Florida, *Proceedings of the Soil and Crop Science Society of Florida*, Vol.40, pp. 4-8.

Geraldson, C.M. (1999). The gradient concept – replacing the soil as the buffer component, *Proceedings of the Soil and Crop Science Society of Florida*, Vol.58, pp. 89-92.

Geraldson, C.M.; Overman, A.J. & Jones, J.P. (1965). Combination of high analysis fertilizers plastic mulch and fumigation for tomato production on old agricultural land, *Proceedings of the Soil and Crop Science Society of Florida*, Vol.25, pp. 18-24.

Gilbert, R.A.; Rice, R.W. & Lentini, R.S. (2008). Characterization of selected mineral soils used for sugarcane production. Florida Cooperation Extension Service SS-AGR-227, Institute of Food and Agricultural Sciences, University of Florida, Gainesville, Florida, USA

Haman, D.Z.; Smajstrla, A.G. & Pitts, D.J. (2005). Efficiencies of irrigation systems used in Florida nurseries, Bulletin 312, Institute of Food and Agricultural Sciences, University of Florida, Gainesville, Florida, USA.

Hanafi, M.M.; Eltaib, S.M.; Ahmad, M.B. & Syed Omar, S.R. (2002). Evaluation of controlled-release compound fertilizers in soil, *Communications in Soil Science and Plant Analysis*, Vol. 33, pp. 1139-1156.

Hartz, T.K. & Hochmuth, G.J. (1996). Fertility management of drip-irrigated vegetables, *HortTechnology*, Vol.6, pp.168-172.

He, Z.L.; Alva, A.K.; Yan, P.; Li, Y.C.; Calvert, D.V.; Stoffella, P.J. & Banks, D.J. (2000). Nitrogen mineralization and transformation from composts and biosolids during field incubation in a sandy soil, *Soil Science*, Vol.165, pp. 161-169.

Hebbar, S.S.; Ramachandrappa, B.K.; Nanjappa, H.V. & Prabhakar, M. (2004). Studies on NPK drip fertigation in field grown tomato (*Lycopersicon esculentum* Mill.), *European Journal Agronomy*, Vol.21, pp. 117-127.

Hochmuth, R. C. (2007). Vegetable growers get irrigation help, *Citrus and Vegetables Magazine*, Vol.71. pp. 28-29.

Hutchinson, C.; Simonne, E.; Solano, P.; Meldrum, J. & Livingston-Way, P. (2003). Testing of controlled release fertilizer programs for seep irrigated Irish potato production, *Journal of Plant Nutrition*, Vol.26, pp. 1709-1723.

Jaber, F.H.; Shukla, S.; Hanlon, E.A.; Stoffella, P.J.; Obreza, T.A. & Bryan, H.H. (2006). Groundwater phosphorus and trace element concentrations from organically amended sandy and calcareous soils of Florida, *Compost Science & Utilization*, Vol.14, pp. 6-15.

Kang, J.; Amoozegar, A.; Hesterberg, D. & Osmond, D.L. (2011). Phosphorus leaching in a sandy soil as affected by organic and inorganic fertilizer sources, *Geoderma*, Vol.161, pp. 194-201.

Kern, J.S. (1992). Geographic patterns of soil water-holding capacity in the contiguous United States, *Soil Science Society of America Journal*, Vol.59, pp. 1126-1133.

Khaleel, R.; Reddy, K.R. & Overcash, M.R. (1981). Changes in soil physical properties due to organic waste application: A review, *Journal of environmental Quality*, Vol.10, pp. 133-141.

Kramer, P.J. & Boyer, J.S. (1995). *Water relations of plant and soils*, Academic Press, ISBN 978-0124250604, San Diego, California, USA

Leinweber, P.; Meissner, R.; Eckhardt, K.U. & Seeger, J. (1999). Management effects on forms of phosphorus in soil and leaching losses, *European Journal of Soil Science*, Vol.50, pp.413-424.

Liu, K.; Zhang, T.Q. & Tan, C.S. (2011). Processing tomato phosphorus utilization and post-harvest soil profile phosphorus as affected by phosphorus and potassium additions and drip irrigation, *Canadian Journal of Soil Science*, Vol.91, pp. 417-425.

Locascio, S. J. (2005). Management of irrigation for vegetables: past, present, and future. *HortTechnology*, Vol.15, pp.482-485.

Lu, P. & O'Connor, G.A. (2001). Biosolids effects on P retention and release in some sandy Florida soils, *Journal of Environmental Quality*, Vol.30, pp.1059-1063.

Muchovej, R.M.; Hanlon, E.A.; McAvoy, E.; Ozores-Hampton, M.; Roka, F.M.; Shukla, S.; Yamataki, H. & Cushman, K. (2005). Management of soil and water for vegetable production in southwest Florida. Florida Cooperation Extension Service SL223, Institute of Food and Agricultural Sciences, University of Florida, Gainesville, Florida, USA

Mylavarapu, R.S. & Zinati, G.M. (2009). Improvement of soil properties using compost for optimum parsley production in sandy soils, *Scientia Horticulturae*, Vol.120, pp. 426-430.

Obreza, T.A. & Collins, M.E. (2002). Common soils used for citrus production in Florida. Florida Cooperation Extension Service SL193, Institute of Food and Agricultural Sciences, University of Florida, Gainesville, Florida, USA

Olson, S.M.; Stall, W.M.; Vallad, G.E.; Webb, S.E.; Taylor, T.G.; Smith, S.A.; Simonne, E.H.; McAvoy, E. & Santos, B.M. (2009). Tomato production in Florida, In: *Vegetable Production Handbook for Florida 2009-2010*, S.M. Olson & E.H. Simonne, (Eds.), 291-312, Institute of Food and Agricultural Sciences, University of Florida, Gainesville, Florida, USA

Pack, J.E.; Hutchinson, C.M. & Simonne, E.H. (2006). Evaluation of controlled-release fertilizers for northeast Florida chip potato production, *Journal of Plant Nutrition*, Vol.29, pp. 1301-1313.

Paramasivam, S. & Alva, A.K. (1997). Leaching of nitrogen forms from controlled-release nitrogen fertilizers, *Communications in Soil Science and Plant Analysis*, Vol.28, pp. 1663-1674.

Parker, C.F. & Sommers, L. E. (1983). Mineralization of nitrogen in sewage sludges, *Journal of Environmental Quality*, Vol.12, pp. 150-156.

Peterson, A.E.; Speth, P.E.; Corey, R.B.; Wright, T.W. & Schlecht, P.L. (1994). Effect of twelve years of liquid digested sludge application on the soil phosphorus level, In: *Sewage sludge: Land utilization and the environment*, C.E. Clapp, (Ed.), 237-247, Soil Science Society of America, Madison, Wisconsin, USA

Sato, S.; Morgan, K.T.; Ozores-Hampton, M. & Simonne, E.H. (2009a). Spatial and temporal distributions in sandy soils with seepage irrigation: I. Ammonium and nitrate, *Soil Science Society of America Journal*, Vol.73, pp. 1044-1052.

Sato, S.; Morgan, K.T.; Ozores-Hampton, M. & Simonne, E.H. (2009b). Spatial and temporal distributions in sandy soils with seepage irrigation: II. Phosphorus and potassium, *Soil Science Society of America Journal*, Vol.73, pp. 1053-1060.

Scholberg, J.M.S.; Parsons, L.R.; Wheaton, T.A.; McNeal, B.L. & Morgan, K.T. (2002). Soil temperature, nitrogen concentration, and residence time affect nitrogen uptake efficiency in citrus. *Journal of Environmental Quality*, Vol.31, pp.579-768.

Shaviv, A. & Mikkelsen, R.L. (1993). Controlled-release fertilizers to increase efficiency of nutrient use and minimize environmental degradation – A review, *Fertilizer Research*, Vol.35, pp. 1-12.

Shirazi, M.A. & Boersma, L. (1982). A unifying quantitative analysis of soil texture. *Soil Science Society of America Journal*, Vol.48, pp. 142-147.

Simonne, E.H. & Dukes, M.D. (2009). Principles and practices of irrigation management for vegetables, In: *Vegetable Production Handbook for Florida 2009-2010*, S.M. Olson & E.H. Simonne, (Eds.), 17-23, Institute of Food and Agricultural Sciences, University of Florida, Gainesville, Florida, USA

Simonne, E.H. & Hutchinson, C.M. (2005). Controlled-release fertilizers for vegetable production in the era of best management practices: teaching new tricks to an old dog, *HortTechnology*, Vol.15, pp. 36-46.

Sims, J.T. (1995). Organic wastes as alternative nitrogen sources, In: *Nitrogen Fertilization in the Environment*, P.E. Bacon (Ed.), Marcel Dekker, New York, USA

Six, J.; Elliott, E.T.; Paustian, K. & Doran, J.W. (1998). Aggregation and soil organic matter accumulation in cultivated and native grassland soils, *Soil Science Society of America Journal*, Vol.62, pp. 1367-1377.

Smajstrla, A.G.; Boman, B.J.; Clark, G.A.; Haman, D.Z.; Harrison, D.S.; Izuno, F.T.; Pitts, D.J. & Zazueta, F.S. (2002). Efficiencies of Florida agricultural irrigation systems, Bulletin 247, Institute of Food and Agricultural Sciences, University of Florida, Gainesville, Florida, USA

Smith, S.R.; Woods, V. & Evans, T.D. (1998). Nitrate dynamics in biosolids-treated soils. I. Influence of biosolids type and soil type, *Bioresource Technology*, Vol.66, pp. 139-149.

Stanley, C.D. & Clark, G.A. (2003). Effect of reduced water table and fertility levels on subirrigated tomato production in southwest Florida. Florida Cooperation Extension Service SL210, Institute of Food and Agricultural Sciences, University of Florida, Gainesville, Florida, USA

Sui, Y.; Thompson, M.L. & Mize, C.W. (1999). Redistribution of biosolids-derived total phosphorus applied to a Mollisol, *Journal Environmental Quality*, Vol.28, pp. 1068-1074.

USDA/NASS (September 2011). Quick Stats 2.0 Beta, Available from http://quickstats.nass.usda.gov

Wang, F.L. & Alva, A.K. (1996). Leaching of nitrogen from slow-release urea sources in sandy soils, *Soil Science Society of America Journal*, Vol.60, pp. 1454-1458.

Wen, G.; Mori, T.; Yamamoto, T.; Chikushi, J. & Inoue, M. (2001). Nitrogen recovery of coated fertilizers and influence on peanut seed quality for peanut plants grown in sandy soil, *Communications in Soil Science and Plant Analysis*, Vol.32, pp. 3121-3140.

Yang, Y.; He, Z.; Stoffella, P.J.; Yang, X.; Graetz, D.A. & Morris, D. (2008). Leaching behavior of phosphorus in sandy soils amended with organic material, *Soil Science*, Vol.173, pp. 257-266.

Zotarelli, L.; Dukes, M. D.; Scholberg, J. M.; Hanselman, T.; Femminella, K.L. & Munoz-Carpena, R. (2008a). Nitrogen and water use efficiency of zucchini squash for a plastic mulch bed system on a sandy soil, *Scientia Horticulturae*, Vol.116, pp.8-16.

Zotarelli, L.; Scholberg, J.M.; Dukes, M. D.; Munoz-Carpena, R. & Icerman, J. (2009). Tomato yield, biomass accumulation, root distribution and water use efficiency on a sandy soil, as affected by nitrogen rate and irrigation scheduling. *Agricultural Water Management*, Vol.96, pp.23-34.

Zotarelli, L.; Scholberg, J.M.; Dukes, M.D. & Muñoz-Carpena, R. (2008b). Fertilizer residence time affects nitrogen uptake efficiency and growth of sweet corn, *Journal of Environment Quality*, Vol.37, pp.1271-1278.

Zvomuya, F.; Rosen, C.J.; Russelle, M.P. & Gupta, S.C. (2003). Nitrate leaching and nitrogen recovery following application of polyolefin-coated urea to potato, *Journal of Environmental Quality*, Vol.32, pp.480-489.

Part 4

Soil Nitrogen Management

8

Strategies for Managing Soil Nitrogen to Prevent Nitrate-N Leaching in Intensive Agriculture System

Liu Zhaohui et al.[*]
*Institute of Agricultural Resources and Environment,
Shandong Academy of Agricultural Sciences Jinan, 250100,
China*

1. Introduction

Nitrogen fertilizer has played a major role in the global food production over the past 60 years. And about 50 percent of total N comes from fertilizer supply. However, fertilizer N has a low efficiency of use in agriculture (10-50 percent for crops grown in the fields). One of the main causes of low efficiency is the large of N by leaching, runoff, ammonia volatilization or denitrification with resulting in pollution of groundwater and atmosphere. With the limitation on arable land area and the demand for more and more food production, the only way is to increase the efficiency of use of fertilizer N. Thus, it is important to know the forms and pathways of N loss and the factors controlling them so that procedures can be developed to minimize the loss and increase N use efficiency (NUE). A conceptual scheme indicates the nitrogen cycle in crop production systems. Annual N input was about 170 Tg, and about half of added N is removed from the field as harvested crop (85 Tg). The remainder of the N, defined as surplus N, either is lost to the environment or accumulates in the soil (Fig.1).

Food demand of the public is the major promotion for rapid development on intensive agriculture, which is becoming a dynamic industry in China. However, with an excess amount of nitrogen from animal manures and commercial fertilizers, many pollutant incidents have been found and reported on nitrate contamination in intensive agriculture especially in greenhouse vegetable production systems (Zhang, 1996; Ju, 2007; Li, 2002; Song, 2009). Environmental and economic concerns have prompted agriculture researchers and producers to seek for more and more efficient strategies for nutrient managements. The present public concerns on nitrate management are focusing on N, which exceeds crop demand and might migrate from agro-ecosystem to groundwater and surface water (Daniel, 1994). Economic considerations in nitrate managements mainly focus on efforts to improve N utilization and reduce costs of N inputs. Based on its necessity for mobility in the soil and risk to environment systems, popular N management thus aims to balance N inputs with

---
[*] Song Xiaozong, Jiang Lihua, Lin Haitao, Xu Yu, Gao Xinhao, Zheng Fuli, Tan Deshui, Wang Mei, Shi Jing and Shen Yuwen
Institute of Agricultural Resources and Environment, Shandong Academy of Agricultural Sciences Jinan, 250100, China

crop requirements, and decrease the nitrogen loss to the groundwater when irrigation or rainfall occurs. This chapter provides an overview of the general role of nitrogen in agro-ecosystems and then discusses how various N management practices can contribute to preventing nitrate-N leaching in intensive agricultural systems.

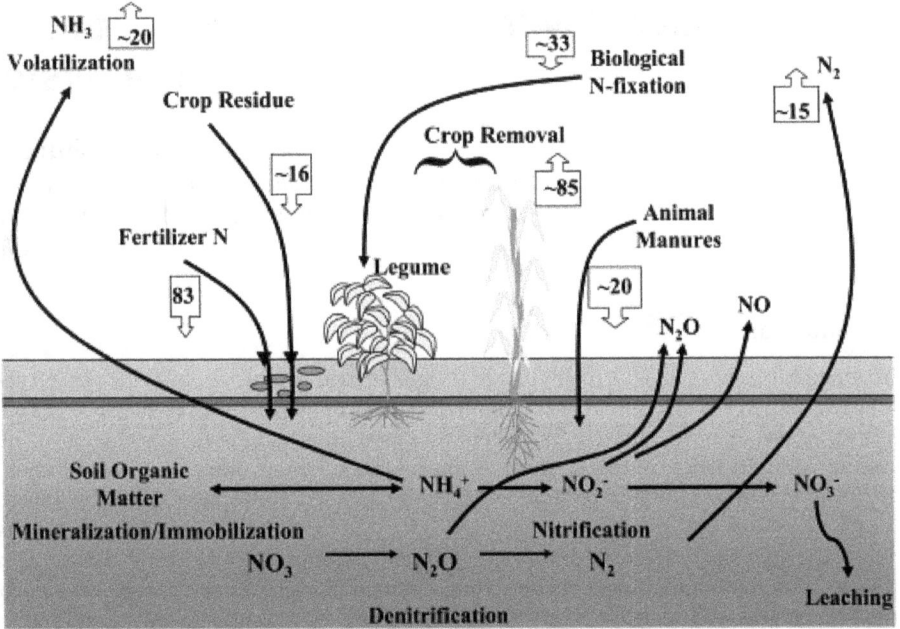

Fig. 1. A conceptual scheme on nitrogen loss in crop production(Tg N). By Mosier, 2004

2. Field conditions for Nitrate-N Leaching

Leaching refers to the movement of N in water moving downward through the soil profile and out of the rooting zone. As the key N form for uptake by most crops, nitrate (NO_3^-) found in soil is usually used to indicate the abundance of N that can be taken up especially in well-aerated soils. As an anion, nitrate usually remains in the soil solution and therefore is relatively free to move with water flows. Drainage of excess water often drives NO_3^- downward through the soil profile and out of the rooting zone, and thus nitrate leaching occurs. Most (usually>95%) of the N in soils is not susceptible to crop uptake until it is exchanged into available N as mineral N form(NO_3^-,NH_4^+) by soil microorganisms under many conditions such as some environmental factors(temperature, moisture availability, aeration status, etc.) and N types or amounts of organic N present in soils. That is to say, the importance of N losses by leaching varies greatly with factors that determine how much and when water flows downward through soils.

Two major inputs, including significant amount of NO_3^- in the soil profile and sufficient precipitation or irrigation water are necessary when nitrate leaching occurs. Substantial N losses occur in systems where mineralization or fertilization results in high concentrations of NO_3^- during periods when leaching is likely. Retained NO_3^- in the soil profile usually comes from many sources such as mineralization of SOM (soil organic matter) crop residues, manures and synthetic fertilizers application. There is also high nitrate leaching potential when rainfall and irrigation events in intensive agricultural systems due to the shallow root systems for cereal and vegetable crops with poor N fixation ability except for cover crop. Flows of excessive water inputs often increase the mobility by moving the soluble N from the soil surface to depths where crops roots cannot uptake and thus that leads to substantial storage capacity for NO_3^--N in the soil profile.

It is difficult for growers to coordinate smoothly NO_3^- leaching control and economic benefits during the crop's active growth period with significant amounts of N input to the soil. Therefore, it is necessary for nutrient managers, soil and environmental scientists to develop the effective strategies to reduce nitrate leaching, which require to know detailed plans on nutrient, water and crop management considering N source, N application method, rate and timing, and others including soil properties and moisture, evapourtranspiration, and crop systems for local conditions and specific sites. Best management practices (BMPs), which reduce N and irrigation inputs without lowering crop yields, are popular and major strategies to reduce nitrate leaching.

2.1 Fertilizer nitrogen management strategies

Applications of animal manures or commercial fertilizers often add more N than that is taken up by crops during a year in systems. The amount of N added for production of many field crops, for example, is the quantities of N expected to be taken up by the crop plus enough extra N to compensate for losses of N expected to occur(Stanford,1973; Bock & Hergert, 1991).

2.1.1 Appropriate N application rate

Excessive N application is generally the universal reason for nitrate leaching. The remainder of N added usually remains in the soil at the end of the cropping season and is vulnerable to leaching. Many literatures reported that nitrate leaching increases rapidly with elevated N application rate (Zvomuya,2003; Guo,2006). Therefore, proper N input is the major consideration for nitrate leaching control. And producers are forced to use a rate that will give them an economic optimum return over the long run. However, most of the current recommendations for the crops requirements are generally at a excessive rate which will result in a marked increase in the loss of N leaching from the greenhouse vegetables (Fig. 2)

It is difficult to establish a nitrogen rate that will be appropriate for every year, since there are several biological reactions that influence the availability of nitrogen for crop use. The economical optimum nitrogen rates varied greatly between different growing seasons even if for the same field. Excess N use in crop production is often identified as a major contributor to NO_3^- enrichment of ground water. Little information is available to show the specific relationships between crop management systems and N fertilizer use on the amounts of NO_3^- lost by leaching. A study was conducted to determine the effect of several cropping systems and N rates, providing a range of N availability to corn (*Zea mays L.*), on soil water NO_3^- concentrations and leaching below the root zone. Four cropping-manure management systems were established in 1993 - 1994 (8-site years) at Arlington, WI, on a Plano silt loam. Ammonium nitrate (0 to 204 kg N ha^{-1} in 34-kg increments) was broadcast at the time of corn planting. The results showed that nitrate N concentrations in the samplers increased as the amount of N applied in excess of the observed EONR increased.

Predicted soil water NO_3^-–N concentration at EONR was 18 mg L^{-1}. Average NO_3^-–N concentrations were <10 mg L^{-1} where fertilizer N rates were >50 kg N ha^{-1} below the EONR and >20 mg L^{-1} where fertilizer N rates were >50 kg N ha^{-1} above the EONR. An end-of-season soil NO_3^- test appears to be capable of evaluating corn N management practices and indicating the amount of excess N fertilizer applied that may be leached from the root zone. These results illustrate the direct relationship between NO_3^- loss by leaching and N application rates that exceed crop needs (Andraski, et al., 2000). Limiting the amount of inorganic N within the soil at the end of a crop's growing season and before the next crop has established an extensive root system is a key factor for reducing N losses. Therefore, although timing, method of N application, and accounting for mineralizable soil N are important for reducing potential NO_3^- leaching, it was concluded that the most important factor was to apply the correct amount of N fertilizer (Power & Schepers 1989).

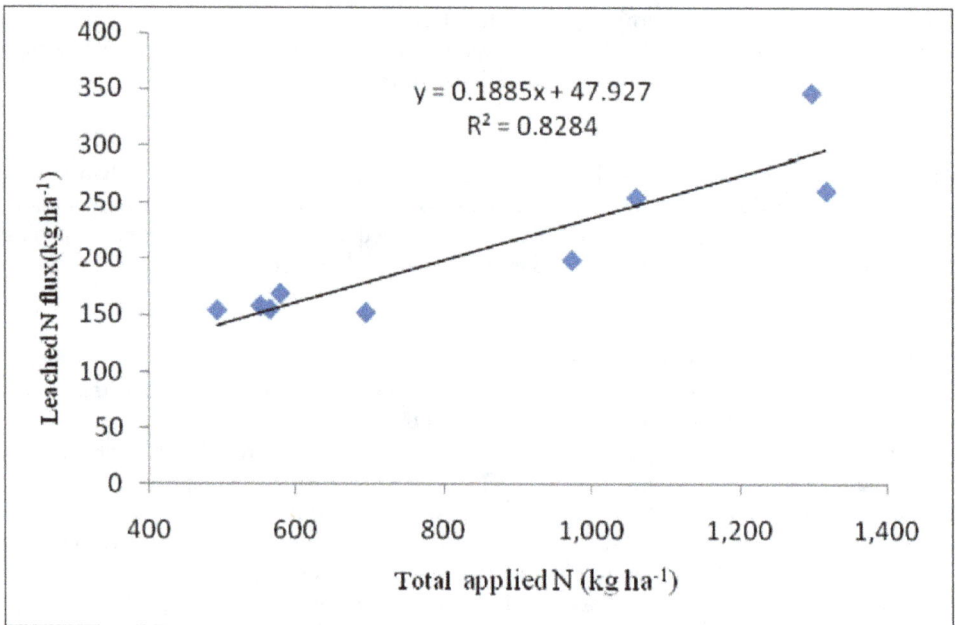

Fig. 2. A linear trend between total applied N and leached N flux.(from Song, 2009)

2.1.2 Timing N application in harmony with crop demand

It is necessary for BMPs development to consider timing of N supply and crop need, i.e., apply N at a proper phase that allows rapid crop uptake. For many crops, cumulative N demand usually follows an S-shaped curve, with slow uptake rate during establishment and an exponential utilization in the vegetative and reproductive phases. Splitting N application is thus recommended by applying N in phase with crop demand, providing high soil-N concentrations at different periods needed for crop growth while minimizing the time with leaching losses risk (Power et al, 1998). And it was reported that decrease in NO_3^- leaching and increase in anticipant yields for potatoes by adopting BMPs (Kelling, 1994; Errebhi, 1998; Waddell, 2000).

In summary, although some of the studies have emphasized the importance of BMPs, they do not provide a complete solution. The effects of these BMPs on reducing NO_3^- leaching are variable, ranging from no effect (Osborne & Curwen, 1990) to 30% reduction (Mechenich & Kraft, 1997). This indicates that BMPs should be carefully evaluated for specific conditions.

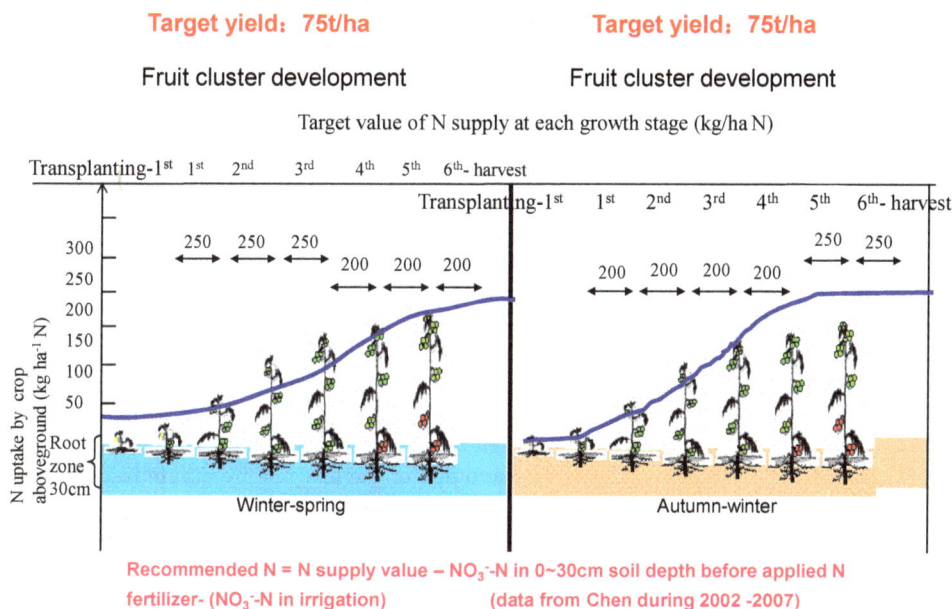

Fig. 3. Optimum N management based on plant N demand and soil N supply for greenhouse tomato cropping system. By Ren, 2009

In recent years in China, unreasonable nitrogen fertilization management in intensive vegetable production region always results in some serious environmental problems, which limit the sustainable development of local vegetable industry. Long-term field experiments were conducted in the six successive greenhouse tomato growing seasons in Shouguang, Shandong province from 2002 to 2007 (Fig.3). Compared with conventional N management, N fertilizer input with site-specific N management averagely reduced by 72% without any significant fruit yield reduction in all seasons. N agronomic efficiency of site-specific N management was 69 kg FW/kg N and value-cost ratio (VCR) was 27.8. According to fruit yield forming and N uptake pattern, the critical period for fertilization was carried out in April and October in winter-spring and autumn-winter growing season, respectively. During the critical periods of fertilization 3-4 events of side-dressing was needed with every 7 or 10 days at a rate of 50 kg N ha^{-1}. With conventional N management, the yearly total N input, including Nmin residue in 0-90cm soil profile before transplanting, N from chemical fertilizer, manure and irrigation water, was 2917 kg N ha^{-1}. However, the apparent N loss was 1816 kg N ha[-1] through leaching, soil fixation, gaseous emission etc. In contrast to conventional N management, N use efficiency in site-specific N management increased by 7% and up to 25% while N input and apparent N loss reduced by 44% and 57% on average, respectively (Ren, 2009).

2.1.3 Slow release fertilizer and Nitrification Inbibitor application

The agronomic and environmental benefits by applying slow-release fertilizer and nitrification inhibitor have been reported with reducing NO_3 leaching and improving NUE (Zerulla, 2001; shaviv, 1993). However, the performance of these newly products depends on climate condition and soil type with temporal and spatial variation.

In contrast to traditional fertilization, nitrate nitrogen content in soil and yield factors of wheat-maize applied slow/controlled release fertilizer with different amounts and rates in North China was analyzed (Fig4, 5). The results indicated that nitrate nitrogen content still maintain high level during late growing period, and the yield traits such as panicle number, 1000 grain weight as well as actual yield keep high level if fertilized according to recommendation at the ratio of 6:4:2 at pre-sowing, reviving and jointing respectively (the treatment named as 100% UD). The formulated slow/controlled release fertilizer (CSR) showed lower nitrate nitrogen content in soil but had no influence on yields compared with 100% UD. CSR showed a positive impact on maize production, e. g. increased fertilizer use efficiency, decreased bare top length. The yield of 80% SCR is 18.3% higher than that of 100% UD (Table 1). Crop could absorb nutrient timely and fully with the application of SCR because the nutrient is released slowly and avoided of the loss risk by nitrate leaching. On the whole, further study on how to optimize SCR fertilization distribution during the wheat and maize growing season was needed for the purpose of increasing economic and environmental benefits simultaneously (Lu et al., 2011).

Recent years, researchers have tried their best to control nitrogenous fertilizer loss and its pollution to environment. The mixed nitrogen nutrition becomes one of the new methods to enhance the effectiveness of nitrogen utilization and reduce the nitrogen loss. In the field conditions, enhanced ammonium nutrition (EAN) by using the nutrification inhibitor in soil becomes a very good way to achieve the mixed nitrogen nutrition. Cotton is sensitive to nitrogen utilization. Meanwhile, due to growing in the hot season, the irrational nitrogenous fertilization utilization will lead to not only the poor cotton growth but also the larger nitrogen loss and environment pollution. With Bt-transgenic cotton 33B as experimental plant and Dicy anodiamide (DCD) as nitrification inhibitor, the effects of nitrogenous fertilizer strategies (including DCD and non-DCD treatment in different nitrogen levels) on the nitrogen accumulation in cotton field soil and cotton functional leaves were discussed. The results showed: 2% DCD treatment enhanced ammonium absorption (increased NH_4^+ - N from 0.70% to 112% in the main stem leaves and from 8.84% to 46.47% in the bearing stem leaves) and restrained nitrate absorption of cotton(decreased NO_3^--N from 0.20% - 22.68% in the main stem leaves and from 0.10% to 28.03% in the bearing stem leaves), the extent of influence is different from one growth stage or nitrogen level from another; at the same time, it reduced the content of rudimental total nitrogen(decreased extent from 0 to 14.39%) and maintained the higher content of ammonium nitrogen(increased extent from 1.11% to 17.83%) in cotton and enhanced the efficiency of nitrogen fertilizer as well as saved nitrogen resource. These above mentioned further showed that it was important to treat cotton with EAN from the physiological and ecological perspectives.

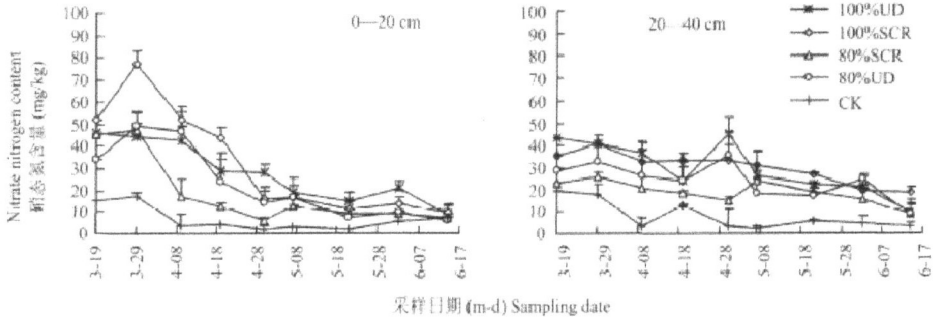

Fig. 4. Comparison of soil nitrate nitrogen content of coated urea, slow-releasing fertilizer and urea under the reduced application rate in winter wheat season

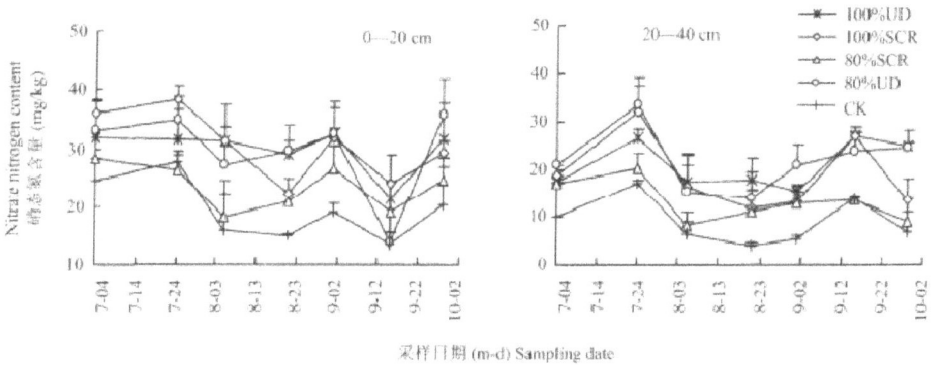

Fig. 5. Comparison of soil nitrate nitrogen content of coated urea, slow-releasing fertilizer and urea under the reduced application rate in summer maize season

Treatments	Winter wheat (kg ha-1)	Summer maize (kg ha-1)
100%UD	3705 a	7045 c
100%SCR	3525 a	7733 ab
80%SCR	3495 a	8336 a
80%UD	3150 a	7671 bc
CK	3225 a	6858 c

Note: Values followed by different letters within a column are significantly different at $P < 0.05$ level. UD- local conventional fertilization; SCR-slow /controlled release fertilization; CK- no fertilization.

Table 1. Difference significance analysis for actual yields of winter wheat-summer maize system in different fertilizing treatments

2.2 Soil management strategies
2.2.1 On-site soil NO3- test
Soil test is defined as "…rapid chemical analyses to the plant-available nutrient status, salinity, and elemental toxicity of a soil ¨a program that includes interpretation, evaluation,

fertilizer and amendment recommendations based on results of chemical analyses and other considerations"(Peck & Soltanpour, 1990). The purpose of soil test is to provide a basic parameter for soil management decisions usually for agricultural systems where yield and quality are the ultimate goals. Also it can be applied for other goals such as human health and environment protection. All modern soil testing programs generally have four basic components: (1) sample collection and handling; (2) sample preparation and analysis; (3) interpretation of analytical results; and (4) recommendations for action. For successful soil test, each component should be conducted properly without no error at each step. Soil test for nitrogen differs greatly between arid and humid regions. In most cases, samples for residual mineral N must be collected to deeper depths (60-200cm) than for standard soil testing (0-20cm) seldom. Due to it, a sample collected from the rooting zone shortly before the start of the growing season and analysed for residual mineral N (NO_3^--N, NH_4^+-N) can precisely determine plant-available N, and thus N inputs are diminished accordingly. Soil samples are often only analysed for NO_3^--N because it is usually the dominant form of inorganic N in most soils in China. Proper handling is critical to avoid changes during storage after sample collection. To avoid these problems, samples extraction of inorganic N should be rapidly done and usually accomplished by shaking some quantitative fresh samples for 30 min to 1 h with a salt solution [e,.g., 2 M KCL, 0.01 $CaCl_2$], followed by filtration. Automated colorimetry is usually used to determine mineral N (NO_3^--N, NH_4^+-N) in soil extraction at present, however, ion chromatography, steam distillation, ion electrodes, and micro-diffusion techniques are also applied for specific needs (Bundy & Meisinger, 1996).

The pre-sidedress soil nitrate test (PSNT) is a valuable method for soil N test and therefore is commonly used in many fields. It is originally developed for corn but now being investigated for wider range of agronomic and vegetable crops (Magdoff, 1984; Bock &Kelley, 1992). The PSNT was described and used to provide a timely monitoring of soil NO_3^--N pool, get guidance for sidedress fertilizer N recommendations and evaluate the ability to nitrate leaching reduction (Bundy, 1994, 1999; Guillard, 1999). It was concluded that PSNT, compared with the conventional N-management system, could reduce fertilizer N, lower nitrate leaching, and diminish the potential for nitrate contamination to groundwater (Durieux, 1995). In China, Field experiments were conducted in a greenhouse of Shouguang city, Shandong province to validate integrated nitrogen management and PSNT techniques in monitoring nitrate dynamic in root zone and corresponding recommendation of sidedressing N fertilizer for greenhouse tomato in spring and autumn seasons in 2004. Considering the target yield level, FW 84 t ha-1, in spring, the rate of N supply (soil NO_3^--N in root zone + N from irrigated water + sidedressed N) were N 300 kg ha-1 of each sidedress at the first, second and the third cluster fruit expanding stage (CFES) , and N 200 kg ha-1 in the later growing stage. Similarly, when the target yield was FW 75 t ha-1, in autumn, the rate of N supply were N 200 kg ha-1 of each sidedressing at the first, second third and the fourth CFES, and N 250 kg ha-1 in the later growing stage in autumn season. Including organic manure application as conventional way, optimized N treatment reduced N application rate by 62% and 78% of total N fertilizer in spring and autumn season, respectively, compared to conventional treatment, because environmental N (N released from organic N pool and N from irrigated water etc.) contributed considerable N to tomato growth. Compared to conventional N treatment, apparent N loss in soil vegetable crop system significantly reduced in optimized N treatment while the yield was the same as that of conventional treatment. It was concluded that integrated nitrogen management, together with PSNT technique, was very useful in increasing nutrient efficiency and reducing the risk of environmental pollution (He et al., 2006). For some open field vegetables, the same results are listed. Trials were conducted in 15 commercial fields in California in 1999-2000 to evaluate the use of presidedress soil nitrate testing (PSNT) to determine sidedress N requirements for lettuce production. In each field a large plot (0.2-1.2 ha) was established in which sidedress N application was based on presidedress soil NO_3^--N concentration. Prior to each sidedress N application scheduled by the cooperating growers, a composite soil sample (top 30 cm) was collected and analyzed for NO_3^--N. No fertilizer was applied in the PSNT plot at that sidedressing if NO_3^--N was >20 mg kg^{-1}; if NO_3^--N was lower than that threshold, only enough N was applied to increase soil available N to 20 mg kg^{-1}. The productivity and N status of PSNT plots were compared to adjacent plots receiving the growers' standard N fertilization. Cooperating growers applied a seasonal average of 257 kg ha^{-1} N, including one to three sidedressings containing 194 kg.ha^{-1} N. Sidedressing based on PSNT decreased total seasonal and sidedress N application by an average of 43% and 57%, respectively. The majority of the N savings achieved with PSNT occurred at the first sidedressing. There was no significant difference between PSNT and grower N management across fields in lettuce yield or postharvest quality. At harvest, PSNT plots had on average 8 mg kg^{-1} lower residual NO_3^--N in the top 90cm of soil than the grower fertilization rate plots, indicating a substantial reduction in subsequent NO_3^--N leaching hazard. It was concluded that PSNT is a reliable management tool that can substantially reduce unnecessary N fertilization in lettuce production (Breschini & Hartz, 2002).

In practice, the PNST actually measures, that is to say, it is not the determination of residual inorganic N, but a field-based expression of the capacity to provide an adequate supply of mineral N during the growing season (i.e., of the soil N mineralization potential). The PNST has been successfully used to evaluated in amounts of field studies in many countries especially in the United States (Magdoff et al.,1990; Meisinger et al.,1992; Sims et al.,1995; Sogbedji et al., 2000). The general approach used to make N recommendation was summarized as follows (Tisdale et al, 1993):

$$N_{fertilizer} = N_{crop} - N_{soil} - (N_{organic\ matter} + N_{previous\ crop} + N_{organic\ waste}) \qquad [1]$$

$N_{fertilizer}$ — amount of N needed from fertilizers, manures, biosolids, etc.
N_{crop} — crop N requirement at realistic yield goal
N_{soil} — residual soil inorganic N
$N_{organic\ matter}$ — N mineralized from soil organic matter
$N_{previous\ crop}$ — residual N available from previous legume crops
$N_{organic\ waste}$ — residual N available from previous organic waste use such as animal manures, biosolids wastewater irrigation, etc.

2.2.2 Tillage

Many literatures showed that tillage alter the soil environment and thus lead to elevated oxidation of SOM and mineralization of soil N (Randall, 1997a). Therefore Chinese farmers conventionally adopt tillage to manage soil N depending on tillage to release N for crop production. Effects of tillage on nitrate leaching control have been demonstrated in studies comparing no-till with conventional tillage in Iowa in USA (Kanwar, 1993; weed, 1996). It was concluded that despite having higher average NO_3^--N concentration in drainage water, tillage had less nitrate leaching losses than no-till under continuous corn systems due to its higher water retention capacity. Ploughless soil tillage impacts on yields and selected soil quality parameters is reviewed from the Scandinavian countries of Denmark, Finland, Norway and Sweden. The success of reduced tillage and direct drilling depends on the crop species as well as on the soil type and the climatic conditions. The best results seem to be obtained on the heaviest clay soils, which is the most difficult soils to prepare with conventional soil tillage methods. Satisfactory yields were obtained after ploughless tillage in winter wheat (*Triticum sp.*), winter oil seed rape (*Brassica sp.*) and late harvested potatoes (*Solanum tuberosum L.*). The influence of crop rotations and preceding crops in ploughless tillage systems for small grain cereals has received relatively little attention. Also, fertilization of reduced tilled crops has received too little attention, but it seems that nitrogen cannot compensate for sub-optimal tillage. One of the most striking effects of ploughless tillage is the increased density of the soil just beneath the depth of tillage. Nutrients and organic matter accumulated near the soil surface after ploughless tillage, and in the long run the soil reaction (pH) declined. Nearly all species of earthworms increased in number in ploughless tillage. The leaching of nitrogen seemed to increase with more intensive cultivation, particularly when carried out in autumn (Rasmussen, 1999).

Effects of tillage on N management have been demonstrated in studies comparing no-till with conventional tillage at several mid-western locations in America. In a long-term Minnesota study (Randall & Iragavarapu, 1995), residual soil NO_3 contents in the 0-1.5-m soil profile using were significantly higher with conventional tillage than N no-till for 5 out of 11 yr and were not significantly different for the other 6 yr. Average flow-weighted NO_3– N concentrations were 13.4 and 12.0 mg L^{-1} for conventional and no-till corn production treatments, respectively. Furthermore, while the no-till treatment had 12% greater subsurface drainage flow than the conventional treatment, NO_3 losses were marginally greater with conventional tillage. Although insignificant, these results suggest a minimal trend toward greater NO_3 losses with conventional tillage in this study. The authors concluded that NO_3 losses through tile drainage depend more on growing-season precipitation than on tillage. Recently, it was also concluded that NO_3 losses from agricultural fields are minimally affected by differences in tillage systems compared with N management practices (Randall & Mulla, 2001).

2.3 Crop management strategies
2.3.1 Introducing cover crops
Except for minimizing N losses, growing cover crops is another well-established method of managing nitrate leaching. The goal is to add crops to capture or recover residual N in the soil after main crop harvest. In recent years, the use of cover crops to reduce nitrate leaching has received much interest in many locations (Delgado, 1998; Logsdon, 2002) in addition to protecting soil from salinization. The choice of cover crops species depends on the cropping system, amount of fallow time, climate and soil type (Meisinger, 1991; Jackson, 1993). In general, shallow-rooted vegetable crops are vulnerable to higher nitrate leaching losses than deep-rooted vegetable crops. Cover crops thus can be introduced and act as scavengers that recover nitrate leached from the precious vegetable crop (Shrestha, 1998), and even reduce the NO_3^- leaching losses that occurs in the next crop (Delgado, 1998; 2001a).

The influence on nitrate leaching of ryegrass (*Lolium perenne L.*) used as a catch crop in spring barley (*Hordeum vulgare L.*) was investigated during three successive years in a lysimeter experiment on a sandy loam soil. Four treatments were included with combinations of time of tillage (November/March) and handling strategy of the aboveground ryegrass biomass (return/removal). Reference plots tilled in March were sown to spring barley alone. The ryegrass reduced nitrate leaching by 1.4–4.3 g N m^{-2} year^{-1} when incorporation took place in November. If incorporation was carried out in March, reductions in nitrate leaching were 2.1–5.6 g N m^{-2} year^{-1}. The herbage cut of ryegrass had accumulated 1.0–2.4 g N m^{-2} year^{-1} and 0.9–2.1 g N m^{-2} year^{-1} in November and March, respectively. Nitrate leaching losses increased with higher rates of N both with and without a catch crop. Grain yield and N uptake of the spring barley were unaffected by a catch crop and the management strategy did not interact with N fertility level. The study showed that growing a ryegrass catch crop repeatedly for three years was effective in reducing nitrate leaching losses, but the retained N did not have any immediate beneficial effect on spring barley grain yield (Thomsen, 2005).

Planting cover crops immediately after harvest or relay with main crop is important as it reduces fallow period and allows enough crop growth to accumulate soil N before winter NO_3 leaching (Fielder & Peel, 1992). For example, a rye cover crop planted on 1, 14 and 30 October in Maryland showed an increase in N accumulation and a decrease in soil NO_3-N with early planting (Staver & Brainsfield, 1998). In a loamy sand soil, over-winter cover crops (e.g., wheat, Triticum aestivum) planted after early potato (Milburn et al., 1997); and wheat, rye, rapeseed (B.napus) seeded after sweet corn and incorporated in spring appeared to be most effective in recycling N to potato crop (Weinert et al., 2002).

A strategy of over seeding cover crops (e.g., oat) after 80 DAE of potato when N uptake is negligible can capture residual fertilizer N or soil N from late season mineralization. Relayed oat crop can capture unutilized N from potato, and rye can capture mineralized N from oat and residual N from potato, if there is any (Bundy & Andraski, 2005). Cover crops can be incorporated in winter just before soil freezes, which can recycle nutrients for succeeding crops.

Guo et al (2008) reported that total N uptake by sweet corn at harvest was 187 kg N ha^{-1} in 2005 and 154 kg N ha^{-1} in 2006 (Fig.6). Shoot N uptake by sweet corn was up to 42 and 56 kg N ha^{-1} from sweet corn transplanting to 21 July in 2005 and 2006. During the later growth stages of sweet corn (from 20 July to the end of the harvest), the amounts of N removed by the sweet corn shoots were 131 and 112 kg N ha^{-1} in 2005 and 2006. After three continuous cucumber growing periods with root zone N management, soil N_{min} was lower at the beginning of the fallow period in 2006 than in 2005. In 2005 no significant difference was found in N_{min} content between sweet corn cropping and the fallow treatment in the top 0.3 m of the soil profile. However, sweet corn cropping evidently depleted N_{min} compared to the fallow treatment in 2006. In both years soil N_{min} at 0.3–0.9 m depth was reduced by sweet corn cropping with root growth and N uptake occurring in contrast to the fallow period. At the sweet corn harvest less N_{min} was retained in the top 1.8 m of the soil profile under sweet corn cropping compared with the fallow period (Fig. 8). Soil N_{min} in the top 1.8 m of the soil profile was reduced by 333 and 304 kg N ha^{-1} with sweet corn cropping compared to the fallow treatment in 2005 and 2006 and soil water content in the 0–1.8 m soil layer increased with or without sweet corn cropping in contrast to the beginning of the fallow period. No significant difference in soil water content was found between N_{mr} and N_{mr+C} treatments after the fallow period (Fig. 7).

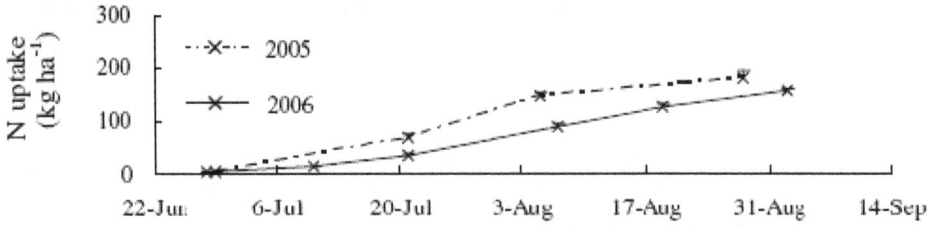

Fig. 6. Total N uptake by sweet corn at harvest. (from Guo et.al., 2008)

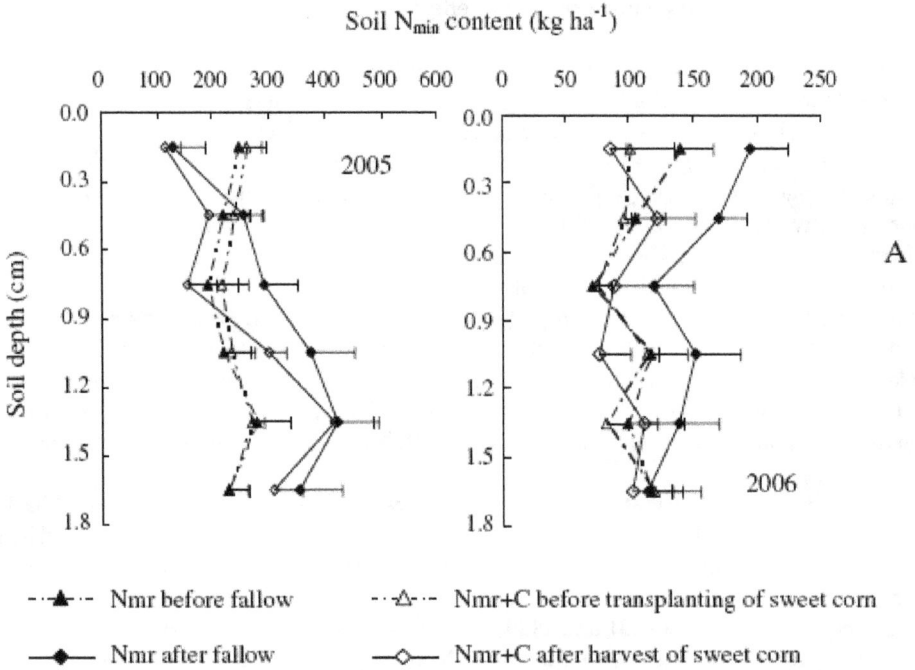

Fig. 7. Soil N_{min} content throughout the top 1.80 m of the soil profile with or without sweet corn as catch crop in the summer fallow period in the greenhouse cucumber cropping system. Bars represent SD of means with three replicates per treatment. (From Guo et.al., 2008)

2.3.2 Manipulating diversified crop rotation

It was showed that change from continuous corn to diversified crop rotation is a better solution for soil residual N by planting crops or varieties with different rooting depths (Ju, 2007). Crop root depths were negatively correlated with NO_3^- leaching, and thus rotations of potato with barley, winter wheat and cover crops were good choices of helping to improve NUE while reducing nitrate leaching (Delgado, 1998; 2007).

Factors contributing to differences in NO_3^- leaching potential for various crop rotations extend beyond fertilizer practices. Interactions between hydrology and tillage are very important because any residual NO_3 that accumulates in the soil profile, whether from N fertilizer or microbial processes, can be leached if it is not assimilated by microbes decomposing crop residue or taken up by another plant. When a crop such as alfalfa depletes profile water content and the amount of precipitation is not sufficient to fully recharge the profile, the leaching potential will be minimal and very little water will be moving into subsurface drainage lines. Differences in residue and root decomposition relationships, as well as soil–plant–water dynamics, among various plant species also influence the leaching potential (Baker & Melvin, 1994; Randall, et al., 1997a; Malpassi, et al., 2000). The rate of N cycling is important because although N-fixing legumes can release large quantities of N to soils over time, organic N derived from plant and microbial residues is not as rapidly available to plants as inorganic N provided by most commercial fertilizers. Additionally, the gradual release of organic N often better synchronizes with subsequent plant needs and microbial population dynamics than point-in-time applications of N fertilizers. The large flush of available N following an inorganic fertilizer N application can often supply more N than can be assimilated by plants and microbes. When this pool is nitrified, large amounts of NO_3 are susceptible to leaching and can potentially contaminate surface and ground water resources.

Better use of soil resources (nutrients and moisture) can also be done by including crops or varieties with different rooting depths (deep and shallow) in a crop rotation (Shrestha & Ladha, 1998). Rooting depths was positively correlated with N use efficiencies and the capability of crops to mine NO_3 from ground irrigation waters (Delgado, 2001a). Crop root depths were negatively correlated with NO_3 leaching. Commercial operations that used cover crops and crops that were rooted more deeply were able to increase the N use efficiency of their farm operations while minimize the amount of residual soil NO_3 in the profile and NO_3 leaching to groundwater (Delgado, et al. 2000; 2006). The deeper rooted crops acted as a biological filter that recovered NO_3 from irrigated groundwater, helping to mine the NO_3 (Delgado et al. 2007). Rotations of potato with barley, winter wheat and cover crops help to increase N use efficiency in the system while minimizing NO_3 leaching (Delgado, 1998). Including alfalfa in a rotation especially in moderate sandy soil is also an effective approach in reducing leaching because of its deep rooting and high water usage (Owens, 1987).

2.3.3 Managing plant residues

As plant tissue is a primary source and sink for C and N, rational management of plant residues can affect N cycling in soils during the growing season and contribute to N immobilization and release in synchronization with crop demand. Plant residue decomposition proceeds depending on the C/N ratio, temperature, water content and other factors. Therefore, this was very important to understand the factors affecting plant residue decomposition and how to manipulated to reduce NO_3 leaching potential with the availability of N to the main crop (Varvel, 1990; Gale, 2000; Dai, 2010).

The amount of crop residue N varies with the crop species, varieties, management practices, climate, and soil. Recovery of fertilizer N in potato is about 50% with current management practices. Distribution of fertilizer N recovery in potato averaged 24% in tubers, 9% in residue, 14% in soil, and 53% leached (Bundy & Andraski, 2005). This suggests that 23% of residues and soil N could be returned to the soil, if properly managed. A study conducted in Canada with cauliflower, red cabbage and spinach residues incorporated in autumn and spring and mulched in autumn showed greater risk of NO_3 leaching with autumn residue handling compared to spring handling (Guerette et al., 2002). Autumn handling of cauliflower residues and both incorporation treatments (spring and autumn) for red cabbage residues contributed significant amounts of N to the following wheat crop (equivalent to 27 to 77 kg N ha^{-1}). Incorporation of crop residue with high carbon to nitrogen ratio should be encouraged to immobilize residual NO_3 left in the root zone (Brinsfield & Staver, 1991).

Effects of different returning amount of maize straw on soil fertility and yield of winter wheat were studied using randomized block design in Loess Plateau in China (Zhang et al., 2010). The results showed that straw returning can increase soil organic matter content and reduce soil total nitrogen loss , enhance the capacity of soil microbial fixing and supplying C and N , increase C/N , and change the distribution of soil microbial community. Higher soil microbial C/N and redistribution of original soil microbial community was propitious to the soil organic transformation and mineralize carbon decomposition, as consequence, improve the soil nutrient supply. The results indicate that under condition of local study area, applying N 138 kg ha^{-1}, combined with returning amount of maize straw 9000 kg ha^{-1} can enhance soil fertility and increase yield by 7.47% significantly (Table 2 and 3).

Items	Treatments					
	Before sowing	ST0	ST6000	ST9000	ST12000	ST15000
BC/TC	1.07±0.03 c	0.76±0.02 d	1.08±0.03 c	1.37±0.04 a	1.14±0.03 b	1.15±0.03 b
BN/TN	2.23±0.07 d	2.82±0.14 c	3.04±0.14 bc	3.82±0.11 a	3.22±0.15 b	2.85±0.09 c

Note: Values followed by different letters within a row are significantly different at $P < 0.05$ level. BC/TC-Microbial biomass C/Total C, BN/TN-Microbial biomass N/Total N.

Table 2. BC/ TC and BN/ TN of soil with different treatments

Treatment	Yield	Increase
ST0	6401.9±38.4 c	
ST6000	6528.7±44.7 b	1.98
ST9000	6880.2±68.8 a	7.47
ST12000	6508.9±45.6 b	1.67
ST15000	6263.3±31.3 d	-2.16

Note: Values followed by different letters within a column are significantly different at $P < 0.05$ level.

Table 3. Wheat yield with straw into field treatment

2.4 Water management strategies

Nitrate leaching is driven by water transport through the soil profile, so good irrigation strategies including proper amount at the right time, are greatly important to N leaching. N management alone cannot effectively reduce NO_3 leaching, while N management scheduling, an important tool for water management, should integrate local factors such as soil moisture, infiltration, texture, crop water use and rainfall. Optimization measures on N and irrigation with frequent but little amounts can decrease NO_3 leaching losses without yield reduction (Saffigna, 1977; Waddle, 2000). Many studies showed that drip irrigation would be a useful choice in reducing N leaching with improving water use efficiency (Waddle, 1999; 2000). This subsurface irrigation system is effective in reducing NO_3 leaching due to its low irrigation amount since it delivers water directly to the root zone where N uptake is greatest (Starr, 2008).

Nitrogen management alone also cannot effectively reduce NO_3 leaching in sandy soils. It is a challenge to supply water to the crops on a sandy soil, which has low water holding capacity, while trying to minimize leaching. Good irrigation strategies (the right amount at the right time) are important as irrigation amount and timing are strongly related to leaching, especially in sandy soils (Cates & Madison, 1994). It was reported that 40% reduction in irrigation amount could help to reduce the risk of NO_3 leaching from potato without affecting yield (Waddell et al., 2000).

A recent study reported that water content within the center of the potato hill, where the greatest densities of roots occur, were greater under drip irrigation than those of sprinkler irrigation (Cooley et al., 2007). Therefore, management strategies targeted at wetting the hill center would likely improve water use efficiency (Starr et al., 2005).

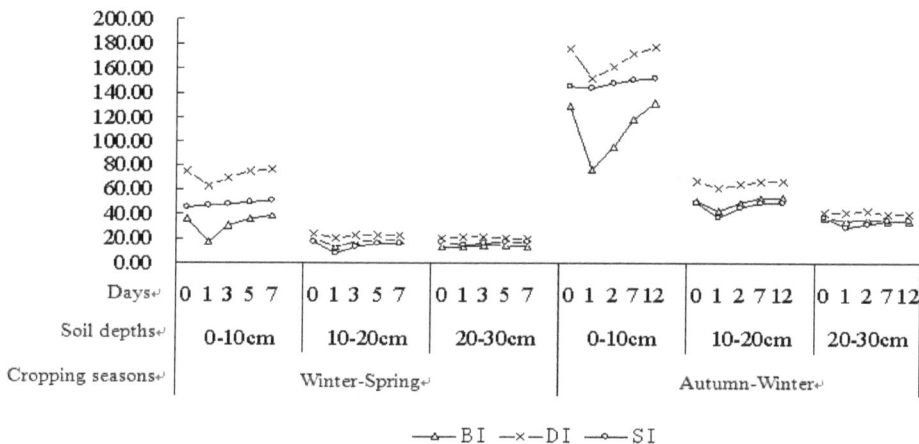

Fig. 8. Effects of different irrigation methods on nitrate nitrogen transport

Kg ha⁻¹

Cropping seasons	BI	DI	SI
Winter-Spring	90.79 a	10.50 b	9.26 b
Autumn-Winter	117.52 a	18.94 b	8.08 c

Table 4. Effects of different irrigation methods on volume nitrate leaching

In order to reveal the effects of different irrigation methods on water distribution and nitrate nitrogen transport in solar greenhouse, border irrigation, drip irrigation and subsurface irrigation were evaluated by using cucumber *Jinyu No.5*(Fig.8). Irrigation water distribution, nitrate leaching, root zone nitrate nitrogen transport, yield and water use efficiency were conducted in the current study (Table 4). The experiments showed that the amount of leaching and evaporation decreased but transpiration increased under drip irrigation and subsurface irrigation. As the results, compared to border irrigation, the above irrigation systems saved water by 25.9% and 32.0%, cucumber yield increased by 11.6% and 15.3% and water use efficiency (WUE) increased by 49.9% and 68.7%, respectively (Table 5). The drip irrigation and subsurface irrigation also reduced the amount of nitrate leaching, and it was important to protect groundwater (Wei et al., 2010).

Cropping seasons	Treatments	Economic yields (Kg ha-1)	WUE$_Y$ (%)
Winter-Spring	BI	110643 a	14.20 c
	DI	123771 b	21.96 b
	SI	128132 a	25.15 a
Autumn-Winter	BI	45526 b	10.31 b
	DI	50564 a	14.79 a
	SI	51858 a	16.20 a

Table 5. Effects of different irrigation methods on economic yield and water use efficiency (WUE$_Y$)

3. Conclusion

Nitrate leaching is considered the major pathway for the loss of N from intensive agriculture systems in China. Therefore, it is of great importance for scientists to seek for efficient and economic strategies on controlling nitrate leaching. Although there is no quick fix for preventing nitrate leaching from the soil profile to groundwater, integrated use of various strategies can decrease nitrate leaching potential significantly by manipulating the N management practices on fertilizers and manures, soil, water and crop. The primary effect

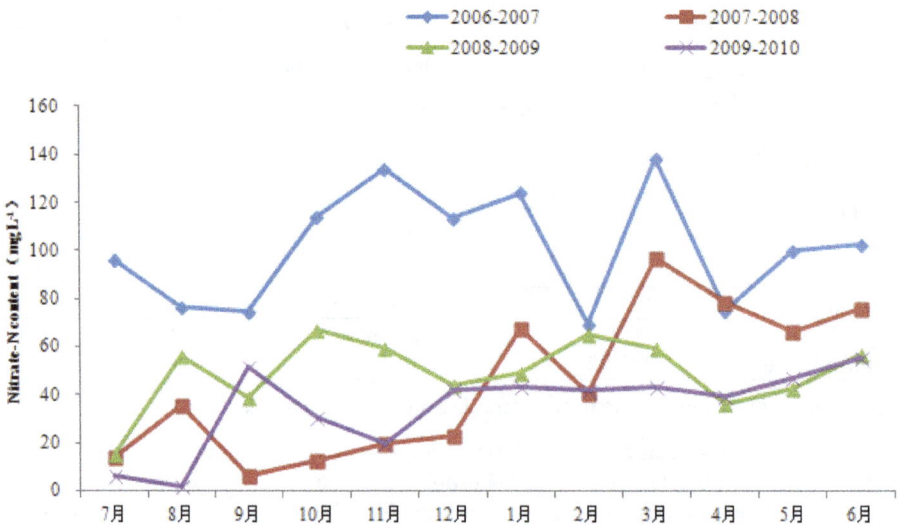

Fig. 9. Nitrate content monthly fluctuation curve within 4 years for a observation well by manipulating the N management practices. by Song, Unpublished data

has been found in a demonstration area (Fig.9). The local groundwater nitrate-N content decreased obviously from initial 101.67 mg L^{-1} to present 35.36mg L^{-1} by a monthly continuous monitoring on a fixed-point observation well within four years. However, that is not all for preventing nitrate contamination to groundwater, the ultimately desirable solution to minimize nitrate threats to water resources, is a better public recognition and implement on the above several different management practices in intensive agricultural production systems.

4. Acknowledgment

This work was financially supported by Special Fund for National Nature Science Foundation of China (40701077), Agro-scientific Research in the Public Interest (201003014) and Non-profit Research Foundation for Agriculture (201103039).

5. References

Andraski, W. ; Bundy, G. & Brye, R. (2000). Crop Management and Corn Nitrogen Rate Effects on Nitrate Leaching. *Journal of environmental quality*, 29(4): 1095-1103, ISSN 0047-2425

Baker, L. & Melvin, W. (1994) .Chemical management, status, and findings.p.27–60. In *Agricultural drainage well research and demonstration project-annual report and project summary*. Iowa Department of Agriculture and Land Stewardship, Des Moines, IA

Bock, R. ; & Hergert. (1991). Fertilizer nitrogen management. pp. 139-164. In R.F. Follett, D. R. Keeney and R.M. CRESE (ed.) *Managing nitrogen for groundwater quality and farm profitability*. Soil Science Society of America. Madison, WI.

Bock, R. ; & Hergert. (1992). *Predicting nitrogen fertilizer needs for corn in humid regions*. Natioanl Fertilizer and Environmental Research Center, Muscle Shaoals, AL.

Breschini, J. & Hartz, K. (2002). Presidedress soil nitrate testing reduces nitrogen fertilizer use and nitrate leaching hazard in lettuce production. *HortScience*, 37(7):1061–1064, ISSN0018-5345

Brinsfield, B. & Staver, W. (1991). Use of cereal grain cover crops for reducing ground water nitrate contamination in the Chesapeake Bay Region. In *Cover crops for clean water*, ed. W.L. Hargrove, pp.79–81. Ankeny: Soil Water Conservation Society

Bundy, G. & Andraski W. (2005). Recovery of fertilizer nitrogen in crop residues and cover crops on an irrigated sandy soil. *Soil Science Society of America Journal*, ISSN 0361-5995, 69: 640–648

Bundy, G. & Meisinger, J. (1994). Nitrogen availability indices. pp. 951–984. In R.W. Weaver et al. (ed.) *Methods of soil analysis*. Part 2 : Microbiological and biochemical properties. Soil Science Society of America, Madison, WI.

Bundy, G., Walters T., & Olness E. (1999). *Evaluation of soil nitrate tests for predicting corn nitrogen response in the North Central Region*. NC Regional Res. Pub. no. 342. Univ. of Wisconsin-Madison.

Cates, K. & Madison, A. (1994). Study of nitrate and atrazine concentration in groundwater from agricultural use on a sandy, irrigated corn field in the lower Wisconsin River valley. WI. *Groundwater Research and Monitoring Projects report*, Project no DNR-81, University of Wisconsin Water Resources Institute

Cooley, T; Lowery, B.; Kelling, A. & Wilner, S. (2007). Water dynamics in drip and overhead sprinkler irrigated potato hills and development of dry zones. *Hydrological Processes*, 21: 2390–2399

Dai, G., Lu, .W. ; Li, K., et al. (2010). Nutrient release characteristic of different crop straws manure. *Transactions of the CSAE*, 26(6): 272-276.

Dana, D.; Douglas, K.; Dan, J., et al. (2002). Nitrogen Management Strategies to Reduce Nitrate Leaching in Tile-Drained Midwestern Soils. *Agronomy Journal*, 94(1):153-171, ISSN 0002-1962.

Daniel C., Sharpley N. & Edwards R., et al. (1994). Minimizing surface water eutrophication from agriculture by phosphorus management. *Journal Soil Water Conservation*, 40: 30-38, ISSN 0022-4561.

Delgado, A., Riggenbach, R.; Sparks, T.; Dillon, A.; Kawanabe, M. & Ristau, J. (2001b). Evaluation of nitrate-nitrogen transport in a potato-barley rotation. *Soil Science Society of America Journal*, 65: 878–883

Delgado, A. (1998). Sequential NLEAP simulations to examine effect of early and late planted winter cover crops on nitrogen dynamics. *Journal of Soil and Water Conservation*, ISSN 0022-4561, 53: 241-244.

Delgado, A. (2001a). Potential use of innovative nutrient management alternatives to increase nutrient use efficiency, reduce losses, and protect soil and water quality. *Special issue Journal of Communication Soil Science and Plant Analysis*. New York: Marcel Dekker Inc.

Delgado, A.; Dillon, A.; Sparks, T. & Samuel, E. (2007). A decade of advances in cover crops: Cover crops with limited irrigation can increase yields, crop quality, and nutrient and water use efficiencies while protecting the environment. *Journal of Soil and Water Conservation*, 62: 110–117.

Delgado, A.; Follett, F. & Shaffer, J. (2000). Simulation of NO3-N dynamics for cropping systems with different rooting depths. *Soil Science Society of America Journal*, 64: 1050–1054.

Delgado, A.; Shaffer, M.; Hu, C.; Lavado, S.; Cueto Wong, J.; Joosse, P.; Li, X.; Rimski-Korsakov, H.; Follett, R.; Colon, W. & Sotomayor, D. (2006). A decade of change in nutrient management requires a new tool: A new nitrogen index. *Journal of Soil and Water Conservation*, 61: 62–71.

Dong, R.; LI, C. & LI, D. (2009). Effects of Nitrogenous Fertilizer Strategies on the Nitrogen Accumulation in Cotton Field Soil and Cotton Functional leaves. *Cotton Science*, ISSN 1002-7807, 21(1) : 51- 56.

Durieux P. ; Brown, J. ; Stewart, J. ; Zhao, Q. ; Jokela, E. & Magdoff, R. (1995). Implications of nitrogen management strategies for nitrate leaching potential: Roles of nitrogen source and fertilizer recommendation system. *Agronomy Journal*, 87:884-887, ISSN 0002-1962.

Errebhi, M.; Rosen, J.; Gupta, C. & Birong, E. (1998a). Potato yield response and nitrate leaching as influenced by nitrogen management. Agron Journal, 90: 10-15.

Fielder, G. & Peel S. (1992). The selection and management of species for cover cropping. *Aspects of Applied Biology*, 30: 283-290, ISSN: 0265-1491.

Gale, J. & Cambardella, A. (2000). Carbon dynamics of surface residue- and root-derived organic matter under simulated no-till. *Soil Science Society of America Journal*, 64:190-195.

Guerette, V.; Desjardins, Y.; Belec, C.; Tremblay, N.; Weier, U. & Scharpf, C. (2002). Nitrogen contribution from mineralization of vegetable crop residues. *Acta Horticulturae*, 571: 95–102, ISSN 0567-7572

Guillard, K, Morris T.F. & Kopp L. (1999). The pre-sidedress soil nitrate test and nitrate leaching from corn. *Journal of Environmental Quality*, 28(6): 1845-1852, ISSN 0047-2425.

Guo, M.; Li, H.; Zhang, Y.; Zhang, X. & Lu, A. (2006). Effects of water table and fertilization management on nitrogen loading to groundwater. *Agricultural Water Management*, 82: 86-98, ISSN 0378-3774, 82: 86-98.

Guo, Y.; Li, L.; Christie, P.; Chen, Q.; Jiang, F. & Zhang, S. (2008). Influence of root zone nitrogen management and a summer catch crop on cucumber yield and soil mineral nitrogen dynamics in intensive production systems. *Plant and Soil*, 313: 55-70, ISSN 0032-079X.

He, F.; Xiao, L.; Li, L., et al. (2006). Integrated nitrogen management in greenhouse tomato production. *Plant Nutrition and Fertilizer Science*, 12 (3) :394 – 399.

IPCC. (2007). *Climate change: The physical science basis contribution of working group I to the fourth assessment report of the intergovernmental panel on climate change.* In eds. Solomon S., Qin D., Manning M., Chen Z., Marquis M., Averyt K.B., Tignor M. and Miller H.L.

Jackson, E.; Wyland, J.; Klein, A.; Smith, F.; Channey, F. & Koike, T. (1993). Winter cover crops can decrease soil nitrate, leaching potential. *California Agriculture*, 47:12-15, ISSN 0008-0845.

Jiang, M. ; Zhang, F. ; Yang, C.; Liu, H.; Song, Z.; Jiang, H. & Zhang, S. (2010). Effects of different models of applying nitrogen fertilizer on yield and quality of tomato and soil fertility in greenhouse. *Plant Nutrition and Fertilizer Science*, 16(1): 158-165, ISSN 1008-505X

Ju, T.; Liu, J.; Pan, R. & Zhang, S. (2007). The fate of 15N-labeled urea and its residual effect in a winter wheat and summer maize cropping rotation system on the North China Plain. *Pedosphere*, 17(1): 52-61, ISSN 1002-0160

Ju, T.; Gao, Q.; Christie, P.; Zhang, S. (2007). Interception of residual nitrate from a calcareous alluvial soil profile on the North China plain by deep-rooted crops: A 15N tracer study. *Environmental Pollution*, 146(2): 534-542, ISSN 0269-7491

Ju, T.; Kou, L.; Christie, P.; Dou, X. & Zhang, S. (2007). Changes in soil environment from excessive application of fertilizers and manures to two contrasting intensive cropping systems on the North China Plain. *Environmental Pollution*, 145(2): 497-506

Kanwar, S.; Stolenberg, E.; Pfeiffer, R.; Karlen, L.; Colvin, S. & Simpkins, W. (1993). Transport of nitrate and pesticides to shallow groundwater systems as affected by tillage and crop rotation practices. pp. 270-273. In Proceedings of the Conference, February 21-24, 1993, Minneapolis, Minnesota, USA. ISBN 10 0935734341

Kelling, A. & Hero, D. (1994). Potato nitrogen management Lessons from 1993. *Proceeding Wisconsin Annual Potato Meeting*, 7: 29-37.

Li, W.; Zhang, M.; Li, H. & Zan, L. (2002). The study of soil nitrate status in fields under plastic house gardening, *Acta Pedologica Sinica*, 39: 283-287, ISSN 0564-3929

Logsdon, D.; Kaspar, C.; Meek, W. & Prueger, H. (2002). Nitrate leaching as influenced by cover crops in large soil monoliths. *Agronomy Journal*, 94: 807-814, ISSN 1435-0645

Lu, L.; Bai, L.; Wang, L.; Wang, H.; Du, J. & Wang, Y. (2011). Efficiency analysis of slow/controlled release fertilizer on wheat-maize in North China. *Plant Nutrition and Fertilizer Science*, 2011, 17(1): 209- 215

Magdoff, R. ; Jokela, E. ; Fox, H. & Griffin F. (1990). A soil test for nitrogen availability in the Northeastern Unites States. *Communications in Soil Science and Plant Analysis*, 21(13-16): 1103-1115, ISSN 0010-3624

Magdoff, R. ; Ross, D. & Amadon J. (1984). A soil test for nitrogen availability to corn. Soil *Soil Science Society of America Journal*, 48: 1301-1304.

Malpassi, N.; Kaspar, C.; Parkin, B.; Cambardella, A. & Nubel, A. (2000). Oat and rye root decomposition effects on nitrogen mineralization. *Soil Science Society of America Journal*, 64:208–215

Mechenich, J. & Kraft, J. (1997). *Contaminant source assessment and management using groundwater flow and contaminant models in the Stevens Point-Whiting-Plover wellhead protection area.* Central Wisconsin Groundwater Center, Cooperative Extension Service, College of Natural Resources, University of Wisconsin-Stevens Point

Meisinger, J. & Delgado, A. (2002). Principles for managing nitrogen leaching. *Journal of Soil and Water Conservation* 57: 485-498.

Meisinger, J. ;Bandel, A.; Angle, S.; O'Keefe, E. & Reynolds, M. (1992). Pre-sidedress soil nitrate test evaluation in Maryland. *Soil Science Society of America Journal*, 56:1527-1532

Meisinger, J.; Hargrove, L.; Mikkelson, L.; Williams, R. & Benson, W. (1991). Effects of cover crops on groundwater quality. In Cover crops for clean water. *Proceedings of international conference on cover crops for clean water*, Tennesse, April 9-11, ed. W.L. Hargrove, 57-68. Ankeny: Soil Water Con. Soc.

Milburn, P.; MacLeod A. & Sanderson B. (1997). Control of fall nitrate leaching from early harvested potatoes on Prince Edward Island. *Canadian Agriculture Engineering*, 39: 263-271, ISSN 0045-432X

Mosier, A. ; Syers, J. & Freney, J. (2004). Agriculture and the nitrogen cycle: assessing the impacts of fertilizer use on food production and the environment. ISBN 1-55963-708-0, pp6.

Osborne, T., & Curwen, D. (1990). Quantifying groundwater quality and productivity effects of agricultural best management practices on irrigated sands. *Proceedings of symposium on agricultural impacts on groundwater quality.* NWWA Groundwater Management, 1: 129–143

Owens, B. (1987). Nitrate leaching losses from monolith lysimeters as influenced by nitrapyrin. *Journal of Environmental Quality*, 16:34–38

Peck, R. & Soltanpour, N. (1990). The principles of soil testing. pp. 1-10. In R. L. Westerman(ed.) *Soil testing and plant analysis*. Soil Science Society of America, Madison, WI.

Power, F. & Schepers, S. (1989). Nitrate contamination of groundwater in North America. *Agriculture, Ecosystems & Environment*, 26:165–187, ISSN 0167-8809

Randall, W. & Iragavarapu, K. (1995). Impact of long-term tillage systems for continuous corn on nitrate leaching to tile drainage. *Journal of Environmental Quality*, 24:360–366

Randall, W. & Mulla, J. (2001). Nitrate nitrogen in surface waters as influenced by climatic conditions and agricultural practices. *Journal of Environmental Quality*, 30:337–344

Randall, W.; Huggins, R.; Russelle, P.; Fuchs, J.; Nelson, W. & Anderson, L. (1997a). Nitrate losses through subsurface tile drainage in conservation reserve program, alfalfa, and row crop systems. *Journal of Environmental Quality*, 26:1240-1247.

Rasmussen, J. (1999). Impact of ploughless soil tillage on yield and soil quality: A Scandinavian review. *Soil and Tillage Research*, 53(1):3-14.

Ren, T. (2009). Agronomic and Environmental Evaluation of Optimizing Nitrogen Management in Greenhouse Tomato Cropping System. Master thesis of Chinese Agricultural University, Beijing.

Saffigna, G. & Keeney, R. (1977). Nitrate and Chloride in groundwater under irrigated agriculture in central Wisconsin. *Ground Water*, 15: 170-177, ISSN 0017-467X

Shaviv, A. & Mikkelsen, L. (1993). Controlled-release fertilizers to increase efficiency of nutrient use and minimize environmental degradation: A review. *Fertilizer Research*, 35(1-2): 1-12, ISSN 0167-1731

Shrestha, K. & Ladha, K. (1998). Nitrate in groundwater and integration of nitrogen-catch crop in rice-sweet pepper cropping system. *Soil Science Society of America Journal*, 62: 1610–1619

Sims, T.; Vasilas, L.; Gartley, L.; Milliken, B. & Green, V. (1995). Evaluation of soil and plant nitrogen tests for maize on manured soils of the Atlantic Coastal Plain. *Agronomy journal*, 87(2): 213-222, ISSN0002-1962

Sogbedji, M.; Es, M. van; Yang, L.; Geohring, D. & Magdoff, R. (2000). Nitrate leaching and nitrogen budget as affected by maize nitrogen rate and soil type. *Journal of Environmental Quality*, 29(6): 1813-1820, ISSN0047-2425

Song, Z. ; Zhao, X. ; Wang, L. & Li, J. (2009). Study of nitrate leaching and nitrogen fate under intensive vegetable production pattern in northern China. *Comptes Rendus Biologies*, 332: 385-392, ISSN 1631-0691

Stanford, G. (1973). The rational for optimum nitrogen fertilization in corn production. *Journal of Environmental Quality*, 2:159-166.

Starr, C.; Cooley, T.; Lowery, B. & Kelling, K. (2005). Soil water fluctuations in a loamy sand under irrigated potato. *Soil Science*,170(2): 77-89, ISSN 0038-075X

Starr, C.; Rowland, D.; Griffin, S. & Olanya, M. (2008). Soil water in relation to irrigation, water uptake and potato yield in a humid climate. *Agricultural Water Management*, 95: 292-300.

Staver, W. & Brainsfield B. (1998). Using winter cover crops to reduce groundwater nitrate contamination in the mid-Atlantic coastal plain. *Journal of Soil and Water Conservation*, 53: 230-240

Thomsen, K. (2005). Nitrate leaching under spring barley is influenced by the presence of a ryegrass catch crop: Results from a lysimeter experiment. *Agriculture, Ecosystems & Environment*, 111:21-29.

Tisdale, L. ; Nelson, L. ; Beaton D. & Havlin L. (1993). *Soil fertility and fertilizers*. 5th Ed. MacMillan Publishers, New York, NY.

Varvel, E. & Peterson, A. (1990). Residual soil nitrogen as affected by continuous, two-year, and four-year crop rotation systems. *Agronomy Journal*, 82: 958-962, ISSN 0002-1962

Waddell, T.; Gupta, C.; Moncrief, F.; Rosen, J. & Steele, D. (2000). Irrigation and nitrogen management impacts on nitrate leaching under potato. Journal of Environmental Quality, 29: 251-261.

Weed, J. & Kanwar, S. (1996). Nitrate and water present in and flowing from root-zone soil. J. Environ. Qual. 25:709-719.

Wei, Y.; Sun, L.; Wang, S. et al. (2010). Effects of different irrigation methods on water distribution and nitrate nitrogen transport of cucumber in greenhouse. *Transactions of the CSAE*, 26(8): 67-72

Weinert, L.; Pan L.; Moneymaker, R.; Santo, S. & Stevens, G. (2002). Nitrogen recycling of nonleguminous winter cover crops to reduce leaching in potato rotations. *Agronomy Journal*, 94: 365–372

Zerulla, W.; Barth, T.; Dressel, J.; Erhardt, K.; Horchler, L.; Pasda, G.; Rädle, M. & Wissemeier, H. (2001). 3,4-Dimethylpyrazole phosphate (DMPP)-a new nitrification inhibitor for agriculture and horticulture. *Nutrient Cycling in Agroecosystem, 60:* 57-64, ISSN 1385-1314

Zhang, J.; Wen, X.; Liao, C.; Liu, Y. (2010). Effects of different amount of maize straw returning on soil fertility and yield of winter wheat. *Plant Nutrition and Fertilizer Science*, 16 (3): 612-619.

Zhang, L.; Tian, X.; Zhang, N. & Li, Q. (1996). Nitrate pollution of groundwater in northern China. *Agriculture, Ecosystems & Environment*, 59: 223-231.

Zvomuya, F.; Rosen, J.; Russelle, P. & Gupta, C. (2003). Nitrate leaching and nitrogen recovery following application of polyolefin-coated urea to potato. *Journal of Environmental Quality*, 32: 480-489.

9

Nitratation Promotion Process for Reducing Nitrogen Losses by N₂O/NO Emissions in the Composting of Livestock Manure

Yasuyuki Fukumoto
Institute of Livestock and Grassland Science,
National Agriculture and Food Research Organization (NARO)
Japan

1. Introduction

The livestock industry already has a large impact on the environment by contributing to desertification, eutrophication, global warming, acidic rain, and so on (Dodd, 1994; Isermann, 1990; Pearson and Stewart, 1993; Steinfeld and Wassenaar, 2007). However, it has been predicted that the global demand for livestock products, such as milk, meat and eggs, will increase because of the growing human population and urbanization. Therefore, for having a sustainable livestock production, it is important to remove environmentally harmful factors from the livestock industry's activities as much as possible.

Animal manure is one of the important contributing factors affecting environmental issues that are caused by livestock activity. Livestock animals excrete huge amounts of manure. For example, a milk cow 600 kg in body weight excretes approximately 18,000 kg of manure, while she produces 7,600 kg of milk during one lactation period. Therefore, livestock production is regarded as an industry with more waste than product. In Japan, approximately 90 Tg of livestock manure is generated annually, which accounts for one-quarter of the nation's total organic waste (Ministry of Agriculture, Forestry and Fisheries 2008). Modern livestock production systems integrate numerous livestock animals in a limited area to take advantage of the efficiency of a small scale operation, and to utilize scientific nutrition and management techniques. However, in such a production system, a huge amount of manure is concentrated in the same area, which leads to serious environmental problems such as diffusion of offensive odor and contamination of underground water if the manure is handled in improper ways (Criado, 1996; Rappert and Muller, 2005). Therefore, proper treatment or handling of manure is important for an environmentally sound livestock production system.

Composting is one of the principal treatment methods of organic waste such as livestock manure. The objectives of composting are to stabilize the biodegradable organic matter in raw wastes, to reduce offensive odors, to kill weed seeds and pathogenic organisms, and finally, to produce a uniform organic fertilizer suitable for land application (Haga et al. 1998). Controlled conditions are important for composting, so as to distinguish it from other natural biological decomposition processes such as rotting and putrefaction (Haga, 1990). After composting, the handling of livestock manure is improved, making it possible to distribute the manure from a limited to a wider area. Therefore, local pollution by livestock manure can be avoided. Moreover, the construction of a sustainable agricultural system is expected to be achieved by enhancing the circulation between the livestock industry and field husbandry via compost of livestock manure.

It is important to improve composting techniques because substantial amounts of harmful gaseous compounds are emitted during the composting process (Fukumoto et al. 2003; Kuroda et al. 1996; Smet et al. 1999). Because of its high nitrogen contents, nitrogenous emission is substantial in livestock manure composting. Those nitrogenous emissions cause not only serious environmental risks, but also a decline in the compost's value as a fertilizer. Ammonia (NH_3) is one of the common nitrogenous emissions, which becomes a main cause of odor coming from the livestock industry (McGinn and Janzen, 1998). Moreover, NH_3 can be the cause of more extensive environmental pollution such as acid rain (Pearson and Stewart, 1993). Due to its high impact on the regional environment, there have been numerous attempts to reduce NH_3 emission from the composting process (Burrows, 2006; Kuroda et al. 2004; la Pagans et al. 2005; Lin, 2008; Yasuda et al. 2009).

Nitrous oxide (N_2O) is also one of the nitrogenous emissions arising from livestock manure composting (Beck-Friis et al. 2001; Czepiel et al. 1996; Osada et al. 2000; Sommer, 2001; Zeman et al. 2002). N_2O is an important greenhouse gas, having a 296-fold stronger effect than carbon dioxide (IPCC, 2007). Additionally, a recent study disclosed that N_2O has now become the largest ozone-depleting substance, surpassing chlorofluorocarbons (Ravishankara et al. 2009). Agriculture is the largest source of anthropogenic N_2O emission (Oenema et al. 2005). Livestock activities in particular contribute to almost two-thirds of all anthropogenic N_2O emissions, and 75-80 percent of agricultural emissions, which are mainly caused by manure (FAO, 2006). However, as of now, the numbers of countermeasures to reduce N_2O emissions from composting are quite few compared with those established for NH_3 emission.

This chapter presents recent knowledge of N_2O reduction from the composting process. The technique developed for reducing N_2O emission is termed the nitratation promotion process. The author will explain effects of the nitratation promotion process on N_2O/NO emissions and nitrogen conservation during swine manure composting. Moreover, the collaboration effect of nitrogen conservation with another countermeasure for reducing NH_3 emission, and the possibility of an adaptation of the nitratation promotion process to other kinds of livestock manure (cattle and poultry) will also be discussed.

2. Abreviations

AOB: Ammonia-oxidizing bacteria
NOB: Nitrite-oxidizing bacteria
AOA: Ammonia-oxidizing archaea
TN: Total nitrogen
DM: Dry matter
WM: Wet matter
OM: Organic matter
NH_3: Ammonia
N_2O: Nitrous oxide
NO: Nitric oxide
NIPRO: Nitratation promotion

MAP: Magnesium ammonium phosphate (struvite)
AGP: Ammonium generation potential

3. N₂O emission from composting

Nitrous oxide is generated via both nitrification and denitrification processes as intermediate products or by-products during the composting process (Fig. 1).

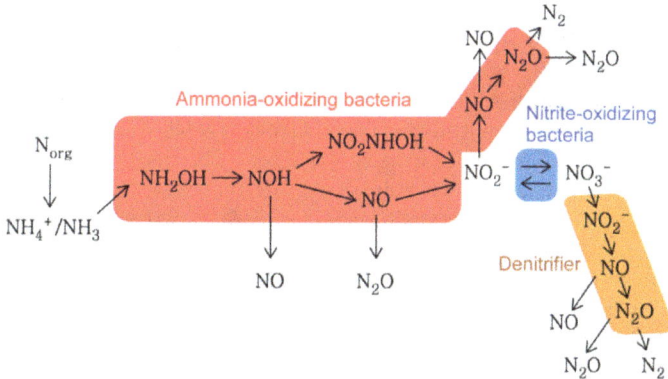

Fig. 1. Transformation of inorganic nitrogen compounds in the composting process

Nitrification is performed by different groups of microbes. Ammonia-oxidizing bacteria (AOB) converts NH_3 into nitrite (NO_2^-) (this reaction is called nitritation), and nitrite-oxidizing bacteria (NOB) converts NO_2^- into nitrate (NO_3^-) (this reaction is called nitratation). Additionally, ammonia-oxidizing archaea (AOA) is thought to have important role in the oxidization of ammonia under various conditions, such as soil, sediments, and seawater (Francis et al. 2005; He et al. 2007; Leininger et al. 2006; Tourna et al. 2011; Wuchter et al. 2006). However, the evidence that AOA activity takes place in the composting process has not yet been confirmed (Maeda et al. 2010b). Nitrification is a prerequisite for N_2O generation from stored manure because little nitrate nitrogen is contained in the manure immediately after excretion. Nitrate produced by nitrification has an opportunity to transfer to an anaerobic portion by disturbance of compost pile (turning). The nitrate is then reduced into dinitrogen gas (N_2) via N_2O by denitrifiers. Inside the compost pile, both aerobic/anaerobic portions coexist, which makes it difficult to estimate the respective contributions of nitrification and denitrification in actual N_2O emissions from composting (Maeda et al. 2011).

It is known that there are several factors affecting N_2O emission from composting, such as moisture content, amount of mixed bedding material, frequency of pile turning, compost pile scale, and so on (Beck-Friis et al. 2001; Fukumoto et al. 2003; Parkinson et al. 2004; Szanto et al. 2007). It is thought that these factors influence the N_2O generation pathway directly or indirectly. As a result, the amount of N_2O emitted from the composting would be decided by a synthetic influence of those factors. Therefore, reduction of gas emission by control of these factors is thought possible if there is a factor largely responsible for gas generation. The relative contributions of nitrification and denitrification to N_2O emission do not become an important issue in the development of countermeasures, because N_2O can be reduced by controlling such factors, whether the factor induces N_2O generation via nitrification or denitrification. The more important issue is to find such a factor during the treatment process. However, when considering how to magnify the adaptable range of a countermeasure, clarifying the mechanism of the N_2O generation would be an important issue that also has large scientific interest.

4. Effect of nitrite accumulation on N_2O emission

One important factor for N_2O emission from the composting is that NO_2^- can be easily found. He et al. (2001) showed that the amount of N_2O emission from food waste composting had increased when NO_2^- was accumulated. Moreover, a good NO_2^- – N_2O correlation has been confirmed under various environmental conditions. Nitrite is an intermediate product of the nitrification process. Generally, NO_2^- is scarcely observed in the natural environment because nitrite oxidizers, such as *Nitrobacter* and *Nitrospira*, oxidize NO_2^- to NO_3^- immediately. However, it is also a fact that NO_2^- accumulation is observed under several environmental conditions (Burns et al. 1996; Corriveau et al. 2010; Silva et al. 2011).

In swine manure composting, notable N_2O emission begins after the thermophilic phase, because nitrifying bacteria cannot be active under thermophilic conditions, such as high temperature, high free ammonia and high organic matter content. After N_2O emission starts, it tends to continue for a long time during the maturation phase of swine manure composting. It has been confirmed that NO_2^- accumulates in the composting material during this period of N_2O emission. Because of the long duration of N_2O emission in the maturation phase, the amount of N_2O emission induced by NO_2^- accumulation accounts for a large portion of the total N_2O emission in swine manure composting. Nitrite accumulation in swine manure composting is due to an inadequate nitrification process, i.e., the growth of indigenous NOB is inhibited while indigenous AOB is active immediately after the thermophlic phase, leading to a lower oxidization rate of NO_2^-. In this case, NO_2^- can be regarded as a critical factor of N_2O generation. Therefore, it is possible to develop a countermeasure for reducing N_2O emission by the regulation of NO_2^- during the composting process.

Two ways of avoiding NO_2^- accumulation can be considered. One method is to use a reagent of nitrification inhibitor, such as nitrapyrin. Its effect as a nitrification inhibitor that reduces N_2O emission from soil has been confirmed in numerous studies (Bhatia et al. 2010; Dittert et al. 2001; Zaman and Blennerhassett, 2010). However, to our knowledge, there have been no studies investigating the effect of nitrification inhibitors on N_2O emission during the composting process. Probably, disadvantages, such as ammonia accumulation and increasing treatment cost, would make it difficult to use in actual practice.

Another way to reduce NO_2^- is to enhance NO_2^- oxidization, i.e., nitratation promotion. It is thought that the nitratation function is hurt or reduced when NO_2^- accumulates during nitrification. Therefore, for nitratation promotion, recovery of the nitratation function is necessary. In swine manure composting, the growth of NOB is inhibited, causing NO_2^- accumulation. There are two ways to recover the NOB growth. One way is to control the environment so that the composting material is suitable for NOB growth, e.g., decreasing the level of free ammonia. On the other hand, a bioremediational technique of adding NOB is also an effective candidate for the recovery of NOB growth. In fact, controlling the compost to be suitable for NOB growth is difficult. Therefore, the authors tried to develop a countermeasure reducing N_2O emission by the addition of a NOB source during composting.

5. Nitratation promotion process

In the first study conducted for reducing N_2O emission from swine manure composting, effects from the addition of a NOB source on NO_2^- accumulation, and the state of nitrifying bacteria and N_2O emission have been investigated using a laboratory-scale apparatus (Fukumoto et al. 2006).

5.1 Materials and methods

Two kinds of NOB sources were used. One was incubated mature swine compost, in which NOB had been concentrated at a high density by an incubation process (NOB density: 10^{11} cell/g). The other NOB source was normal mature swine compost (NOB density: 10^6 cell/g). These materials were added to the composting swine manure after the thermophilic phase because NOB activity is strongly restricted under high temperature conditions. Gas emission was monitored continuously using an infrared photoacoustic detector.

5.2 Results and discussion

After the addition of an NOB source to the composting swine manure, the establishment of an NOB population at the cell density of 10^5-10^7 cell/g in the composting material was confirmed, while the absence of NOB continued for 5-6 weeks after the start of AOB growth in the composting bin without the addition of NOB (control) (Fig. 2).

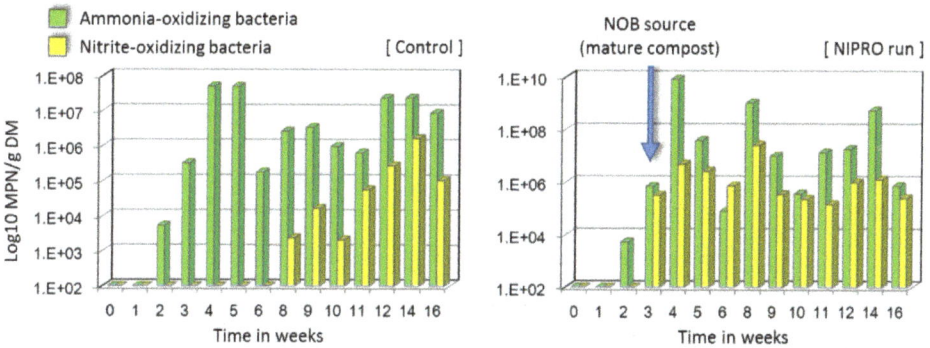

Fig. 2. Changes in the cell number of nitrifying bacteria during swine manure composting. Arrow indicates addition of NOB source of mature swine compost

Because NOB was absent, NO_2^- accumulated for a long duration in the control (precisely, the absence of NOB resulted in the cell number of NOB being under the detection limit of 10^2 cell/g). On the other hand, a prolonged NO_2^- accumulation was not observed in the composting with an NOB source addition (Fig. 3).

Moreover, the effect on NO_2^- avoidance was also obtained by the addition of normal mature compost, which meant that no specific treatment, such as incubation of NOB, was necessary for preparation of the NOB source. The required cell number of NOB for the oxidation of NO_2^- to NO_3^- seems to be more than 10^5 cell/g compost. Blouin et al. (1990) showed a similar result regarding the required cell number for complete NO_2^- oxidization in swine manure. However, because the population size of NOB grows after a source addition, it is not always necessary for NOB to exceed 10^5 cells per gram compost at the time of the NOB source addition. The pattern of N_2O emission from swine manure composting agreed well with changes in the NO_2^- concentration (Fig. 4).

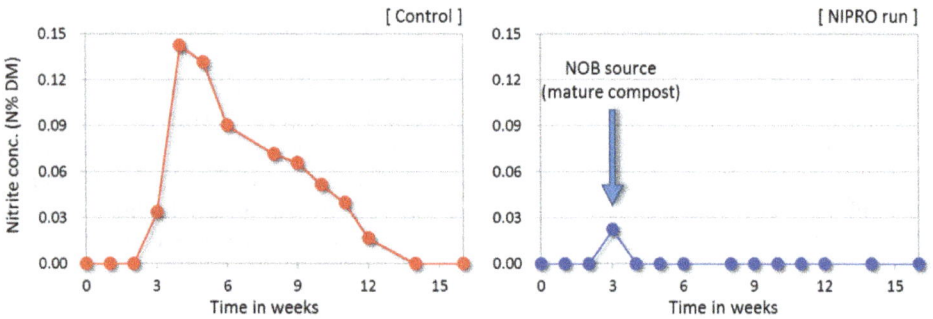

Fig. 3. Changes in concentration of nitrite nitrogen during swine manure composting. Arrow indicates addition of NOB source of mature swine compost

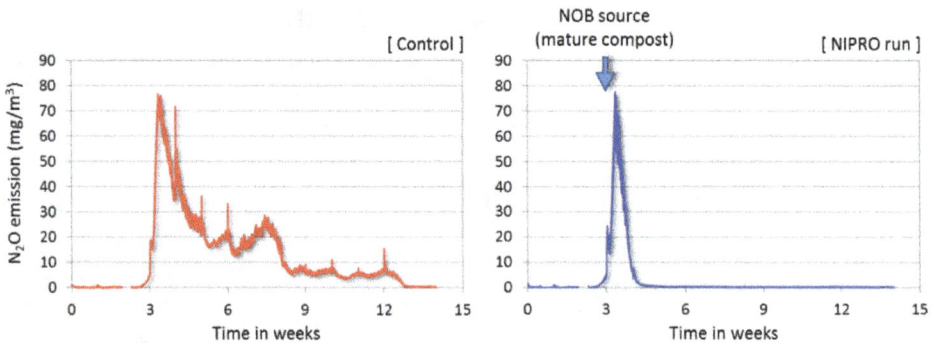

Fig. 4. Emission patterns of N_2O during swine manure composting. Arrow indicates addition of NOB source of mature swine compost

Therefore, the duration of NO_2^- accumulation affects the amount of N_2O emission. In the normal composting (control), the N_2O emission continued until NO_2^- disappeared. As a consequence, the amount of N_2O emission throughout the composting became large (N_2O emission rate: 88.5 g N_2O-N/kg $TN_{initial}$). On the other hand, because of its shortened duration of NO_2^- accumulation, the amount of N_2O emission was reduced in the compost with an NOB source addition (N_2O emission rate: 17.5-20.2 g N_2O-N/kg $TN_{initial}$). The rate of N_2O emission decrease by the addition of an NOB source was calculated as 77-80% in this study (Table 1).

	Total emission, mg N	Emission rate, g N/kg $TN_{initial}$
Control	10,351	88.5
MSC addition	2,046	17.5
cul-MSC addition	2,362	20.2

$TN_{initial}$, initial total nitrogen; MSC, mature swine compost; *cul*-MSC, cultured MSC.

Table 1. Total N_2O emission and its emission rate during swine manure composting

Summarizing our published and unpublished data from the laboratory-scale composting experiments, the average decrease in the rate of N_2O emission by the addition of an NOB source during swine manure composting was calculated to be 60%, and, therefore, this technique appeared to allow a quantitative reduction of N_2O emission in the composting process. The schematic of this technique (nitratation promotion (NIPRO) process) is shown in Fig. 5.

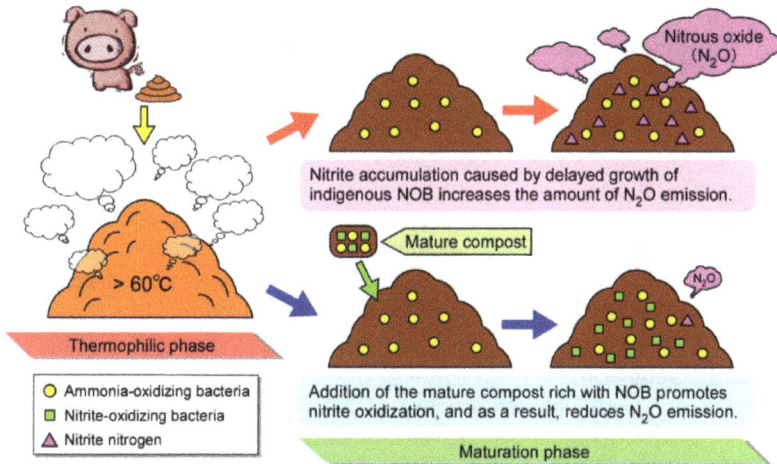

Fig. 5. Schematic of nitratation promotion process for reducing N_2O emission during composting

6. Nitrogen conservation

The nitrogen that had avoided being lost as an N_2O emission by applying the NIPRO process seemed to be preserved in the form of nitrate nitrogen in the compost product. Fig. 6 shows the changes in the NH_4^+, NO_2^- and NO_3^- contents during swine manure composting. During the thermophilic phase (0-3 week), NH_4^+ accounted for most of the inorganic nitrogen. After the start of nitrification, NO_2^-/NO_3^- nitrogen began to increase, while NH_4^+ declined. However, a remarkable difference in NO_3^- increase between the two treatments was observed. In the composting without an NOB addition (control), NO_3^- content increased slowly. In particular, the rate of NO_3^- increase became slower during NO_2^- accumulation. On the other hand, NO_3^- in the composting with the NIPRO process increased quickly after the addition of an NOB source, and then it reached a concentration in the final product higher than that of the control. Therefore, it was hypothesized that the NIPRO process contributes to nitrogen conservation.

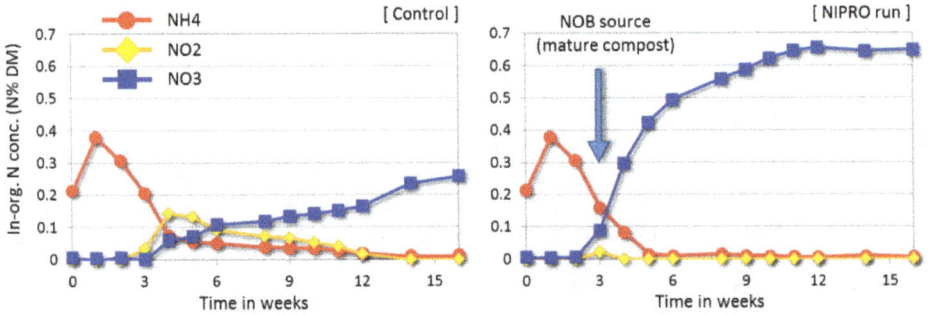

Fig. 6. Changes in concentration of inorganic nitrogen compounds during swine manure composting. Arrow indicates addition of NOB source of mature swine compost

To quantify the effect of the NIPRO process on nitrogen conservation, a nitrogen balance in swine manure composting was investigated (Fukumoto and Inubushi, 2009).

6.1 Materials and methods

Fresh swine manure was mixed with sawdust to make it suitable for aerobic decomposition. Sixteen kilograms of mixture as wet matter (WM) was piled inside a laboratory-scale apparatus. In the NIPRO run, 500 g WM of mature swine compost was added at turning on day 18 to avoid decreasing the number of NOB added at high temperature.

6.2 Results and discussion

In this composting experiment, the amount of N_2O emission was reduced by 70% by applying the NIPRO process, while there was little difference in NH_3 emission (Fig. 7).

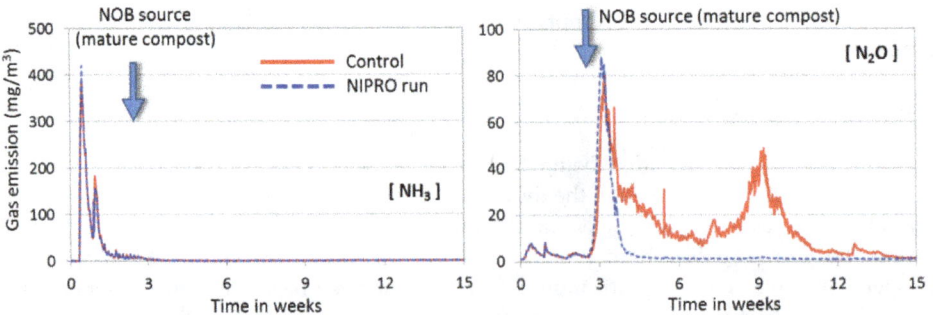

Fig. 7. Emission patterns of NH_3 and N_2O during swine manure composting. Arrow indicates addition of NOB source of mature swine compost

At the start of this composting experiment, approximately 130 g of nitrogen was contained in the initial compost pile of 16 kg (WM). The total mass of nitrogen emitted as N_2O in the control and the NIPRO run were 12.3 g and 4.0 g, respectively, and the amount of NH_3-N emission, which occurred mostly in the thermophilic phase, was 13.0 g in both runs. Therefore, the amounts of nitrogen loss by NH_3 and N_2O emission in the control and the NIPRO run were calculated as 25.2 g and 16.9 g, respectively. However, total nitrogen loss in the control and the NIPRO run were 36.8 g and 17.6 g, respectively (Fig. 8). Nitrogen loss mainly occurred after the thermophilic phase, with the exception of NH_3 and N_2O, and its magnitude became very small in the NIPRO run process that prevented prolonged NO_2^- accumulation. Therefore, it is considered that the unexplained nitrogen loss is expanded by NO_2^- accumulation during the composting, and that the effect of the NIPRO process on nitrogen conservation has a possibility to become larger than expected.

When the composition of nitrogen components in the final compost product between the control and the NIPRO run was compared, the organic nitrogen content made little difference between the runs, but NO_3^- nitrogen, which is a fast release fertilizer, increased greatly in the NIPRO run (Table 2).

Run	Elapsed time (d)	FW (kg)	MC (%)	Nitrogen compounds (gN/kg DM)			
				TN	NH_4^+	NO_2^-	NO_3^-
Control	0	16.00	64.6	23.4	2.0	0.0	0.1
	110	7.77	53.2	24.0	0.1	0.0	1.5
NIPRO	0	16.00	64.6	23.4	2.0	0.0	0.1
	110	8.04	51.3	28.6	0.1	0.0	4.3

FW, fresh weight; MC, moisture content; DM, dry matter; TN, total nitrogen.

Table 2. Properties of initial/final compost material in the swine manure composting

Kester et al. (1997) reported that higher NO_2^- concentrations enhanced both N_2O and nitric oxide (NO) emissions in the continuous cultures of nitrifiers and denitrifiers. Therefore, the emission of NO was measured during swine manure composting to clarify the components of the unknown nitrogen emissions. As a result of the measurement, it was revealed that significant NO emission begins after the thermophilic phase and is enhanced by NO_2^- accumulation during the composting, as with N_2O. Therefore, NO emission will also be reduced by applying the NIPRO (Fig. 9). However, the portion of NO emission from the total nitrogen loss tended to be small compared with NH_3 and N_2O emissions, even in the control, especially when the moisture content of the composting material was high. In one example, the level of nitrogen loss as an NO emission was only one-tenth the magnitude of the N_2O emission in our composting experiment. However, there was also a case in which the amount of NO emission had become half of the N_2O emission under comparatively dry conditions (Fukumoto et al. 2011a). It is known that moisture content is an important factor related to NO emission (del Prado et al. 2006), and these results seemed to have reflected it. Information concerning NO emissions from composting is limited (Hao and Chang, 2001), though it exerts a strong impact on chemical and physical processes in the atmosphere.

Therefore, further research is warranted to assess the environmental risk caused by NO emission from livestock activity.

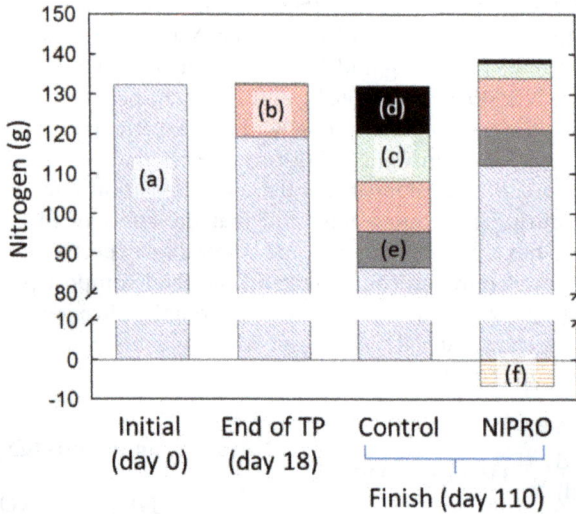

Fig. 8. Nitrogen mass balance over 110 days of swine manure composting. TP, thermophilic phase of composting; (a), N in the compost material; (b), N loss as NH_3 emission; (c), N loss as N_2O emission; (d), N loss as other emissions; (e), N loss as sample; (f), N in the added MSC for NIPRO process

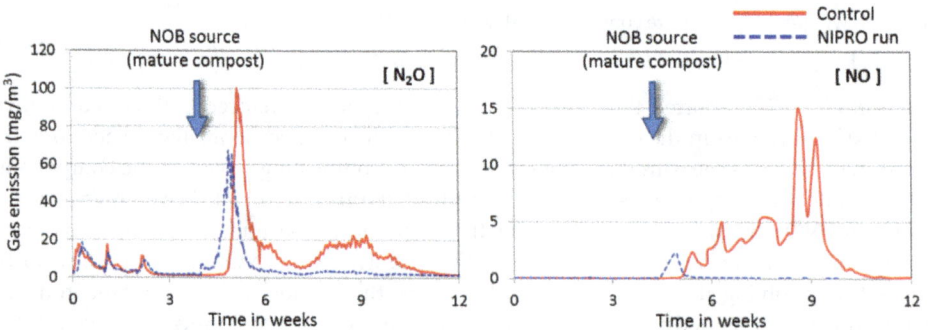

Fig. 9. Emission patterns of N_2O and NO during swine manure composting. Arrow indicates addition of NOB source of mature swine compost

7. Collaboration with struvite crystallization

Ammonia emission during the thermophilic phase is thought to be a principal cause of nitrogen loss during composting, though N_2O, NO, and other emissions during the maturation phase have an important role in nitrogen loss. Therefore, reducing NH_3 emission is necessary for nitrogen conservation in composting. Recently, struvite crystallization has been considered to be one effective countermeasure for reducing NH_3 emission (Jeong and Kim, 2001). Struvite is crystallized magnesium ammonium phosphate (MAP, $MgNH_4PO_4 \bullet 6H_2O$), which is formed according to the following equation:

$$HPO_4^{2-} + NH_4^+ + Mg^{2+} + OH^- + 6H_2O \rightarrow MgNH_4PO_4 \bullet 6H_2O \text{ (Struvite)} + H_2O$$

In the composting process, struvite crystallization used for reducing NH_3 emission is prompted by the addition of magnesium (Mg) and phosphate (PO_4) salts. The struvite crystallization reaction is accelerated when the pH is between 8 and 9. During the thermophilic phase, the pH of composting material generally rises to over 8 by the increase of NH_4^+ nitrogen, which is generated by the decomposition of organic nitrogen. Therefore, the adjustment of pH is not necessary. The effect of struvite crystallization on reducing NH_3 emission has been confirmed in the composting of food waste, and of poultry and swine manure. Struvite is a valuable slow-release fertilizer. The ingredients of struvite are released under acidic conditions. Nitrate nitrogen that is generally contained in mature compost is a fast-release type fertilizer; therefore, the struvite crystallization is thought to add a new and different value to the compost product, as well as reducing environmental risks. As reported in several papers, the reagent addition of struvite crystallization has a negative effect on the composting microorganisms decomposing in organic matter. Therefore, there is a possibility that nitratation promotion would be affected by the reagent addition of struvite crystallization in the case that these two countermeasures are applied simultaneously.

7.1 Materials and methods

To quantify the combined effect of struvite crystallization (MAP) and the NIPRO process on the reduction of nitrogenous emissions, laboratory- and mid-scale composting experiments of swine manure were conducted (Fukumoto et al. 2011a). The dose of Mg and PO_4 sources is an important issue related to struvite crystallization in composting (Jeong and Hwang, 2005; Lee et al. 2009). A higher dose of Mg and PO_4 could reduce the amount of nitrogen loss; however, there would be adverse effects on the decomposing organic matter and treatment costs would be increased. In our study, the respective doses of Mg and P placed into 0.045 and 0.030 mol/kg of raw feces was decided according to our past study, considering the balance of three factors (N conservation, degradation of organic matter and cost). The reagents for struvite crystallization were added at the start of composting, and then the mature swine compost (NOB source) was added after the thermophilic phase for nitratation promotion.

7.2 Results and discussion

By the addition of Mg and PO_4 salts, the amount of NH_3 emission could be reduced by 25-43% compared with the control. To confirm the struvite formation, the amount of nitrogen fixed in struvite crystals was measured according to the procedure of Tanahashi et al. (2010). As the result of these analyses, the amount of struvite nitrogen contents in the final product of sole struvite crystallization treatment became larger than those in the control. Therefore, it was confirmed that struvite crystallization had been enhanced by the addition of reagents during swine manure composting. However, in the treatment of two combined countermeasures (MAP and NIPRO), the struvite nitrogen content had become lower than those in the control despite the addition of Mg and PO_4 salts. To evaluate the effect reagent addition has on struvite crystallization, changes in the struvite nitrogen content were investigated. During the thermophilic phase (0-28 days), the struvite nitrogen content in the treatment of two combined countermeasures had changed more than those in the control, which indicated that the struvite crystallization was enhanced by the addition of reagents. However, after the addition of mature swine compost for nitratation promotion, the struvite nitrogen content was suddenly decreased, and became lower than that of the control. During the thermophilic phase, the material pH rose to over 8 due to the high NH_4^+ contents. However, the material pH generally declined due to the start of nitrification. Particularly, because more NO_3^- nitrogen is accumulated in the compost product by the NIPRO process, the material pH tended to become lower than what was found in normal composting. Due to the lower pH, the struvite crystals formed during the thermophilic phase are considered to have been dissolved again during the maturation phase (Fig. 10).

NOB source
(mature compost)

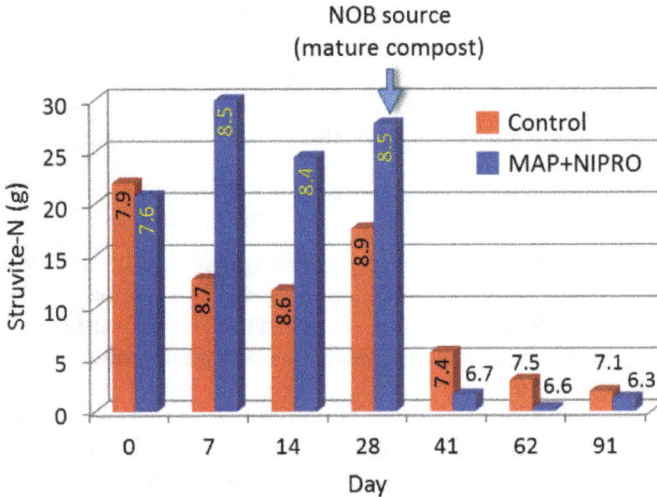

Fig. 10. Changes in the struvite nitrogen content during swine manure composting. Arrow indicates addition of NOB source of mature swine compost. Numbers on bars show the material pH

On the other hand, the effect of struvite crystallization on reducing N_2O emission was small (decreasing rate of N_2O was 10%). When the NIPRO process was applied simultaneously, the amount of N_2O emission could be reduced by 52-80%. However, the struvite crystallization showed a reducing effect on NO and other nitrogenous emissions during the maturation phase, as well as the NIPRO process. Conserving nitrogen in the form of struvite crystals has a benefit for stable nitrogen conservation because microbes cannot use the nitrogen in the struvite before it is dissolved. Therefore, it is considered that the struvite crystallization had a reduction effect on nitrogen losses during the maturation phase (Fig. 11).

The amount of total nitrogen loss was reduced by 60% by applying the two combined countermeasures, opposed to 50% by the struvite crystallization alone. No adverse effects from adding reagents to the growth of NOB were observed. However, the NIPRO process dissolved struvite crystals due to a decline in pH. Therefore, the effectiveness of struvite as a slow-release fertilizer cannot be expected when the struvite crystallization is applied with the NIPRO process.

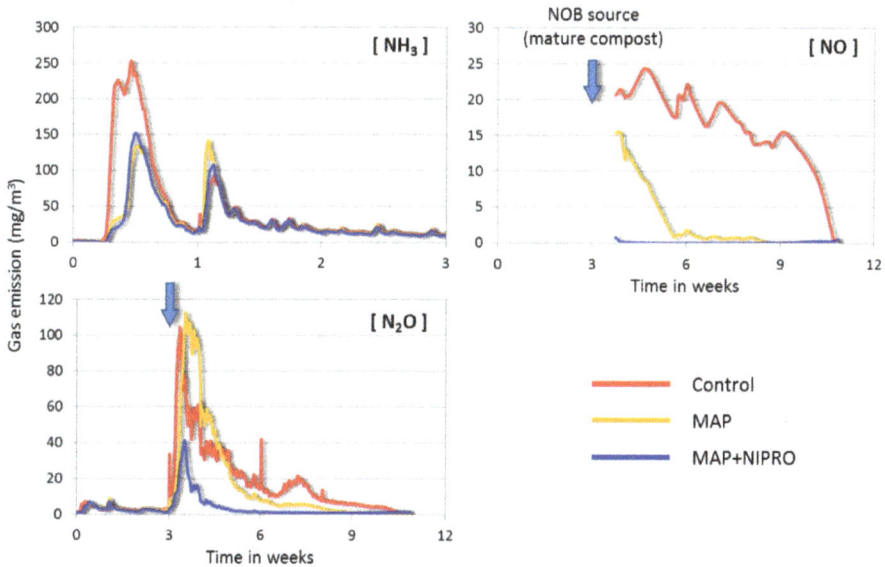

Fig. 11. Patterns of NH_3, N_2O and NO emissions during swine manure composting. Arrows indicate addition of NOB source of mature swine compost

8. Cattle and poultry manure composting

In the sections above, all composting experiments were conducted using swine manure. This is because swine manure has a natural property that is suitable for applying the NIPRO process. A precondition for showing the maximum effect of the NIPRO process is that N_2O induced by NO_2^- accumulation contributes to a large portion of its total emission. In swine manure composting, because the decreasing rate of N_2O emission by applying the NIPRO process is comparatively high (average decreasing rate is 60%), the portion of N_2O induced by NO_2^- accumulation is considered to be large. However, such information is inapplicable in the composting of cattle and poultry manure. Thereupon, to inspect the adaptability of the NIPRO process, composting experiments of cattle and poultry manure, and incubation tests for the quantification of ammonium generation potential (AGP) in respective animal manure, were conducted (Fukumoto et al. 2011b).

8.1 Materials and methods

In composting experiment of cattle manure, the house-shaped dynamic chamber system, which was constructed in a former study (Fukumoto et al. 2011a), was used. On the other hand, the laboratory-scale composting apparatus was used for poultry manure composting.

To estimate the maximum ammonium generation potential (AGP), incubation tests of respective livestock manure were conducted. The mixed solution of fresh manure (3-5% w/v) and nitrification inhibitor was agitated continuously at 25°C in the dark, and the concentration of NH_4^+-N in the solution was periodically measured until the rising curve became a plateau. The AGP was calculated as the amount of generated NH_4^+-N per gram of organic matter

8.2 Results and discussion

In cattle manure composting, the trends of nitrogen transition and N_2O emission were different from those of swine manure composting (Fig. 12). Most N_2O in cattle manure composting tends to be generated in the thermophilic phase, whereas it occurs in the maturation phase in swine manure composting. Similar results of N_2O emission in cattle manure composting were also observed in other studies (Maeda et al. 2010a; Maeda et al. 2010b). Moreover, the delayed growth of indigenous NOB, which is a cause of prolonged NO_2^- accumulation in swine manure composting, was not observed in cattle manure composting. Therefore, it is considered that the activity of nitrification, which becomes the starting point of N_2O generation from livestock manure, is high even during the thermophilic phase in cattle manure composting. The level of NH_4^+-N in cattle manure is usually lower than that of swine manure, which also indicates that the inhibitory effect of free ammonia on nitrification is low. A high NH_4^+-N level is thought to be one of the causes of delayed growth of indigenous NOB in swine manure composting. The quick recovery of indigenous NOB, i.e., complete nitrification, in cattle manure composting seems to be due to the lower influence of inhibitors, such as free ammonia. Therefore, the prolonged NO_2^- accumulation after the start of nitrification did not occur in cattle manure composting, which indicates that the NIPRO process may be not suitable for cattle manure composting.

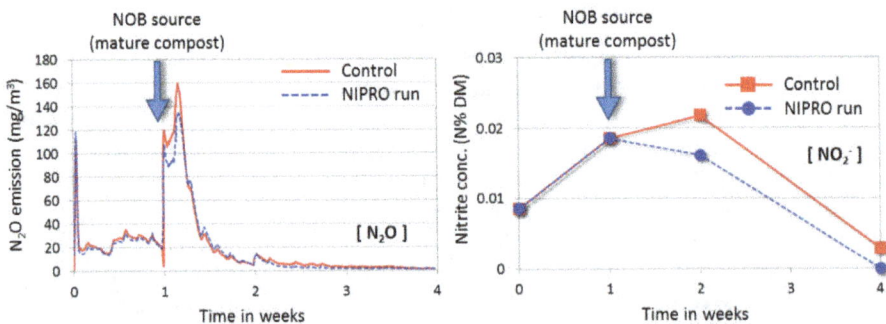

Fig. 12. Emission patterns of N_2O and changes in the concentration of nitrite nitrogen during cattle manure composting. Arrow indicates addition of NOB source of mature bovine compost

With the poultry manure composting, no obvious nitrification activity could be observed even in the long maturation phase (over 300 days). The excretion mechanism of poultry is different from other livestock animals. The bird excretes urine together with feces in the form of uric acid, making nitrogen content in the manure extremely high. Therefore, it is considered that nitrification was completely inhibited by a high concentration of free ammonia in poultry manure composting. In fact, any growth of nitrifiers was not observed in this study, except for some temporary detections of AOB. Because nitrification could not be initiated, no significant N_2O emissions were confirmed. Therefore, it is considered that the NIPRO process, which reduces the amount of N_2O generated via inadequate nitrification, has no positive effect on poultry manure composting.

These three kinds of livestock manure each showed characteristic nitrogen transitions and N_2O emission patterns. It is considered that the respective unique nitrogen turnover of each livestock depends on the amount of active nitrogen generated via decomposition of manure. Therefore, as one of the factors affecting nitrogen turnover, the AGP of fresh livestock manure was investigated (Fig. 13).

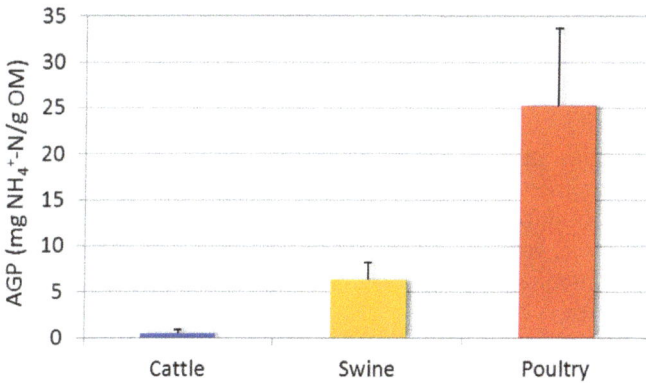

Fig. 13. Ammonium generation potential (AGP) of cattle, swine and poultry manure. OM, organic matter. Error bars indicate standard deviation

From the results of the incubation tests, poultry manure showed the highest AGP value among the three kinds of manure, as expected. On the other hand, the lowest AGP value was observed in cattle manure, which was approximately one-fiftieth the value of poultry manure. Swine manure showed an intermediate value between cattle and poultry manure.

The AGP value is thought to affect nitrification. In poultry manure composting, large amounts of ammonium are generated, which inhibits the activity of nitrification completely (Anthonisen et al. 1976). On the other hand, because of the low AGP, the effect of inhibitory factors, such as free ammonia, would be small in cattle manure composting, which would lead to the quick recovery of complete nitrification. Therefore, in cattle or poultry manure composting, it is thought that prolonged NO_2^- accumulation scarcely occurs, although when it does, it is for completely different reasons. This indicates that the quantity of N_2O emission induced by NO_2^- accumulation is small. Therefore, the effect of the NIPRO process on reducing N_2O emission would be slight in cattle and poultry manure composting. On the other hand, the level of AGP in swine manure might be suitable for NO_2^- accumulation. That is, it is considered that the AGP of this level does not inhibit the activity of the first half of nitrification (i.e., nitritation), but would have a serious influence on the latter half of nitrification (i.e., nitratation) because the tolerance of NOB on inhibition factors, such as free ammonia, is lower than that of AOB. It is thought that the AGP would become one of the decisive materials when choosing to apply the NIPRO process to the composting of organic wastes.

9. Conclusions

The present report showed the effect of the nitratation promotion (NIPRO) process on nitrogenous emissions during the composting of livestock manure. The NIPRO process can reduce the nitrogenous emissions induced by NO_2^- accumulation by the addition of an NOB source such as mature compost. In swine manure composting, a remarkable reducing effect on N_2O, NO and other nitrogenous emissions (probably N_2) by applying the NIPRO process, was confirmed. As a result, more nitrogen could be preserved in the final compost product in the form of NO_3^- nitrogen, which is expected to improve the compost's value as a fertilizer. Moreover, the NIPRO process can collaborate with the struvite crystallization which reduces NH_3 emission in the thermophilic phase. However, it was revealed that struvite crystals formed during the thermophilic phase have been dissolved during the maturation phase due to a pH decline induced by NO_3^- accumulation in the case applying the NIPRO process. In cattle or poultry manure composting, it seemed to be difficult to apply the NIPRO process for reducing nitrogen losses, because prolonged NO_2^- accumulation would be difficult to maintain in those manure composts. Before putting these findings to practical use, however, some issues still remain. For instance, how to decide the timing of NOB addition, and how to evaluate effects and outcomes in an actual case. Therefore, further study is indeed necessary.

10. Acknowledgement

The works constituting this chapter have been supported by the Ministry of Agriculture, Forestry and Fisheries of Japan, the Ministry of the Environment of Japan, and the Ministry of Education, Culture, Sports, Science and Technology of Japan.

11. References

Anthonisen, A.C., Loehr, R.C., Prakasam, T.B.S. & Srinath, E.G. (1976). Inhibition of nitrification by ammonia and nitrous-acid. *Journal Water Pollution Control Federation,* 48, 835-852.
Beck-Friis, B., Smårs, S., Jönsson, H. & Kirchmann, H. (2001). Gaseous emissions of carbon dioxide, ammonia and nitrous oxide from organic household waste in a compost reactor under different temperature regimes. *Journal of Agricultural Engineering Research,* 78, 423-430.
Bhatia, A., Sasmal, S., Jain, N., Pathak, H., Kumar, R. & Singh, A. (2010). Mitigating nitrous oxide emission from soil under conventional and no-tillage in wheat using nitrification inhibitors. *Agriculture Ecosystems & Environment,* 136, 247-253.
Blouin, M., Bisaillon, J.-G., Beaudet, R. & Ishaque, M. (1990). Nitrification of swine waste. *Canadian Journal of Microbiology,* 36, 273-278.

Burns, L.C., Stevens, R.J. & Laughlin, R.J. (1996). Production of nitrite in soil by simultaneous nitrification and denitrification. *Soil Biology & Biochemistry*, 28, 609-616.

Burrows, S. (2006). The chemistry of mushroom composts. II.-Nitrogen changes during the composting and cropping processes. *Journal of the Science of Food and Agriculture*, 2, 403-410.

Corriveau, J., van Bochove, E., Savard, M.M., Cluis, D. & Paradis, D. (2010). Occurrence of high in-stream nitrite levels in a temperate region agricultural watershed. *Water Air and Soil Pollution*, 206, 335-347.

Criado, S.R. (1996). Considerations on main factors which take part in nitrate contamination of ground water in Spain with relationship to other EU countries. *Fertilizer Research*, 43, 203-207.

Czepiel, P., Douglas, E., Harriss, R. & Crill, P. (1996). Measurements of N₂O from composted organic wastes. *Environmental Science & Technology*, 30, 2519-2525.

del Prado, A., Merino, P., Estavillo, J.M., Pinto, M. & Gonzalez-Murua, C. (2006). N₂O and NO emissions from different N sources and under a range of soil water contents. *Nutrient Cycling in Agroecosystems*, 74, 229-243.

Dittert, K., Bol, R., King, R., Chadwick, D. & Hatch, D. (2001). Use of a novel nitrification inhibitor to reduce nitrous oxide emission from N-15-labelled dairy slurry injected into soil. *Rapid Communications in Mass Spectrometry*, 15, 1291-1296.

Dodd, J.L. (1994). Desertification and degradation in sub-Saharan Africa - the role of livestock. *Bioscience*, 44, 28-34.

FAO (2006). Livestock's role in climate change and air pollution. in: H. Steinfeld, P. Gerber, T. Wassenaar, V. Castel, M. Rosales & C. de Haan (Eds.), Livestock's long shadow. LEAD Virtual Research and Development Centre, Rome, Italy, pp. 102-123.

Francis, C.A., Roberts, K.J., Beman, J.M., Santoro, A.E. & Oakley, B.B. (2005). Ubiquity and diversity of ammonia-oxidizing archaea in water columns and sediments of the ocean. *Proceedings of the National Academy of Sciences of the United States of America*, 102, 14683-14688.

Fukumoto, Y. & Inubushi, K. (2009). Effect of nitrite accumulation on nitrous oxide emission and total nitrogen loss during swine manure composting. *Soil Science and Plant Nutrition*, 55, 428-434.

Fukumoto, Y., Osada, T., Hanajima, D. & Haga, K. (2003). Patterns and quantities of NH₃, N₂O and CH₄ emissions during swine manure composting without forced aeration–effect of compost pile scale. *Bioresource Technology*, 89, 109-114.

Fukumoto, Y., Suzuki, K., Kuroda, K., Waki, M. & Yasuda, T. (2011a). Effects of struvite formation and nitratation promotion on nitrogenous emissions such as NH₃, N₂O and NO during swine manure composting. *Bioresource Technology*, 102, 1468-1474.

Fukumoto, Y., Suzuki, K., Osada, T., Kuroda, K., Hanajima, D., Yasuda, T. & Haga, K. (2006). Reduction of nitrous oxide emission from pig manure composting by addition of nitrite-oxidizing bacteria. *Environmental Science & Technology*, 40, 6787-6791.

Fukumoto, Y., Waki, M. & Yasuda, T (2011b). Characteristics of nitrogen transition and N₂O generation in the composting of cattle, swine and poultry manure (22IOS4-04), *Proceedings of CIGR International Symposium on "Sustainable Bioproduction – Water, Energy, and Food"*, Tokyo, Japan, 19-23 September 2011.

Haga, K. (1990). Production of compost from organic wastes. *FFTC (Food and Fertilizer Technology Center)/ ASPAC Extension Bulletin*, 311, 1-18.

Haga, K., Osada, T., Harada, Y., Izawa, T. & Nishimura, Y. (1998). Constituents of the anaerobic portion occurring in the pile during composting of cattle wastes. *The Journal of the Society of Agricultural Structures, Japan*, 29, 125-130.

Hao, X. & Chang, C. (2001). Gaseous NO, NO_2, and NH_3 loss during cattle feedlot manure composting. *Phyton-Annales Rei Botanicae*, 41, 81-93.

He, J., Shen, J., Zhang, L., Zhu, Y., Zheng, Y., Xu, M. & Di, H.J. (2007). Quantitative analyses of the abundance and composition of ammonia-oxidizing bacteria and ammonia-oxidizing archaea of a Chinese upland red soil under long-term fertilization practices. *Environmental Microbiology*, 9, 2364-2374.

He, Y., Inamori, Y., Mizuochi, M., Kong, H., Iwami, N. & Sun, T. (2001). Nitrous oxide emissions from aerated composting of organic waste. *Environmental Science & Technology*, 35, 2347-2351.

IPCC (2007). Fourth Assessment Report - Climate Change 2007: The Physical Science Basis.

Isermann, K. (1990). Share of agriculture in nitrogen and phosphorus emissions into the surface waters of western-Europe against the background of their eutrophication. *Fertilizer Research*, 26, 253-269.

Jeong, Y.K. & Hwang, S.J. (2005). Optimum doses of Mg and P salts for precipitating ammonia into struvite crystals in aerobic composting. *Bioresource Technology*, 96, 1-6.

Jeong, Y.K. & Kim, J.S. (2001). A new method for conservation of nitrogen in aerobic composting process. *Bioresource Technology*, 79, 129-133.

Kester, R.A., de Boer, W. & Laanbroek, H.J. (1997). Production of NO and N_2O by pure cultures of nitrifying and denitrifying bacteria during changes in aeration. *Applied and Environmental Microbiology*, 63, 3872-3877.

Kuroda, K., Hanajima, D., Fukumoto, Y., Suzuki, K., Kawamoto, S., Shima, J. & Haga, K. (2004). Isolation of thermophilic ammonium-tolerant bacterium and its application to reduce ammonia emission during composting of animal wastes. *Bioscience, Biotechnology, and Biochemistry*, 68, 286-292.

Kuroda, K., Osada, T., Yonaga, M., Kanematu, A., Nitta, T., Mouri, S. & Kojima, T. (1996). Emissions of malodorous compounds and greenhouse gases from composting swine feces. *Bioresource Technology*, 56, 265-271.

la Pagans, E., Font, X. & Sanchez, A. (2005). Biofiltration for ammonia removal from composting exhaust gases. *Chemical Engineering Journal*, 113, 105-110.

Lee, J.E., Rahman, M.M. & Ra, C.S. (2009). Dose effects of Mg and PO_4 sources on the composting of swine manure. *Journal of Hazardous Materials*, 169, 801-807.

Leininger, S., Urich, T., Schloter, M., Schwark, L., Qi, J., Nicol, G.W., Prosser, J.I., Schuster, S.C. & Schleper, C. (2006). Archaea predominate among ammonia-oxidizing prokaryotes in soils. *Nature*, 442, 806-809.

Lin, C. (2008). A negative-pressure aeration system for composting food wastes. *Bioresource Technology*, 99, 7651-7656.

Maeda, K., Hanajima, D., Toyoda, S., Yoshida, N., Morioka, R. & Osada, T. (2011). Microbiology of nitrogen cycle in animal manure compost. *Microbial Biotechnology*, DOI: 10.1111/j.1751-7915.2010.00236.x.

Maeda, K., Morioka, R., Hanajima, D. & Osada, T. (2010a). The impact of using mature compost on nitrous oxide emission and the denitrifier community in the cattle manure composting process. *Microbial Ecology*, 59, 25-36.

Maeda, K., Toyoda, S., Shimojima, R., Osada, T., Hanajima, D., Morioka, R. & Yoshida, N. (2010b). Source of nitrous oxide emissions during the cow manure composting process as revealed by isotopomer analysis of and amoA abundance in betaproteobacterial ammonia-oxidizing bacteria. *Applied and Environmental Microbiology*, 76, 1555-1562.

McGinn, S.M. & Janzen, H.H. (1998). Ammonia sources in agriculture and their measurement. *Canadian Journal of Soil Science*, 78, 139-148.

Ministry of Agriculture, Forestry and Fisheries, Japn (2008). Statistics of Livestock Industry in 2008.

Oenema, O., Wrage, N., Velthof, G.L., van Groenigen, J.W., Dolfing, J. & Kuikman, P.J. (2005). Trends in global nitrous oxide emissions from animal production systems. *Nutrient Cycling in Agroecosystems*, 72, 51-65.

Parkinson, R., Gibbs, P., Burchett, S. & Misselbrook, T. (2004). Effect of turning regime and seasonal weather conditions on nitrogen and phosphorus losses during aerobic composting of cattle manure. *Bioresource Technology*, 91, 171-178.

Pearson, J. & Stewart, G.R. (1993). The deposition of atmospheric ammonia and its effects on plants. *New Phytologist* 125, 283-305.

Rappert, S. & Muller, R. (2005). Odor compounds in waste gas emissions from agricultural operations and food industries. *Waste Management*, 25, 887-907.

Ravishankara, A.R., Daniel, J.S. & Portmann, R.W. (2009). Nitrous oxide (N₂O): the dominant ozone-depleting substance emitted in the 21st century. *Science* 326, 123-125.

Silva, C.D., Cuervo-Lopez, F.M., Gomez, J. & Texier, A.C. (2011). Nitrite effect on ammonium and nitrite oxidizing processes in a nitrifying sludge. *World Journal of Microbiology & Biotechnology*, 27, 1241-1245.

Smet, E., Van Langenhove, H. & De Bo, I. (1999). The emission of volatile compounds during the aerobic and the combined anaerobic/aerobic composting of biowaste. *Atmospheric Environment*, 33, 1295-1303.

Sommer, S.G. (2001). Effect of composting on nutrient loss and nitrogen availability of cattle deep litter. *European Journal of Agronomy*, 14, 123-133.

Steinfeld, H. & Wassenaar, T. (2007). The role of livestock production in carbon and nitrogen cycles. *Annual Review of Environment and Resources*, 32, 271-294.

Szanto, G.L., Hamelers, H.V.M., Rulkens, W.H. & Veeken, A.H.M. (2007). NH₃, N₂O and CH₄ emissions during passively aerated composting of straw-rich pig manure. *Bioresource Technology*, 98, 2659-2670.

Tanahashi, T., Yano, H., Itou, H. & Oyanagi, W. (2010). Magnesium ammonium phosphate in cattle and swine manure composts and an extraction method for its evaluation. *Japanese Journal of Soil Science and Plant Nutrition*, 81, 329-335.

Tourna, M., Stieglmeier, M., Spang, A., Konneke, M., Schintlmeister, A., Urich, T., Engel, M., Schloter, M., Wagner, M., Richter, A. & Schleper, C. (2011). Nitrososphaera viennensis, an ammonia oxidizing archaeon from soil. *Proceedings of the National Academy of Sciences of the United States of America*, 108, 8420-8425.

Wuchter, C., Abbas, B., Coolen, M.J.L., Herfort, L., van Bleijswijk, J., Timmers, P., Strous, M., Teira, E., Herndl, G.J., Middelburg, J.J., Schouten, S. & Damste, J.S.S. (2006). Archaeal nitrification in the ocean. *Proceedings of the National Academy of Sciences of the United States of America*, 103, 12317-12322.

Yasuda, T., Kuroda, K., Fukumoto, Y., Hanajima, D. & Suzuki, K. (2009). Evaluation of full-scale biofilter with rockwool mixture treating ammonia gas from livestock manure composting. *Bioresource Technology*, 100, 1568-1572.

Zaman, M. & Blennerhassett, J.D. (2010). Effects of the different rates of urease and nitrification inhibitors on gaseous emissions of ammonia and nitrous oxide, nitrate leaching and pasture production from urine patches in an intensive grazed pasture system. *Agriculture Ecosystems & Environment*, 136, 236-246.

Zeman, C., Depken, D. & Rich, M. (2002). Research on how the composting process impacts greenhouse gas emissions and global warming. *Compost Science & Utilization*, 10, 72-86.

Part 5

Soil Salinity

10

Effect of Salinity on Soil Microorganisms

Celia Maria Maganhotto de Souza Silva and Elisabeth Francisconi Fay
Embrapa Environment
Brazil

1. Introduction

Whilst the majority of countries have criteria to evaluate the quality of the air and water, the same does not occur for the quality of the soil. Traditionally soil quality is associated with productivity, but recently it has been defined in terms of sustainability, that is, the capacity of the soil to absorb, store and recycle water, minerals and energy such that the production of the crops can be maximized and environmental degradation minimized. Thus preservation of soil quality is a critical factor for environmental sustainability.

A significant decline in soil quality has occurred throughout the entire world as a result of adverse changes in its physical, chemical and biological properties. According to Steer (1998), in the last decades of the last century, about 2 billion of the 8.7 billion agricultural lands, permanent pastures, forests and wild native lands were degraded. The global grain production growth rate fell from 3% in the seventies to 1.3% in the period from 1983-1993, and one of the main reasons for this decline was inadequate soil and water management. Inventories carried out on the soil productive capacity in the last decade indicated that 40% of the degradations of arable land were induced by man as a result of soil erosion, atmospheric pollution, intensive cultivation, over-grazing, deforestation, salinization and desertification (Oldeman, 1994).

Soil degradation processes constitute a serious problem on a worldwide basis, with significant environmental, social and economic consequences. As the world population increases, so does the need to protect the soil as a vital resource, particularly for food production, .

The soil is a dynamic medium, constituting the habitat of abundant biodiversity, with unique genetic patterns where one can find the greatest amount and variety of living organisms, which serve as a nutrient reservoir. One gram of soil in good conditions can contain 600 million bacteria belonging to 15,000 or 20,000 different species. These values decrease to 1 million bacteria encompassing from 5000 to 8000 species in desert soils (Informativo Capebe, 2010). Depending on the amount of organic matter present in the soil, the biological activity eliminates pathogenic agents, decomposes organic matter and other pollutants into simpler components (frequently less noxious), and contributes to maintaining the physical and biochemical properties required for soil fertility and structure.

However, soil is not an inexhaustible resource and consciousness of this, allied to knowledge of the need to maintain or increase the capacity of this agro-ecosystem, directing its multiple functions in an adequate way, is increasing, as also changes in the overall perception of its importance as an environmental component.

As an open system, the soil is dynamic and in constant interaction with the atmosphere, the hydrosphere, the biosphere and the lithosphere. Depending on the intensity with which these factors act, soil can present differentiated characteristics which define its potentialities for exploitation by man. Its structure is defined as the aggregation of primary particles in compound particles, separate from adjacent aggregates. However its structure implies in an arrangement of the primary particles (sand, silt, clay) which, by way of cementing agents, can group together forming aggregates with certain structural patterns, which necessarily include porous space.

Alterations in the chemical conditions of cultivated soils, such as the concentrations and types of ions in solution in the soil, variations in pH and in the critical flocculation concentration of the particles, can cause modifications in the dispersion of the clay fraction, degrading the original soil fraction. The sodium ion, being monovalent, increases the width of the diffuse double layer on the surface of the clays, reducing the attractive forces between them with a consequent increase in particle dispersion. The consequence of this dispersion of the clay is also shown by a reduction in stability of the soil aggregates, which are thus easily transported by rain or irrigation (Almeida Neto, 2007).

Although soil structure is not considered as a plant growth factor, it exerts an influence on the air and water supplies to the roots of the crops, on nutrient availability, on the penetration and development of the roots, as also on the movement of the soil macro-fauna. It also influences the loss of agrochemicals by way of erosion and leaching and can have considerable importance on the negative environmental impact of some agricultural practices. According to Machado (2002), inadequate soil use has been causing a gradual loss in their productive capacity.

The main threats to soil are erosion, mineralization of the organic matter, reduction in bio-diversity, contamination, water proofing, compacting, salinization, and the degrading effects of floods and landslides. Soil degradation produces deterioration of the plant covering and the hydric resources. In addition, by mean of a series of physical, chemical, biological and hydrological processes, it causes destruction of both the biological potential of the land and of their use to sustain the population connected to it.

2. Soil salinization

Soil salinity is part of the natural ecosystem in arid and semi-arid regions and an increasing problem in agricultural soils the world over. In temperate, moist climates salinity occurs on a smaller scale, principally in salt water marshes, at the side of highways and in salty effluent discharges (Pathak & Rao, 1998; Keren, 2000; Qadir et al, 2000; Wichern, 2006).

Salinization consists of an accumulation of water soluble salts in the soil. These salts include the ions potassium (K^+), magnesium (Mg^{2+}), calcium (Ca^{2+}), chloride (Cl^-), sulfate (SO_4^{2-}), carbonate (CO_3^{2-}), bicarbonate (HCO_3^-) and sodium (Na^+). Sodium accumulation is also called sodification. High sodium contents result in destruction of the soil structure which, due to a lack of oxygen, becomes incapable of assuring plant growth and animal life.

Salt affects crop germination and density, as also vegetative development, reducing productivity and, in the most serious cases, leading to generalized plant death, limiting nutrient absorption and reducing the quality of the available water. For example, elevated salinity weakens plants due to the increase in osmotic pressure and the toxic effect of the salts. In addition, salinization affects the metabolism of the organisms present in the soil, drastically reducing soil fertility and increasing water proofing of the deeper layers, impeding cultivation of the land. In an indirect way, soil salinization can adversely affect plant growth, due to destruction of the soil structure and its consequent compacting. This occurs due to a dispersion of the clay particles caused by substitution of the calcium (Ca^{+2}) and magnesium (Mg^{+2}) ions present in the complex by sodium (Na^+), resulting in an increase in soil sodicity, that is, in the percentage of exchangeable sodium (PES), which, in the last instance, is the main factor responsible for the deterioration of the physical properties of salt-affected soils (sodic, or alkaline, and saline-sodic). Saline soils show the following physical-hydric characteristics: low permeability, low hydraulic conductivity and aggregate instability (Freire, 2009).

On a world wide scale, the production by approximately 400 million hectares of arable land is being severely restricted by salinity (Bot et al., 2000). In the European Union salinization affects about 1 million hectares, mainly in the Mediterranean countries, constituting one of the main causes of desertification (Iannetta & Colonna, 2011). The most affected soils are situated in Hungary, Romania, Greece, Italy and the Iberian Peninsula (Agricultura...., 2011). About 8.1 million hectares are salinized in India, of which 3.1 million are in coastal regions (Triphati et al., 2007). In Nordic countries, the use of salt to remove ice from highways produces localized salinization phenomena (Agricultura...., 2011). Considering the increasing temperatures and decrease in pluviosity which have characterized climates in recent years, the salinization problem has increased.

Salinization results from natural or anthropogenic factors, constituting a process of soil degradation which, in some cases, is responsible for irreparable losses in their productive capacity, with great extensions of arable land becoming sterile.

2.1 Natural soil salinization and sodification factors

The natural factors influencing soil salinity are:

- geological phenomena which increase the salts concentration in groundwater and consequently in the soil;
- natural factors capable of bringing groundwater containing elevated salt contents to the surface;
- infiltration of groundwater in below sea-level zones (micro-depressions with reduced or absent drainage);
- drainage of waters from zones with geological substrates capable of liberating large amounts of salts;
- action of winds, which, in coastal zones, can transport moderate amounts of salts to the interior.

The weathering of primary minerals (which make up the rocks or the original soil material) is the indirect source of nearly all the salts present in soils, although there are only a few cases in which this results in sufficient accumulation of salt (primary or pedogenetic salinization) to form saline soils. The areas of lands salinized by such natural processes, from which salt-affected soils could arise, such as Planossolo Solódico, Solonetz Solorizado, Solonchack Solonétzico, do not increase so drastically when compared to the increasing growth intensity of the extension of land salinized by anthropic activity.

In general, the salinization process occurs in soils situated in regions of low rainfall and which have a water-bearing stratum near the surface. In coastal zones, salinization could be associated with the over-exploitation of groundwaters due to the demand induced by increased urbanization, or by industry and agriculture. The over-extraction of groundwaters can result in a lowering of the normal water-bearing stratum levels, leading to the intrusion of sea water.

2.2 Secondary factors leading to soil salinization and sodification

The most influential anthropogenic factors are:

- irrigation with water containing elevated salt contents;
- rise in phreatic water level due to human activities (infiltration of water from unlined channels and reservoirs, irregular distribution of irrigation water, deficient irrigation practices, inadequate drainage);
- use of fertilizers and other production factors, namely for intensive agriculture in land with low permeability and reduced possibilities for leaching;
- irrigation with residual waters with high salt contents;
- elimination of residual waters with high salt contents by way of the soil;
- contamination of the soil with industrial water and sub-products with high salt contents.

Soils affected by salts commonly appear in irrigated areas due to inadequate management of the irrigation and other practices, such that important extensions of fertile, arable land are becoming more and more saline. This is due to management practices that do not aim at conserving the productive capacity of the soil, such as, for example, the non-existence of an efficient drainage system, the use of inadequate quality water in inadequate amounts, and also the incorrect and excessive use of chemical fertilizers.

Irrigation is an ancient agricultural practice, widely used throughout the world, principally in tropical regions where hot, dry climates prevail, such as, for example, the semi-arid region in northeastern Brazil, where the evapotranspiration rate exceeds the rainfall throughout the better part of the year. In these areas, where there is not sufficient water available to supply the hydric needs of the crops throughout the whole vegetative cycle, irrigation assumes a fundamentally important role in order to guarantee good agricultural harvests. Since all natural waters contain variable amounts of soluble salts, be they of meteoric (rain), surface (rivers, lakes, dams, etc.) or subterranean (aquifers) origin, the application of water to the soil by irrigation implies necessarily in the addition of salts to their profile.

Thus salinization of a soil depends on the quality of the water used for irrigation, on the existence and level of natural and/or artificial drainage of the soil, on the depth of the water-bearing stratum and on the original concentration of salt in the soil profile. The basic principle to avoid soil salinization is to maintain the equilibrium between the amount of salt provided to the soil by irrigation and the amount of salt removed by drainage. In arid or highly ventilated climes, evaporation of the water enriches the soil with solutes, increasing the danger of salinization. In the same way, soils with limited permeability tend to concentrate salts. In irrigated zones, the low rainfall, elevated evapotranspiration rates and the soil structure impede leaching of the salts, which accumulate in the surface layers.

Estimates by FAO indicate that of the 250 million hectares of irrigated land in the world, approximately 50% already show salinization and soil saturation problems, and 10 million hectares are abandoned annually due to these problems (CODEVASF, 2011).

The excessive amounts of salts provided by irrigation waters can have adverse effects on the chemical and physical properties of the soils and on their biological processes (Garcia & Hernandez, 1996; Rietz & Haynes, 2003; Tejada &Gonzalez, 2005). These effects include mineralization of the carbon and nitrogen and the enzymatic activity, which is crucial for the decomposition of organic matter and liberation of the nutrients necessary for sustainability of the production (Azam & Ifzal, 2006; Wong et al., 2008). In addition, the agricultural practices can increase or reduce the microbial population, thus altering the activity, source and persistence of the enzymes in the soil (Parham et al., 2003).

Organic fertilizers are considered useful for crops due to their nutritive value, principally of nitrogen (N), and for their merits in improving the physical properties of the soil (Jackson & Bertsch, 2001; Garbarino et al, 2003), but their salt content is usually ignored, which could prejudice plant growth and soil quality after continued application. The flow of nitrogen (N) and phosphorus (P) from the application of animal manure is considered to contribute to non-precise pollution (Parker, 2000; Anderson & Xia; 2001; Ekholm et al, 2005; Allen et al, 2006). Salinity is considered as a non-precise source of pollution, but the secondary salinity induced by the application of organic fertilizers has not been considered as particularly worrying. Li-Xian et al. (2007) evaluated the effect of applying poultry manure and its ionic composition on soil salinization. The authors showed that the increase or decrease in the concentration of a determined ion in the soil depended on its concentration in the manure, the application rate, the removal of the crops and to leaching. The ionic composition of the soil salinity changed according to the types and doses of the fertilizers used and to their applications. The results also showed that, even in humid regions, the potential for the risk of secondary soil salinization exists with the successive application of animal manure.

3. The effect of salinity on the soil microorganisms

The microbial communities of the soil perform a fundamental role in cycling nutrients, in the volume of organic matter in the soil and in maintaining plant productivity. Thus it is important to understand the microbial response to environmental stress, such as high concentrations of heavy metals of salts, fire and the water content of the soil. Stress can be detrimental for sensitive microorganisms and decrease the activity of surviving cells, due to the metabolic load imposed by the need for stress tolerance mechanisms (Schimel et al, 2007; Yuan et al., 2007, Ibekwe et al., 2010; Chowdhury, 2011). In a dry hot climate, the low humidity and soil salinity are the most stressful factors for the soil microbial flora, and frequently occur simultaneously.

Saline stress can gain importance, especially in agricultural soils where the high salinity may be a result of irrigation practices and the application of chemical fertilizers. Research has been carried out on naturally saline soils, and the detrimental influence of salinity on the microbial soil communities and their activities reported in the majority of studies (Batra & Manna, 1997; Zahran, 1997; Rietz & Haynes, 2003; Sardinha et al., 2003).

The effect is always more pronounced in the rhizosphere according to the increase in water absorption by the plants due to transpiration. The simple explanation for this is that life in high salt concentrations has a high bio-energetic taxation, since the microorganisms need to maintain osmotic equilibrium between the cytoplasm and the surrounding medium, excluding sodium ions from inside the cell. As a result, energy sufficient for osmo-adaptation is required (Oren, 2002; Jiang et al, 2007).

3.1 Fungi

The composition of the microbial community may be affected by salinity (Pankhurst et al., 2001; Gros et al., 2003; Gennari et al., 2007; Llamas et al., 2008; Chowdhury et al., 2011) since the microbial genotypes differ in their tolerance of a low osmotic potential (Mandeel, 2006; Llamas et al., 2008). In fungi, a low osmotic potential decreases spore germination and the growth of hyphae and changes the morphology (Juniper & Abbott, 2006) and gene expression (Liang et al., 2007), resulting in the formation of spores with thick walls (Mandeel, 2006).

Fungi have been reported to be more sensitive to osmotic stress than bacteria (Pankhurst et al., 2001; Sardinha et al., 2003; Wichern et al., 2006). There is a significant reduction in the total fungal count in soils salinized with different concentrations of sodium chloride. Similarly, with an increase in the salinity level to above 5%, the total count of bacteria and actinobacteria was drastically reduced (Omar et al., 1994). Van Bruggen & Semenov (2000) reported that on a long-term basis there is a decrease in the genetic diversity of fungi as a result of stress. On the other hand, Killham (1994) mentioned that the filamentous fungi are highly tolerant of hydric stress. However they have to deal with the increase in osmotic pressure and may therefore change their physiology (Killham, 1994) and morphology in response to this (Zahran, 1997). Two strategies used by microorganisms to adapt to osmotic stress were described by Killham (1994), both of which result in an accumulation of solutes in the cell to counteract the increase in osmotic pressure. One is the selective exclusion of the solute incorporated (for example, Na+, Cl-), thus accumulating the ions necessary for metabolism (for example, NH4+). The other cell adaptation mechanism is the production of organic compounds that will antagonize the concentration gradient between the soil solution and the cell cytoplasm. This adaptation finally results in a physiologically more active microbial community, and, in consequence, reduced substrate use efficiency. However these mechanisms are known for single microorganisms, but little has been studied at the community level.

According to Oren (2001) and Hagemann (2011), while sensitive cells are damaged by the low osmotic potential, some microorganisms can adapt by accumulating osmolytes (including amino acids in bacteria and polyols in fungi), that help retain water (Beales, 2004). Nevertheless the synthesis of osmolytes requires large amounts of energy: 30 to 110 ATPs, when compared to the 30 ATPs required to synthesize the cell wall (Oren, 1999), representing a significant metabolic responsibility for the microorganisms, and reduces the energy available for growth.

In order to better understand what happens to the microbial biomass and its activity in saline soils, one must also consider the water potential (osmotic potential + matrix potential), especially the low water content when the salt concentration in the soil solution increases. Since the water content changes, the microorganisms will be subject to different osmotic and water potentials, even though the modifications in the electrical conductivity (EC) measurement are small. Thus the EC is an indicator of little importance with respect to microbial stress in saline environments. According to Chowdhury et al. (2011), microorganisms have two strategies to respond to the water potential. A decrease in this potential to up to -2MPa damages a proportion of the microbial population, but the remaining microorganisms will adapt themselves and be active. For lower water potentials, the adaption mechanisms are not sufficient and, although the microorganisms survive, they do so with reduced activity per unit of biomass. However more studies are required in different soils and, in particular, in saline soils, in order to discern which effects can be generalized.

Considering the forecast for an increase in saline and sodic areas, an understanding of the effects of salinity and sodicity on the soil carbon (C) stock and flow is fundamental for environmental management. Wong et al. (2008) evaluated the effects of salinity and sodicity on the microbial biomass and on soil respiration, under controlled conditions, submitting perturbed soil samples to leaching after receiving different salts concentrations. The highest soil respiration rates were observed in soils with low salinity, and the lowest in soils with medium salinity, whilst the microbial biomass was greater in the treatments with high salinity and lower in those with low salinity. According to the authors, the results can be attributed to a greater availability of substrate in high salt concentrations, or by an increase in the dispersion of the aggregates of soil or from the dissolution or hydrolysis of the organic material in the soil, which can compensate, at least in part, the stress to which the microbial population is submitted in high salt concentrations. The apparent disparity between the evolution of respiration and that of the biomass could be due to a change induced in the microbial population from one dominated by more active microorganisms to one dominated by less active microorganisms.

The microbial biomass is an important labile fraction of the soil organic matter, functioning both as an agent of transformation and recycling of the organic matter and soil nutrients, as also of a source of nutrients for the plants. It is also a potential source of enzymes in the soil. High salinity reduces the microbial biomass (Tripathi et al., 2006; Wichern et al., 2006), affects amino acid capture and protein synthesis (Norbek & Blomberg, 1998) and respiration (Laura, 1974; Pathak & Rao, 1998; Gennari et al., 2007) and causes increases and decreases in C and N mineralization (Pathak & Rao, 1998; Wichern et al., 2006).

Since the soil organic matter, and consequently, the biomass and microbial activity, are generally more relevant in the first few centimeters at the surface of the soil, salinization close to the surface can significantly affect a series of microbiologically mediated processes. This is a considerable problem, since the microbial processes of the soil control its ecological functions and fertility.

The availability of nutrients for plants is regulated by the rhizospheric microbial activity. Thus any factor affecting this community and its functions influences the availability of nutrients and growth of the plants. One of the microbial responses playing a significant role in plant growth is the internal recycling of nitrogen (N) by way of immobilization and re-mineralization. In the majority of studies, the immobilization of NH_4^+-N is reported as being quicker than that of NO_3^--N, whilst the re-mineralization of the N immobilized in NH_4^+ is slower than that immobilized in NO_3^- (Herrmann et al., 2005). However, little has been reported about immobilization/re-mineralization in the two forms of N under conditions of salinity. Since nitrification is more or less inhibited in the presence of salts (Laura, 1977; Sethi et al., 1993) resulting in an accumulation of NH_4^+-N, the cycling of the two forms of N will have a significant impact on the dynamics and availability of N for the plants. According to Azam & Ifzal (2006) the presence of NaCl retards the N immobilization process. Both re-mineralization and nitrification were significantly retarded in the presence of NaCl, maximum inhibition occurring with 4000 mg NaCl kg^{-1} of soil. The inhibitory effect of NaCl on N re-mineralization was relatively higher in soils treated with NH_4^+. The results of this study suggest greater sensitivity to NaCl by microorganisms that have assimilated NO_3^-. However, N re-mineralization in the population that had assimilated NO_3^- was less affected by salinity when compared to the population that had assimilated NH_4^+.

3.2 Effect on enzymatic activities

Since the greater part of soil biochemical transformations are dependent on or related to the presence of enzymes, an evaluation of their activities could be useful to indicate if a soil is adequately carrying out the processes closely connected to its quality.

Soil enzymes carry out a fundamental role in the ecosystems, acting as catalysts of various reactions that result in the decomposition of organic residues, cycling of nutrients and the formation of organic matter in the soil, in addition to taking part in intercellular metabolic reactions responsible for the functioning and maintenance of living beings, quite apart from their biotechnological potential, with various applications in the industrial and environmental areas. They generally originate from microorganisms, but can also have animal and vegetable origins.

Amongst the diverse soil enzymes, dehydrogenase, β-glucosidase, urease and the phosphatases are important in the transformation of different nutrients for plants. The activity of dehydrogenase reflects the total oxidative capacity of the microbial biomass (Nannipieri et al., 1990) and is involved in the central aspect of metabolism. β-glucosidase is an important enzyme in the land carbon cycle, in the production of glucose, which constitutes an important energy source for the microbial mass (Tabatabai, 1994). Thus, the determination of β-glucosidase activity, amongst other hydrolytic enzyme activities, has been suggested as a good indicator of soil quality (Dick et al., 1996). The phosphatases play an important role in the transformation of organic phosphorus into inorganic forms more appropriate for plants. Phosphorus (P) is one of the essential nutrients for a plant, and the greater part of soil phosphorus occurs in the organic form.

Urease predominates amongst the enzymes involved in the N cycle of the soil (Tabatabai & Bremner, 1972; Cookson, 1999). It catalyzes the hydrolysis of urea into ammonia or the ammonium ion, depending on the pH of the soil and carbon dioxide. Urease and catalase are the enzymes responsible for the decomposition of vegetable residues. The activity of these enzymes transforms the residue into humus, which is then completely decomposed into the free nutrients (Ahmad & Khan, 1988). On the other hand, amylase hydrolyzes the polysaccharides, converting them into simpler constituents. The activity of this enzyme is associated with high productivity of the crops (Ahmad & Khan, 1988).

Under laboratory conditions, salinity influenced soil enzyme activity negatively, although the degree of inhibition varied according to the enzyme analyzed and the nature and amount of soil added (Frankenberger & Bingham, 1982). Dehydrogenase activity was severely inhibited whereas the hydrolases showed a milder degree of inhibition. The reduction of enzyme activity in saline soils could be due to the osmotic dehydration of the microbial cells that liberate intracellular enzymes, which become vulnerable to the attack by soil proteases, with a consequent decrease in enzyme activity. The salting-out effect modifies the ionic conformation of the protein-enzyme active site, and specific ionic toxicity causes a nutritional imbalance for microbial growth and subsequent enzyme synthesis (Frankenberger & Bingham, 1982). Ahmad & Khan (1988) and Rietz & Haynes (2003) obtained similar results. According to Rietz & Haynes (2003) the increase in salinity due to an influx of salty water under controlled conditions, decreased the carbon content of the soil microbial biomass and enzymes. Other researchers, for example Omar et al. (1994) and

Jialiang (2008) also indicated the effects of soil salinity on the carbon of microbial biomass and on enzyme activity. Garcia & Hernandez (1996) and Ghollarata & Raiesi (2007) showed that an increase in soil salinity inhibited the enzyme activities of benzoyl argininamide alkaline phosphatase and β-glucosidase, and also microbial respiration. Invertase and urease activities were also severely reduced by an increasing concentration of sodium chloride (NaCl) during incubation. In addition, the effect was inhibitory of nitrate reductase in the majority of the treatments (Omar et al., 1994). On comparing the enzyme activities of saline soil with those of normal soil, Ahmed & Khan (1988) also observed a decline in amylase, catalase, phosphatase and urease activities with increasing salinity.

Controlled conditions (laboratory) do not usually reflect the natural situation prevailing in coastal region soils, where the salinity varies temporally. Tripathi et al. (2006; 2007) studied the influence of the salinity of arable soils in Indian coastal regions on the microbial biomass and the following enzyme activities: dehydrogenase, β-glucosidase, urease, and acid and alkaline phosphatases, in three different seasons of the year. The microbial and biochemical parameters were adversely affected by the salinity, and the most extreme situation occurred in the summer. Of the enzymes studied, the activity of dehydrogenase was the most affected.

Another particular ecosystem is the mangrove swamps, areas restricted to zones between coastal seas and islands in tropical regions, associated with estuaries, bays and lagoons in places protected from the impact of waves, where the salinity is between 5 and 30%, but can reach 90% (Museu do Una, 2010). This is a highly degraded natural environment for a variety of reasons, amongst which the discharge of domestic and industrial effluent. Variations in the salinity of this environment can affect the retention of the pollutants and the microbiological responses as a function of the discharge of effluent. On investigating such areas, Tam (1998) observed that the addition of effluent to mangrove swamps, independent of their salinity, stimulated microbial growth and increased the activities of the enzymes dehydrogenase and alkaline phosphatase. According to the author these effects were due to supplementation with additional carbon sources and other nutrients, provided by the effluent.

4. Recovery of saline soils

The low productivity of saline soils can be attributed not only to their toxicity due to the salt or to the damage caused by excessive amounts of soluble salts, but also to low soil fertility. The fertility problems are usually evidenced by a lack of organic matter and of available mineral nutrients, especially N and P (Shi et al., 1994). These soils are also usually characterized by a reduction in the activities of some key soil enzymes, such as urease and phosphatase (Shi et al., 1994, Yuan et al., 1997), which are associated with biological transformations and the bioavailability of N and P. The adverse effect of soil salinity on crops depends both on their tolerance and on other factors with important roles in the selection of the natural soil microbial flora during salinization, such as: soil composition, organic matter, pH, heavy metals, water and oxygen availability (Ross et al., 2000).

4.1 Organic amendents

Recently various organic supplements, such as ground coverings, manures and compounds, have been investigated for their efficiency in recovering saline soils. It has been shown that the application of organic matter can accelerate the leaching of NaCl, decrease the percentage of exchangeable sodium and the electrical conductivity, and increase water filtration, the water holding capacity and aggregate stability (El-Shakweer et al., 1998).

In soils affected by salts and showing low productivity, the adoption of adequate agricultural practices is of fundamental importance for the success of their exploitation, including modifications in the organic fertilization (Garcia et al., 2000). For example, supplementation with organic matter improves the quality of saline soil and neutralizes the negative effects of the salt, since the microorganisms profit from the greater availability of substrate and can thus deal better with the high salinity. Supplementation also leads to a differentiation in the soil microbial community, with bacteria dominating the surface of the substrate (Wichern et al., 2006).

The application of decomposing cow manure, straw or decomposing stable manure significantly increased the productivities of rice, wheat, barley and sorghum, cultivated in saline soils (Swarup, 1985; Gaffar et al., 1992; Aich et al., 1997). The incorporation of sewage sludge and the epicarp-mesocarp of the almond tree fruit into saline soil increased the N, P and K concentrations in the soil and in tomato fruits (Gomez et al., 1992), and the iron and manganese concentrations in rice (Swarup, 1985). In contrast, the addition of stable manure reduced the sodium adsorption ratio (SAR) (Gaffar et al., 1992). Tejada & Gonzalez (2005) showed that an increase in the organic matter content of saline soils increased the soil structural stability and density and, consequently, the microbial biomass.

The C/N ratio is an extremely important property in the decomposition of organic matter by microorganisms, and for this reason, the organic matter added to saline soils performs an important role in the positive effect on the microbial activity and enzyme activities.

The incorporation of rice straw, swine excrement or rice straw plus swine excrement significantly increased the activities of urease and phosphatase and the rate of respiration of the soil (Liang et al., 2003), coinciding with previous reports on the incorporation of other organic matters in saline soils (Blagodatsky & Richter, 1998; Luo & Sun, 1994).

Other organic residues with differentiated chemical compositions were studied by Tejada et al. (2006): one a compound obtained from a cotton de-stoner and the other non-composted chicken manure, in two different doses. The application of both in the doses studied, under dry climate conditions, improved the physical, chemical and biological properties of the saline soil. These organic treatments also favor the appearance of spontaneous vegetation, which protects the soil and contributes to its correction. The alterations tested improved the soil structure, reducing the percentage of exchangeable sodium and promoting an increase in various enzyme activities. However, whereas the cotton compound had a greater effect on the physical properties of the soil and the percentage of exchangeable sodium, the chicken manure mainly increased the soil enzyme activity.

However, the excessive use of organic manure should be avoided, especially in areas flooded for long periods, in order to reduce the risk of toxic effects from reduced intermediates, which accumulate from the anaerobic decomposition of organic manure (Liang et al., 2003).

The use of residues as a soil corrective or conditioner is an economically and environmentally interesting practice, and coconut powder stands out amongst the organic materials that could be used to recover saline soils, since it is abundant in the Northeastern region of Brazil due to the great consumption of coconuts. It represents a solution for the use of discarded coconut shells, which are constituted of one fraction of fibers and another known as powder, which is aggregated to the fibers. Silva Junior et al. (2009) evaluated the basal respiration of soil incubated with different concentrations of coconut powder, submitted to different levels of salinity. The incorporation of organic matter increased the amount of C-CO$_2$ mineralization, even at high salinity levels. It also caused a reduction in the negative effect of salinity on microbial activity.

4.2 Plant remediation strategies

Another solution used for the recovery of saline soils in agricultural systems or salinized, abandoned areas is the use of plants (Hatton & Nulsen, 1999). The use of plants to remediate saline and sodic soils is a low-cost, emergent method, but with little acceptance due to its low profitability. However, some farmers have improved the salinity condition of their soils by planting salt-tolerant trees (Marcar et al., 1995) or forage shrubs (Barrett-Lennard & Malcolm, 1995; Porto et al., 2006).

Various plant species (halophytic plants) grow naturally at the coast and in salinized areas, and can survive in salt concentrations equal or greater to that of sea water. The compartmenting of the ions in the vacuoles, the accumulation of compatible solutes in the cytoplasm and the presence of genes for salt tolerance, confer salt resistance on these plants (Gorham, 1995). The re-vegetation of salinized areas with halophytic plants is an example of pro-active phyto-remediation (Porto et al., 2001; 2006).

The introduction of the halophyte *Glycyrrhiza glabra* in the recovery of saline soils and restoration of the subsequent crop systems in irrigated agriculture has been demonstrated in various studies (Mihailova, 1966; Kerbabaev, 1971; Pauzner, 1971). From results presented by Ravindran et al., (2007), the authors concluded that of the six vegetable species evaluated, *Suaeda maritima* and *Sesuvium portulacastrum* L. exhibited a greater accumulation of salt in their tissues and greater salt reduction in the soil. Rabhi et al. (2009) compared *Sesuvium portulacastrum* L. (Aizoaceae) with two other native halophytes: *A. indicum* and *S. fruticosa* with respect to their abilities to desalinate saline soils. The authors showed that of the three species studied, *Sesuvium portulacastrum* L. was the most convenient for use in the leaching of salts from the rhizospere in arid and semi-arid regions, where the rainfall is low.

In the same way, the creation of high productivity plant forage systems by establishing palatable halophytic plants showed that it was possible to remediate a saline/sodic soil and provide extra income for the farmers at the same time (Hyder, 1981; Helalia et al., 1992; Dagar et al., 2004). Species of the saltbushes *Atriplex* have been used both as forage and to rehabilitate degraded areas (dunes, salt-mines, saline soils). They are dominant in many arid and semi-arid regions of the world, particularly in environments combining relatively high soil salinity with aridity (Ortiz-Dorda et al., 2005).

Quantifying the recovery of the biological activity of soils remediated with plants has been the focus of few studies. Silva et al. (2008) evaluated the effect of irrigation with the desalination waste from pink tilapia production tanks, on the chemical and microbiological properties of soils cultivated with *Atriplex nummularia* Lindl. The authors found that although the irrigation with saline waste affected the physical and chemical properties of the soil, cultivation of the halophyte favored microbial activity. Similar results were obtained by Pereira et al. (2004) studying soil cultivated for three years with *Atriplex nummularia* Lindl., and irrigated with saline waste. In the dry periods, the values for pH, electrical conductivity, fluorescein diacetate hydrolytic activity and alkaline phosphatase activity were higher than in other areas. However, a negative correlation was observed between the values for microbial carbon and the metabolic quotient.

Carvava et al. (2005) studied the influence of the following eight halophytes: *Asteriscus maritimus* (L.) Less, *Arthrocnemum macrostachyum* (Moric.) Moris, *Frankenia corymbosa* Desf., *Halimione portulacoides* (L.) Aellen, *Limonium cossonianum* O. Kuntze, *Limonium caesium* (Girard) O. Kuntze, *Lygeum spartum* L., and *Suaeda vera* Forsskål ex J.F. Gmelin, on the microbiological and biochemical properties of the rhizospere and aggregate stability of a saline soil. There was good correlation between the enzyme activities, the C of the microbial biomass, colonization of the roots of the eight halophytes and the levels of stable aggregates. The results also showed that the microbial activity and the soil properties related to the microbial activity, as also the aggregate stability, were determined by the type of halophytic species. The modifications in microbial activity caused by the vegetation were also related to the variation in the activities of protease, phosphatase, urease and β-glucosidase (Ceccanti & Garcia, 1994). In the case of the halophytes *Arthrocnemum macrostachyum* and *Sarcocornia fruticosa*, when grown in a salty swampy area contaminated with metals, whose roots were colonized by arbuscular mycorrhizal fungi, it was found that the salinity and heavy metals negatively affected the degree of colonization by fungi and some of the parameters indicating microbial activity, such as dehydrogenase, urease, protease, phosphatase and β-glucosidase (Carrasco et al., 2006).

4.3 Consequences of remediation activities

The principal objective of the recovery of soils affected by salts is to reduce the concentration of soluble salts and of exchangeable sodium in the soil profile, to a level that does not prejudice the development of crops. A decrease in the degree of salinity involves the process of dissolution and consequent removal by percolation water, whereas a decrease in the exchangeable sodium content involves its displacement from the exchange complex by calcium before the leaching process. Since it is of low cost and relatively abundant in many parts of the world, plaster is the corrective most used to recover sodic and saline-sodic soils (Oad et al., 2002; Barros et al., 2004; Gharaibeh et al., 2009). The substituted sodium is leached from the radicular zone by way of excess irrigation, a process that demands an adequate flow of water through the soil (Qadir & Oster, 2004; Qadir et al, 2006).

There are reports in the literature that the efficiency of washing the radicular zone was higher when irrigation was carried out by dripping rather than by other methods (Bresler et al., 1982). The key question in the recovery of soils affected by salts using irrigation by dripping, is that a reasonable irrigation regime must be carried out to guarantee not only normal growth of the crops, but also an excess of water to leach out the salts. Recently, Kang et al. (2008) reported the recovery of heavy-textured saline soils using irrigation by dripping. However, the alterations in the soil properties during recovery are still not well defined. The physical, chemical and biological alterations occurring in a saline soil during the recovery process with a corn crop and irrigation by dripping were reported by Tan & Kang (2009). The results showed that the soil density in the first 0-20 cm decreased from 1.71 g cm^{-3} to 1.44 g cm^{-3} after three years of rehabilitation. The water content in the saturated soil of the 0-10 cm layer increased from 20.3 to 30.2%. Both the soil salinity and pH value decreased significantly after three years of recovery. The organic matter contents reduced, whereas the total nitrogen, total phosphorus and total potassium tended to increase after cultivation and irrigation. The amount of bacteria, actinobacteria and fungi increased according to the number of years of rehabilitation, with a tendency for a homogenous distribution in the soil profile. The activities of urease and alkaline phosphatase also increased, but the activity of invertase altered little.

Lin et al. (2006) observed that the bacteria, actinobacteria and fungi increased 2.3, 4.3 and 71 times, respectively, by planting *Suaeda salsa* L. and irrigating by dripping in coastal saline soil. There was a reduction in soil salinity and improvement in fertility.

The low solubility of Ca^{2+} during remediation could limit its efficiency, and thus the possibility of using it with microorganisms is being explored so as to provide more active Ca^{2+} from plaster. Experiments carried out with blue-green algae and plaster resulted in greater solubility of the plaster, thus providing recovery of the sodic soils (Subhashini & Kaushik, 1981). However, Syed et al. (2003) reported successful experiments in the recovery of saline-sodic soils when a mixture of different microorganisms was applied, without the prior application of plaster.

Sahina et al. (2011) studied the effect of microbial application in four different saline-sodic soils with saturated hydraulic conductivity, and treated with plaster. Suspensions of three fungal isolates (*Aspergillus* spp. FS 9, 11 and *Alternaria* spp. FS 8) and two bacterial strains (*Bacillus subtilis* OSU 142 and M3 *Bacillus megaterium*) were mixed with the leaching water of the soil treated with plaster, and subsequently applied to the soil columns. The measurement of the saturated hydraulic conductivity of the soil columns after treatment, indicated that it increased significantly ($P<0.01$) in the saline-sodic soils after application of the microorganisms. The data suggest that the microorganisms tested could have the potential to help improve water circulation through saline soils.

Carter (1986) suggested that the addition of plaster caused a decrease in microbial activity, but tended to increase the microbial biomass in the soil. The effect of salinity on the carbon dynamics, with respect to the accumulation or loss of C is not well documented. The rate of C accumulation or loss depends on the balance between the amount entering and leaving. The entrance of C depends on the plant and the accumulation of biomass, when the organic carbon levels of the soil are dominated by the deposition of vegetable and root residues. The entrance of C into saline soils decreases with the decline in growth of the vegetation, due to the direct effect of the toxic ions and of the increase in osmotic potential, and the indirect effect of the structural decline of the sodic soil.

Wong et al. (2009) investigated the flow of C in saline soils to which plaster and organic matter were added. The microbial biomass was lower in the untreated saline soil, but the effect of adding plaster was insignificant. The accumulated respiration was greater in the soils receiving the organic supplement, whereas the \addition of plaster decreased the accumulated respiration rate when compared to the addition of organic matter and of organic matter and plaster. The lowest respiration rate and microbial biomass was attributed to the soil with the lowest rate of organic carbon, resulting from the little or absent entrance of C into the soil due to the high salinity, responsible for the lack or absence of vegetation. There was an increase in respiration and in microbial biomass with the addition of organic matter, independent of the adverse environmental conditions of the soil. The results suggest that the microbial biomass of a hibernating population of salt tolerant microorganisms was present, and multiplied quickly when substrate became available.

5. Conclusions

According to estimates made by the United Nations, the population projected for 2050 is one of 8.9 billion inhabitants, which would exacerbate the challenges of agriculture to meet the food demands of this population. In the past the main directive was to increase the potential for food cultivation and its productivity in the field. These days the paradigm has changed, and now demands that the increase in productivity must be accompanied by sustainable management. Sustainable agriculture involves the management of agricultural resources respecting human needs, the maintenance of environmental quality, and the conservation of natural resources for the future. Soil salinity is widely reported as the main agricultural problem, particularly in irrigated agriculture, and approximately 20% of arable land and 50% of agricultural land in the world are under saline stress. According to statistics from UNESCO and FAO, the area of saline soils in the world is of 9.5 x 10^7 km^2. Salinity causes, directly or indirectly, a harmful influence on the maintenance of soil quality, since it affects the physical, chemical and biological properties of the soil. Research has reported the detrimental influence of salinity on the microbial communities of the soil and their activities. Thus the recovery of soils affected by salt has an important role in the sense of mitigating the pressure, especially in agricultural areas. Saline soil is an important land resource for agriculture.

6. References

Agricultura sustentável e conservação dos solos. Processos de degradação do solo. Salinização e sodificação. Ficha informativa nº 4. Disponível em < http://soco.jrc.ec.europa.eu/documents/PTFactSheet-04.pdf> Acesso em 05/06/2011.

Ahmad, I. & Khan, K.M. (1988). Studies on enzymes activity in normal and saline soils. *Pakistan Journal Agricultural Research*, Vol. 9, No.4, pp. 506-508, ISSN 0251-0480.

Ahmad, Z.; Yahiro, Y.; Kai, H. & Harada, T. (1973). Transformation of the organic nitrogen becoming decomposable due to the drying of soil. *Soil Science Plant Nutrition*, Vol. 19, No 4, (December, 1973), pp. 287–298, ISSN: 0038-0768.

Aich, A.C.; Ahmed, A.H.M. & Mandal, R. (1997). Impact of organic matter, lime and gypsum on grain yield of wheat in salt affected soils irrigated with different grades of brackish water. *Research*, Vol. 10, No (1– 2), (December, 1997), pp. 79– 84, ISSN: 0970-5767.

Allen, S.C.; Fair, V.D.; Graetz, D.A.; SHIBU, J. & Ramachandran N.P.K. (2006). Phosphorus loss from organic versus inorganic fertilizers used in alley cropping on a Florida Ultisol. Agriculture, Ecosystems & Environment, Vol. 117, No 4, (December, 2006), pp. 290–298, ISSN: 0167-8809.

Almeida Neto, O.B. de. (2007). Dispersão da argila e condutividade hidráulica em solos com diferentes mineralogías, lixiviados com soluções salino-sódicas. Tese de Doutorado. Universidade Federal de Viçosa. Viçosa, MG, Brasil.

Anderson, R. & Xia, L.Z. (2001). Agronomic measures of P, Q/I parameters and lysimeter-collectable P in subsurface soil horizons of a long-term slurry experiment. *Chemosphere*, Vol. 42, No 2, (January, 2001), pp. 171–178, ISSN: 0045-6535.

Azam, F. & Ifzal, M. (2006). Microbial populations immobilizing NH_4^+-N and NO_3^--N differ in their sensitivity to sodium chloride salinity in soil. *Soil Biology & Biochemistry*, Vol. 38, No 8, (August, 2006), pp. 2491–2494, ISSN: 0038-0717.

Barrett-Lennard, E.G. & Malcolm, C.V. (1995). *Saltland Pastures in Australia – A Practical Guide*. Department of Agriculture Western Australia ISBN: 1 920860 07 X, Perth, Western Australia.

Barros, M. de F.C.; Fontes, M.P.F.; Alvarez, V.H. & Ruiz, H.A. (2004). Recuperação de solos afetados por sais pela aplicação de gesso de jazida e calcário no Nordeste do Brasil. *Revista Brasileira de Engenharia Agrícola e Ambiental*, Vol. 8, No 1, (Janeiro-Abril, 2004), pp.59-64, ISSN: 1415-4366.

Batra, L. & Manna, M.C. (1997). Dehydrogenase activity and microbial biomass carbon in salt affected soils of semiarid and arid regions. *Arid Soil Research and Rehabilitation*, Vol. 11, No 3, (Available online: January, 2009), pp. 295–303, ISSN 0890-3069.

Beales, N. (2004). Adaptation of microorganisms to cold temperatures, weak acid preservatives, low pH, and osmotic stress: a review. *Comprehensive Reviews in Food Science and Food Safety*, Vol. 3, No 1, (January, 2004), pp. 1-20, ISSN 1541-4337.

Blagodatsky, S.A. & Richter, O. (1998). Microbial growth in soil and nitrogen turnover: a theoretical model considering the activity state of microorganisms. *Soil Biology & Biochemistry*, Vol. 30, No 13, (November, 1998), pp. 1743–1755, ISSN 0038-0717.

Bot, A.; Nachtergaele, F. & Young, A. (2000). Land resource potential and constraints at regional and country levels. Rome: FAO, 2000. (FAO. World Soil Resources Report, 90).

Bresler E.; McNeal, B. L. & Carter, D L. (1982). *Saline and Sodic Soils: Principles-Dynamics-Modeling*. Springer-Verlag, ISBN: 3-540-11120-4, Berlin Heidelberg.

Caravaca, F.; Alguacil, M.M.; Torres, P. & Roldán, A. (2005). Plant type mediates rhizospheric microbial activities and soil aggregation in a semiarid Mediterranean salt marsh. *Geoderma*, Vol. 124, No 3-4, (February 2005), pp. 375-382, ISSN: 0016-7061.

Carrasco, L.; Caravaca, F.; Alvarez-Rogel, J. & Roldan, A. (2006). Microbial processes in the rhizosphere soil of a heavy metals-contaminated Mediterranean salt marsh:A facilitating role of AM fungi. *Chemosphere*, Vol. 64, No 1, (June, 2006), pp. 104–111, ISSN: 0045-6535.

Carter, M.R. (1986). Microbial biomass and mineralizable nitrogen in Solonetzic soils: influence of gypsum and lime amendments. *Soil Biology & Biochemistry*, Vol. 18, No 5, (1986) pp. 531–537, ISSN: 0038-0717.

Ceccanti, B. & Garcia, C. (1994). Coupled chemical and biochemical methodologies to characterize a composting process and the humic substances. In: *Humic Substances in the Global Environment and its Implication on Human Health*, Senesi, N. & Miano, T. (Eds.), pp. 1279-1285, Elsevier, ISBN:10- 0444895930, New York, USA.

Chowdhury, N.; Marschner, P. & Burns, R.G. (2011). Soil microbial activity and community composition: impact of changes in matric and osmotic potential. *Siol Biology and Biochemistry*, Vol. 43, No 6, (June, 2011), pp. 1229-1236, ISSN: 0038-0717.

CODEVASF - Salinização do solo. Disponível em < http://www.codevasf.gov.br/programas_acoes/irrigacao/salinizacao-do-solo> Acesso em 10/03/2011.

Cookson, P. 1999. Special variation in soil urease activity around irrigated date palms. *Arid Soil Research and Rehabilitation*, Vol. 13, No 2, (Available online: November, 2010), pp. 155–169, ISSN: 0890-3069.

Dagar, J.C.; Tomar, O.S.; Kumar, Y. & Yadav, R.K. (2004). Growing three aromatic grasses in different alkali soils in semi-arid regions of northern India. *Land Degradation and Development*, Vol. 15, No 2, (March/April, 2004), pp. 143–151, ISSN: 1085-3278.

Dick, R.P., Breakwell, D.P. & Turco, R.F. (1996). Soil enzyme activities and biodiversity measurements as integrative microbiological indicators, In: *Methods for Assessing Soil Quality Special Publication No. 49*, Doran, J.W. & Jones, A.J. (Eds.), pp. 247–271, Soil Science Society America, ISBN-10: 0891188266, Madison, WI, USA.

Ekholm, P.; Turtola, E.; Gronroos, J.; Sauri, P. & Ylivainio, K. (2005). Phosphorus loss from different farming systems estimated from soil surface phosphorus balance, Agriculture, Ecosystems & Environment, Vol. 110, No 3-4, (November, 2005), pp. 266–278, ISSN: 0167-8809.

El-Shakweer, M.H.A.; El-Sayad, E.A. & Ejes, M.S.A. (1998). Soil and plant analysis as a guide for interpretation of the improvement efficiency of organic conditioners added to different soils in Egypt. *Communications in Soil Science and Plant Analysis*, Vol. 29, No 11-14, (Available online: November, 2008), pp. 2067–2088, ISSN: 0010-3624.

Frankenberger, W.T., Bingham, F.T. (1982). Influence of salinity on soil enzyme activities. *Soil Science Society of America Journal*, Vol. 46, No 6, (November-December, 1982), pp. 1173–1177, ISSN 0361-5995.

Freire, E. de A.; Laime, E.M.M.; Navilta, V. do N.; Lima, V. L. de & Santos, J.S. dos. (2009). Análise dos riscos de salinidade do solo do perímetro irrigado de Forquilha, Ceará. *Revista Educação Agrícola Superior*, Vol. 24, No 2, (December, 2009), pp.62-66, ISSN: 0101-756 X.

Gaffar, M.O.; Ibrahim, Y.M. & Wahab, D.A.A. (1992). Effect of farmyard manure and sand on the performance of sorghum and sodicity of soils. *Journal of the Indian Society of Soil Science*, Vol. 40, No 3, (September, 1992), pp. 540– 543, ISSN : 0019-638X.

Garbarino Jr.; Bednar A.J.; Rutherford, D. W.; Beyer, R.S. & Wershaw, R.L. (2003). Environmental fate of roxarsone in poultry litter. I. Degradation of roxarsone during composting. *Environmental Science & Technology*, Vol. 37, No 8, (Available online: March, 2003), pp. 1509–14, ISSN 0013-936X.

Garcıa, C. & Hernandez, T. (1996). Influence of salinity on the biological and biochemical activity of a calciorthird soil. *Plant and Soil*, Vol. 178, No. 2, (1996), pp. 255–263, ISSN: 0032-079X.

Garcia, C.; Hernandez, T.; Pascual, J.A.; Moreno, J.L. & Ros, M. (2000). Microbial activity in soils of SE Spain exposed to degradation and desertification processes. Strategies for their rehabilitation. In *Research and Perspectives of Soil Enzymology in Spain*. Garcia, C., Hernandez, M.T. (Eds.), pp. 93–143, CEBAS-CSIC, ISBN: 84-605-9821-7, Murcia, Spain.

Gennari, M., Abbate, C., La Porta, V., Baglieri, A. & Cignetti, A. (2007). Microbial response to Na_2SO_4 additions in a volcanic soil. *Arid Land Research and Management*, Vol. 21, No 3, (Available online: June, 2007), pp. 211-227, ISSN: 1532-4982.

Gharaibeh, M.A.; Eltaif, N.I. & Shunnar, O.F. (2009). Leaching and reclamation of calcareous saline–sodic soil by moderately saline and moderate-SAR water using gypsum and calcium chloride. *Journal of Plant Nutrition Soil Science*, Vol. 172, No 5, (Available online: May, 2009), pp. 713–719, ISSN: 1532-4982.

Gomez, I..; Navarro, P.J. & Mataix, J. (1992). The influence of saline irrigation and organic waste fertilization on the mineral content (N, P, K, Na, Ca and Mg) of tomatoes. *Journal of the Science of Food and Agriculture*, Vol. 59, No 4, (published online: September, 2006) pp. 483–487, ISSN: 0022-5142.

Gorham, J. (1995). Mechanism of salt tolerance of halophytes. In: *Halophytes and biosaline agriculture*, Choukr-Allah, R.; Malcolm, C.V. & Hamdy, A. (Eds.), pp. 207-233, Marcel Dekker, ISBN-10: 0824796640, New York, USA.

Gros, R.; Poly, F.; Jocteur-Monrozier, L. & Faivre, P. (2003). Plant and soil microbial community responses to solid waste leachates diffusion on grassland. *Plant and Soil*, Vol. 255, No 2, (March, 2003), pp. 445-455, ISSN: 0032-079X.

Hagemann, M. (2011). Molecular biology of cyanobacterial salt acclimation. *Fems Microbiology Reviews*, Vol. 35, No 1, (January, 2011), pp. 87-123, ISSN: 0168-6445.

Hatton, T.J. & Nulsen, R.A. (1999). Towards achieving functional ecosystem mimicry with respect to water cycling in southern Australian agriculture. Agroforestry Systems, Vol. 45, No 1-3, (1999), pp.203–214, ISSN: 1572-9680.

Helalia, A.M.; El-Amir, S.; Abou-Zeid, S.T. & Zaghloul, K.F. (1992). Bio-reclamation of salinesodic soil by Amshot grass in northern Egypt. *Soil and Tillage Research*, Vol. 22, No 1-2, (January, 1992), pp.109–115, ISSN: 01671987.

Herrmann, A.; Witter, E. & Katterer, T. (2005). A method to assess whether 'preferential use' occurs after 15N ammonium addition; implication for the 15N isotope dilution technique. *Soil Biology & Biochemistry*, Vol. 37, No 1, (January, 2005), pp. 183–186, ISSN: 0038-0717.

Iannetta, M. & Colonna, N. Salinização. Lucinda, Série B, No. 3. Disponível em < http://geografia.fcsh.unl.pt/lucinda/Leaflets/B3_Leaflet_PT.pdf> Acesso em 05 de junho de 2011.

Ibekwe, A.M.; Poss, J.A.; Grattan, S.R.; Grieve, C.M. & Suarez, D. (2010). Bacterial diversity in cucumber (Cucumis sativus) rhizosphere in response to salinity, soil pH, and boron. *Soil Biology & Biochemistry*, Vol. 42, No 4, (April, 2010), pp. 567-575, ISSN 0038-0717.

Informativo CAPEBE 06/12/2010 - Um patrimônio chamado solo. Disponível em >http://www.capebe.org.br/informativo.php?id=355> Acesso em 13 de junho de 2011.

Jackson, B.P. & Bertsch, P.M. (2001). Determination of arsenic speciation in poultry wastes by IC-ICP-MS. *Environmental Science & Technology*, Vol. 35, No 24, (November, 2001), pp. 4868–4873, ISSN 0013-936X.

Jialiang, L: (2008).Research on the effect of saline wetland from pulp wastewater irrigation in Yellow River Dleta, Doctoral dissertation, Ocean University of China, Chine.

Jiang, H.; Dong, H.; Yu, B.; Liu, X.; Li, Y.; Ji, S. & Zhang, C.L. (2007). Microbial response to salinity change in Lake Chaka, a hypersaline lake on Tibetan plateau. *Environmental Microbiology*, Vol. 9, No 10, (July, 2007), pp. 2603-2621, ISSN: 1462-2920.

Juniper, S., Abbott, L.K. (2006). Soil salinity delays germination and limits growth of hyphae from propagules of arbuscular mycorrhizal fungi. *Mycorrhiza*, Vol. 16, No 5, (July, 2006), pp. 371-379, ISSN: 1432-1890.

Kang, Y.; Wan, S.; Jiao, Y.; Tan, J. & Sun, Z. (2008). Saline soil salinity and water management with tensiometer under drip irrigation. In: *Symposia on the Fifth Annual Meeting of Agricultural Land and Water Engineering of Chinese Society of Agricultural Engineering*, pp. 124-131, Beijing

Keren, R. (2000). Salinity. In: *Handbook of Soil Science*, Sumner, M.E. (Ed.), pp. G3–G25, CRC Press, ISBN: 9780849331367, Boca Raton.

Killham, K. (1994). *Soil Ecology* (1), Cambridge University Press, ISBN: 0 521 43521 8, United Kingdom.

Laura, R.D. (1974). Effects of neutral salts on carbon and nitrogen mineralization of organic matter in soil. *Plant and Soil*, Vol. 41, No 1, (1974), pp. 113-127, ISSN: 0032-079X.

Laura, R.D. (1977). Salinity and nitrogen mineralization in soil. *Soil Biology & Biochemistry*, Vol. 9, No 5, pp. 333–336, ISSN: 0038-0717.

Liang, Y.; Chen, H.; Tang, M.J. & Shen, S.H. (2007). Proteome analysis of an ectomycorrhizal fungus *Boletus edulis* under salt shock. *Mycological Research*, Vol. 111, No 8, (August, 2007), pp. 939-946, ISSN: 0953-7562.

Liang, Y.C.; Yang, Y.F.; Yang, C.G.; Shen, Q.Q.; Zhou, J.M. & Yang, L.Z. (2003). Soil enzymatic activity and growth of rice and barley as influenced by organic matter in an anthropogenic soil. *Geoderma*, Vol. 115, No 1-2, (July, 2003), pp. 149–160, ISSN: 0016-7061.

Lin, X. Z.; Chen, K. S.; He, P.Q.; Shen, J.H. & Huang X. H. (2006). The effects of *Suaeda salsa* L. planting on the soil microflora in coastal saline soil. *Acta Ecologica Sinica*, Vol. 26, No 3, (March, 2006), pp. 801-807 (in Chinese), ISSN: 1872-2032.

Li-Xian, Y.; Guo-Liang, L.; Shi-Hua, T.; Gavin, S. & Zhao-Huan, H. (2007). Salinity of animal manure and potential risk of secondary soil salinization through successive manure application. *Science of the Total Environment*, Vol. 383, No 1-3, (June, 2007), pp. 106–114, ISSN 0048-9697.

Llamas, D.P., Gonzales, M.D., Gonzales, C.I., Lopez, G.R., Marquina, J.C. (2008). Effects of water potential on spore germination and viability of *Fusarium* species. *Journal of Industrial Microbiology & Biotechnology*, Vol. 35, No 11, (November, 2008), pp. 1411-1418, ISSN: 1367-5435.

Luo, A. & SUN, X. (1994). Effect of organic manure on the biological activities associated with insoluble phosphorus release in a blue purple paddy soil. *Communications in Soil Science and Plant Analisys*, Vol. 25, No 13-14, (Available online: Nov 2008), pp. 2513– 2522, ISSN: 0010-3624.

Machado, R. E. (2002). Simulação de escoamento e de produção de sedimentos em uma microbacia hidrográfica utilizando técnicas de modelagem e geoprocessamento. 154p. Tese (Doutorado em Agronomia) – Escola Superior de Agricultura Luiz de Queiroz, Piracicaba, Brasil.

Ghollaratta, M. & Raiesi, F. (2007). The adverse effects of soil salinization on the growth of *Trifolium alexandrinum* L. and associated microbial and biochemical properties in a soil from Iran. *Soil Biology & Biochemistry*, Vol. 39, No 7, (July, 2007), pp. 1699- 1702, ISSN: 0038-0717.

Mandeel, Q.A. (2006). Biodiversity of the genus Fusarium in saline soil habitats. *Journal of Basic Microbiology*, Vol. 46, No 6, (December, 2006), pp. 480-494, ISSN: 0233-111X.

Marcar, N.; Crawford, D.; Leppert, P.; Jovanovic, T.; Floyd, R. & Farrow, R. (1995). Trees for Saltland: In: *A Guide to Selecting Native Species for Australia*. CSIRO Press, ISBN: 0 643 05819 2, Melbourne,Australia.

Museu do Una. Manguezais e estuário. Disponível em < http://www.museudouna.com.br/mangue.htm> Acesso em 05/06/2011.

Nannipieri, P.; Gregos, S. & Ceccanti, B. (1990). Ecological significance of the biological activity in soil. In: *Soil Biochemistry, vol. 6*, Bollag, J.M. & Stotzy, G. (Eds.), pp. 293–355, ISBN: 0-8247-8232-1,Marcel Dekker, New York, USA.

Oad, F.C.; Samo, M.A.; Soomro, A.; Oad, D.L.; Oad, N.L. & Siyal, A.G. (2002). Amelioration of salt affected soils. *Pakistan Journal of Applied Sciences*, Vol. 2, No 1, (January, 2002), pp.1–9, ISSN: 1607-8926.

Oldeman, I. R. (1994). The global extent of soil degradation. In: *Soil resilience and sustainable land use*. D. J. Greenland & I. Szabolcs, (Ed.), pp. 99-118, CAB International, ISBN-10: 0851988717, Wallingford, England.

Omar, S.A.; Abdel-Sater M.A.; Khallil, A.M.; Abdalla, M.H. (1994). Growth and enzyme activities of fungi and bacteria in soil salinized with sodium chloride. *Folia Microbiologica*, Vol.39, No 1, (1994), pp. 23-28, ISSN: 0015-5632

Oren, A. (1999). Bioenergetic aspects of halophilism. *Microbiology and Molecular Biology Reviews*, Vol. 63, No 2, (June, 1999), pp. 334-348, ISSN: 1098-5557.

Oren, A. (2001). The bioenergetic basis for the decrease in metabolic diversity at increasing salt concentrations: implication of the functioning of salt lake ecosystems. *Hidrobiología*, Vol. 466, No 1-3, pp. 61-72, ISSN: 0073-2087.

Oren, A. (2002). Molecular ecology of extremely halophilic archaea and bacteria. *FEMS Microbiology Ecology*, Vol. 39, No 1, (January, 2002), pp. 1–7, ISSN: 0095-3628.

Ortiz-Dorda, J.; Martinez-Mora, C.; Correal, E.; Simon, B. & Cenis, J.L. (2005). Genetic structure of *Atriplex halimus* populations in the mediterranean basin. *Annals of Botany*, Vol. 95, No 5, pp. 827–834, ISSN: 0305-7364.

Pankhurst, C.E.; Yu, S., Hawke, B.G. & Harch, B.D. (2001). Capacity of fatty acid profiles and substrate utilization patters to describe differences in soil microbial communities associated with increased salinity or alkalinity at three locations in South Australia. *Biology and Fertility of Soils*, Vol. 33, No 3, (March, 2001), pp. 204–217, ISSN: 0178-2762.

Parker, D. (2000). Controlling agricultural nonpoint water pollution: costs of implementing the Maryland Water Quality Improvement Act of 1998. *Agricultural Economics*, Vol. 24, No 1, (August, 2005), pp. 23–31, ISSN: 1574-0862.

Pathak, H., Rao, D.L.N. (1998). Carbon and nitrogen mineralisation from added organic matter in saline and alkali soils. *Soil Biology & Biochemistry*, Vol. 30, No 6, (June, 1998), pp. 695-702, ISSN: 0038-0717.

Pereira, S.V.; Martinez, C.R.; Porto, E.R.; Oliveira, B.R.B. & Maia, L.C. (2004). Atividade microbiana em solo do Semi-Árido sob cultivo de *Atriplex nummularia*. Pesquisa Agropecuária Brasileira, Vol. 39, No 8, (Agosto, 2004), pp. 757-762, ISSN: 0100-204X.

Porto, E.R.; Amorim, M.C.C. de; Dutra, M.T.; Paulino, R.V.; Brito, L.T. de L. & Matos, A.N.B. (2006). Rendimento da *Atriplex nummularia* irrigada com efluentes da criação de tilápia em rejeito da dessalinização de água. *Revista Brasileira de Engenharia Agrícola e Ambiental*, Vol. 10, No 1, (Janeiro-março, 2006), pp. 97-103, ISSN: 1415-4366.

Porto, E.R.; Amorim, M.C.C. de & Silva Junior, L.G. de A. Uso de rejeito da dessalinização de água salobra para irrigação da erva-sal (*Atriplex nummularia*). (2001). *Revista Brasileira de Engenharia Agrícola e Ambiental*, Vol. 5, No 1, (Janeiro-Abril, 2001), pp.111-114, ISSN: 1415-4366.

Qadir, M., Ghafoor, A., Murtaza, G. (2000). Amelioration strategies for saline soils: a review. *Land Degradation & Development*, Vol. 11, No 6, (January, 2001), pp. 501-521, ISSN: 1099-145X.

Qadir, M.; Noble, A.D., Schubert, S.; Thomas, R.J. & Arslan, A. (2006). Sodicity-induced land degradation and its sustainable management: problems and prospects. *Land Degradation & Development*, Vol. 17, No 6, (May, 2006), pp. 661-676, ISSN: 1099-145X.

Qadir, M. & Oster, J.D. (2004). Crop and irrigation management strategies for saline-sodic soils and waters aimed at environmentally sustainable agriculture. *The Science of the total Environment*, Vol. 323, No 1-3, (May, 2004), pp. 1-19, ISSN: 0048-9697.

Rabhi, M.; Hafsi, C.; Lakhdar, A.; Hajji, S.; Barhoumi, Z.; Hamrouni, M.H.; Abdelly, C. & Smaoui, A. (2009). Evaluation of the capacity of three halophytes to desalinize their rhizosphere as grown on saline soils under nonleaching conditions. *African Journal of Ecology*, Vol. 47, No 4, (August, 2009), pp. 463-468, ISSN: 0141-6707.

Ravindran, K.C.; Venkatesan, K.; Balakrishnan, V.; Chellappan, K.P. & Balasubramanian, T. (2007). Restoration of saline land by halophytes for Indian soils. *Soil Biology & Biochemistry*, Vol. 39, No 10, (October, 2007), pp. 2661-2664, ISSN 0038-0717.

Rietz, D.N., Haynes, R.J., 2003. Effects of irrigation induced salinity and sodicity on soil microbial activity. *Soil Biology & Biochemistry*, Vol. 35, No 6, (June, 2003), pp. 845-854, ISSN 0038-0717.

Ross, I.L.; Alami, Y.; Harvey, P.R.; Achouak, W. & Ryder, M.H. (2000). Genetic diversity and biological control activity of novel species of closely related pseudomonads isolated from wheat field soils in South Australia. *Applied and Environmental Microbiology*, Vol. 66, No 4, (April, 2000), pp. 1609-1616, ISSN: 0099-2240.

Sahin, U.; Eroglu, S. & Sahin, F. Microbial application with gypsum increases the saturated hydraulic conductivity of saline-sodic soils. (2011). *Applied Soil Ecology*, Vol. 48, No 2, (June, 2011), pp. 247-250, ISSN: 0929-1393.

Sardinha, M.; Muller, T.; Schmeisky, H. & Joergensen, R.G. (2003). Microbial performance in soils along a salinity gradient under acidic conditions. *Applied Soil Ecology*, Vol. 23, No 3, (July, 2003), pp. 237-244, ISSN: 0929-1393.

Schimel, J.P.; Balser, T.C. & Wallenstein, M. (2007). Microbial stress response physiology and its implications for ecosystem function. *Ecology*, Vol. 88, No 6, (June, 2007), pp. 1386-1394, ISSN: 0012-9658.

Sethi, V.; Kaushik, A. & Kaushik, C.P. (1993). Salt stress effects on soil urease activity, nitrifier and Azotobacter populations in gram rhizosphere. *Tropical Ecology*, Vol. 34, No 2, (December, 1993), pp. 189-198, ISSN: 0564-3295.

Shi, W.; Cheng, M.; Li, C. & Ma, G. (1994). Effect of Cl_ 1 on behavior of fertilizer nitrogen, number of microorganisms and enzyme activities in soils. *Pedosphere*, Vol. 4, No 4, (December, 1994), pp. 357- 364, ISSN: 1002-0160.

Silva Júnior, J.M.T.da; Tavares, R. de C; Mendes Filho, P.F. & Gomes, V.F.F. (2009). Efeitos de níveis de salinidade sobre a atividade microbiana de um Argissolo Amarelo incubado com diferentes adubos orgânicos. *Revista Brasileira de Ciências Agrárias,* Vol. 4, No 4, (out-dez, 2009), pp. 378-382, ISSN: 1981-1160

Silva, C.M.M.S.; Vieira, R.F. & Oliveira, P.R. de. (2008). Salinidade, sodicidade e propriedades microbiológicas de Argissolo cultivado com erva-sal e irrigado com rejeito salino. Pesquisa Agropecuária Brasileira, Vol. 43, No 10, (Out, 2008), pp. 1389-1396, ISSN 0100-204X .

Sterr, A. Making development sustainable. (1998). *Advanced Geo-Ecology,* Vol. 31, pp. 857-865, SSN: 0145 8752.

Subhashini, D. & Kaushik, B.D. (1981). Amelioration of sodic soils with blue–green algae. *Australian Journal Soil Research,* Vol. 19, No 3, (1981), pp. 361–366, ISSN: 0004-9573.

Swarup, A. (1985). Yield and nutrition of rice as influenced by pre-submergence and amendments in a highly sodic soil. Journal of the Indian Society of Soil Science, Vol. 33, No 2, (April-June, 1985), pp. 352– 357, ISSN: 0019-638X.

Syed, A., Satou, N. & Higa, T. (2003). Mechanisms of effective microorganisms (EM) in removing salt from saline soils. In: *13th Annual West Coast Conference on Contaminated Soils, Sediments and Water* , Mission Valley Marriott, San Diego, CA, USA, March 2003.

Tabatabai, M.A., (1994). Soil enzymes. In *Methods of Soil Analysis. Part 2: Microbial and Biochemical Properties,* Weaver, R.W., Angel, J.S., Bottomley, P.S. (Eds.), pp. 775–833, Soil Science Society America, ISBN 10 089118810X, Madison, WI, USA.

Tabatabai, M.A. & Bremner, J.M. (1972). Assay of urease activity in soils. *Soil Biology & Biochemistry,* Vol. 4, No 4, (November,1972), pp. 479–487, ISSN: 0038-0717.

Tam, N.F.Y. (1998). Effects of wastewater discharge on microbial populations and enzyme activities in mangrove soils. *Environmental Pollution,* Vol. 102, No 2-3, (August, 1998), pp. 233-242, ISSN: 02697491.

Tan, J. & Kang, Y. (2009). Changes in Soil Properties Under the Influences of Cropping and Drip Irrigation During the Reclamation of Severe Salt-Affected Soils. *Agricultural Sciences in China,* 2009, Vol. 8, No 10, (October, 2009), pp. 1228-1237, ISSN: 1671-2927.

Tejada, M. & Gonzalez, J.L. (2005). Beet vinasse applied to wheat under dry land conditions affects soil properties and yield. *European Journal of Agronomy,* Vol. 23, No 4, (December, 2005), pp. 336-347, ISSN: 1161-0301.

Tejada, M.; Garcia, C.; Gonzalez, J.L. & Hernandez, M.T. (2006). Use of organic amendment as a strategy for saline soil remediation: Influence on the physical, chemical and biological properties of soil. *Soil Biology & Biochemistry,* Vol. 38, No 6, (June, 2006), pp. 1413–1421, ISSN 0038-0717.

Tripathi, S.; Kumari, S.; Chakraborty, A.; Gupta, A.; Chakrabarti, K. & Bandyapadhyay, B.K. 2006. Microbial biomass and its activities in salt-affected soils. *Biology and Fertility of Soils,* Vol. 42, No 3, (February, 2006), pp. 273–277, ISSN: 0178-2762.

Tripathi, S.; Chakrabarty, A.; Chakrabarti, K. & Bandyopadhyay, B.K. (2007). Enzyme activities and microbial biomass in coastal soils of India. *Soil Biology &Biochemistry,* Vol. 39, No 11, (November, 2007), pp. 2840–2848, ISSN 0038-0717.

Van Bruggen, A.H.C. & Semenov, A.M. (2000). In search of biological indicators for soil health and disease suppression. *Applied Soil Ecology,* Vol. 15, No 1, (August, 2000), pp. 13–24, ISSN: 0929-1393.

Wichern, J.; Wichern, F. & Joergensen, R.G. (2006). Impacto of salinity on soil microbial communities and the decomposition of maize in acidic soils. *Geoderma*, Vol. 137, No 1-2, (December, 2006), pp. 100-108, ISSN: 0016-7061.

Wong, V.N.L.; Dalal, R.C. & Greene, R. S. B. (2008). Salinity and sodicity effects on respiration and microbial biomass of soil. *Biology and Fertility of Soils*, Vol. 44, No 7, (August, 2008), pp. 943-953, ISSN: 0178-2762.

Wong, V.N.L.; Dalal, R.C. & Greene, R. S. B. (2009). Carbon dynamics of sodic and saline soils following gypsum and organic material additions: A laboratory incubation. *Applied Soil Ecology*, Vol. 41, No 1, (January, 2009), pp. 29– 40, ISSN: 0929-1393.

Yuan, L.; Huang, J. & Yu, S. (1997). Responses of nitrogen and related enzyme activities to fertilization in rhizosphere of wheat. *Pedosphere*, Vol. 7, No 2, (June, 1997), pp. 141–148, ISSN: 1002-0160.

Yuan, B-C; Li, Z-Z; Liu, H; Gao, M. & Zhang, Y-Y. (2007). Microbial biomass and activity in salt affected soils under arid conditions. *Applied Soil Ecology*, Vol. 35, No 2, (February, 2007), pp. 319-328, ISSN: 0929-1393.

Zahran, Z. (1997). Diversity, adaptation and activity of the bacterial flora in saline environments. *Biology and Fertility of Soils*, Vol. 25, No 3, (September, 2007), pp. 211–223, ISSN: 0178-2762.

Part 6

Soil Pollution Management

11

Restoration of Cadmium (Cd) Pollution Soils by Use of Weeds

Masaru Ogasawara
Weed Science Center (WSC), Utsunomiya University,
350 Mine-machi, Utsunomiya
Japan

1. Introduction

Soil contamination by heavy metals such as cadmium (Cd), copper and mercury has become a big concern particularly in metal plating plants, mining sites and surrounding areas as well as residential area and farmlands in the river downstream region neighboring these facilities. In some cases, heavy metals in soils leach into river water and then diffuse onto farmlands with irrigation, resulting in relatively low levels of heavy metals being spread into wider areas rather than being localized in high concentrations (Asami 1972). Therefore, when civil engineering methods including addition of topsoil and removal of contaminated soils are used for remediation of heavy metal contaminated soils, a large amount of uncontaminated fresh soil and large disposal areas are required, which creates a bottle neck for the remediation. Thus, the development of a new remediation technology to replace the conventional civil engineering technology is needed. The remediation technology for soils contaminated with harmful substances using plants is called phyto-remediation, and its potential has demonstrated by many researchers (Elizabeth 2005). However, this new remediation technology is in the initial stage even at present, except for some cases, e. g., the remediation of oil-spilled soil using Italian ryegrass (Kaimi *et al.* 2006) and that of Cd-contaminated paddy soil using rice plants (Honma *et al.* 2009) are currently being conducted, because no appropriate remediation plant and no remediation systems have been found and developed. Thus, weed species, possessing high adaptability to environment, have been pointed out as a suitable plant for soil remediation. Although research on phyto-remediation using weeds has just begun and there are many issues which need to be resolved, this remediation technique is expected to become a valuable technology for the alleviation of heavy metal contaminated soils in the near future.

This chapter focuses on the potential of weed for remediation of Cd-contaminated soils. In order to better understand phyto-remediation by weeds, the rationale of using weeds for Cd remediation and the biological characteristics of weeds are explained. Herbaceous plants are classified into several groups such as crops, grasses, weeds and wild plants. Crops are plants that require artificial protection such as pest control, fertilization, watering and etc., ; on the other hand, weeds can thrive under severe growth conditions. For example, asiatic plantain (*Plantago asiatica*) in highly compacted soil areas, crabgrass (*Digitaria cilliari*) in dry regions, annual bluegrass (*Poa annua*) in cool wet regions, field horsetail (*Equisetum arvense*) in acidic soils, saltbush (*Atriplex subcordata*) in salt-accumulation areas, and broomsedge

(*Andropogon virginicu*) in phosphate-deficient soils can grow vigorously under these conditions (Takematsu and Ichizen 1987, 1993, 1997). For plants, not only Cd contaminated soil but also various environmental factors such as low temperature, aridity, low sunlight (shade), nutrient deficiency (infertile soil), poorly drained soil, competitions between plants for water, nutrition, and light, and allelochemicals generated from plants are regarded as adverse growth conditions. Therefore, remediation plant (plant using for restoration of Cd pollution soils) must be able to grow under these adverse weather and soil conditions. On the other hand, several methods are considered in phyto-remediation, and the capability of remediation plants are depending on the approach considered. Unlike organic compounds, Cd cannot be degraded; therefore, absorption (phyto-extraction) and fixation (phyto-stabilization) are the most effective methods proposed for Cd-remediation. Phyto-extraction is the chemical removal method of Cd by absorption through the roots and accumulation in shoots, followed by plant harvesting. Phyto-stabilization is a method of retaining Cd on the adjacent surface of plant roots. Mulching, which prevents the run-off of Cd contaminated soil into the surrounding non-polluted area by the root system extending in soils, can also be considered for phyto-remediation technologies. Particularly in slopes, mulching with plants may be prior to Cd extraction from the contaminated soils.

Among several screening studies on the remediation plants conducted, it is demonstrated that *Athyrium yokoscense* is highly tolerant to heavy metals (*Nishizono et al.* 1987); however, its biomass is extremely small to be valuable as a remediation plant. The plant species best suited for phyto-extraction require the ability to accumulate large amounts of Cd in their shoots, extension of their roots into soil, rapid growth and a long growing period. The plant species suitable for mulching require the ability to extend their roots into soil, a high LAI (leaf area index) value; plants with a high LAI value can reduce the physical strength of rainfall to scour soil, large biomass, rapid growth and long growing period. Furthermore, it is important that whether seeds and vegetative reproductive organs such as rhizomes and tubers can be inexpensively supplied in large quantities for the remediation plant. As compared with crops, weeds are generally superior in several points such as environment adaptability and Cd tolerance and accumulation, however, are inferior in seed supply (Table-1). Thus, in case when seeds of remediation plants (weeds) are difficult to obtain, top soil (seed bank) of non-pollution areas where weeds are densely grown, are available. As well as Cd tolerant weed species, Cd sensitive weed species are also useful for Cd remediation (Table-2); e. g., results of phyto-remediation at the pollution area can be evaluated by distribution and biomass of the Cd sensitive weed species such as *Arenaria serphylliforia*, *Geranium carolinianum* and *Phseolus aureas*.

Factors	weeds	Crops
Cd tolerance	Superior	Inferior
Cd accumulation	Superior	Inferior
Adaptability to environment *	Superior	Inferior
Pest tolerance (disease, insect)	Superior	Inferior
Seed or seedling supply	Inferior	Superior
Growth speed	Superior	Inferior

*: drought, cool, shade, wet, salinity and infertile land tolerance

Table 1. Comparison of weeds and crops as a Cd remediation plants

Family name	Scientific name	I_{50} value (mg Cd kg^{-1})*
Onagraceae	Epilobium angustifolium	1.5
Geraniaceae	Geranium carolinianum	1.3
Leguminosae	Phaseolus angularis (cv. Tanba Dinagon)	2.0
	Phaseolus aureus	0.8
	Trifolium fragiferum	1.7
Cruciferae	Rorippa cantoniensis	2.3
Caryophyllaceae	Arenaria serpyllifolia	1.1
	Stellaria alsine var. undulata	1.6
	Stellaria graminea	1.9
Polygonaceae	Antenoron filiforme	1.4
Compositae	Eclipta prostrata	1.5
	Sonchus asper	1.8

* Susuceptible weed species having ≤ 2 of I$_{50}$ value (mg Cd kg^{-1}): the concentrations of Cd causing a 50% reduction in fresh weight of shoot to that of the untreated plant grown under sand culture conditions

Table 2. Susceptible weed species to Cd

2. Sensitivities of weeds to Cd

About 6,000 plant species are accounted as a weed in the world. Cd sensitivities weeds vary depending on species (Fig. 1 and Fig. 2). In this section, Cd sensitivity and Cd content of weeds are referred.

It is reported that Arabidopsis halleri and Thlaspi caerulescens are highly tolerant to Cd (Bert et al. 2003; Brown et al. 1995); however, Hibiscus cannabinus, Portulaca oleracea, Xanthium strumarium, Amsinckia barbata, Anthoxathium odoratum, Arthoraxon hipidus, Digitaria ciliaris, Echinochloa crus-galli var. praticola, Lolium multiflorum, Panicum bisulcatum, Paspalum dilatatum, Poa pratensis and Setaria faberi are have also been reported to have a high tolerance to Cd (I$_{50}$: > 30 mg Cd Kg^{-1}) (Abe et al. 2006) (Table-3). When the weed habitat is considered, E. crus-galli var. praticola, D. ciliaris, S. faberi and P. dilatatum, A. odoratum, L. multiflorum and P. pratensis are the most suitable for saturated, semi-arid, warm and cold conditions, correspondingly. When the weed biomass is considered, X. strumarium growing up 1–2 m is the most suitable for Cd extraction. When the life-span of weed is considered, A. odoratum, P. bisulcatum, P. dilatatum, P. thunbergii, and P. pratensis are perennials, and particularly P. pratensis is a long-lived grass that can survive more than a few decades; therefore, phyto-remediation will be proceeded continuously once perennial weed species such as P. pratensis is introduced into the Cd pollution area.

On the other hand, there are great variations on Cd tolerance among weed species. However, the tolerance can be predicted when they belong to the same family. Abe et al. (2006) reported that Gramineae and Compositae weed species were torelant, while Leguminoseae and Cruciferae weed species were sensitive according to the pot tests conducted

Fig. 1. Susceptibilty of *Echinochloa crus-galli* var. *crus-galli* to Cd

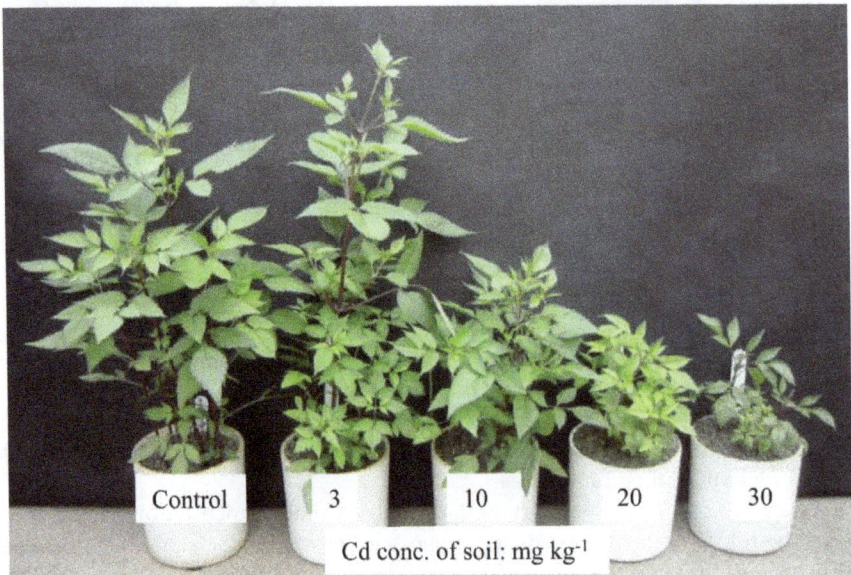

Fig. 2. Susceptibilty of *Bidens frondo*sa to Cd

Family name	Scientific name
Malvaceae	Hibiscus cannabinus
	Malva sylvestris
Caryophyllaceae	Silene ameria
Portulaceae	Portulaca oleracea
Phytolaccaceae	Phytolacca americana
Compositae	Xanthium atrumarium
Boraginaceae	Amsinckia barbata
Gramineae	Anthoxanthum odoratum
	Anthraxon hispidus
	Digitaria ciliaris
	Echinochloa crus-galli var. pracilola
	Lolium multiflorum
	Panicum bisulcatum
	Pasplum dilatatum
	Poa pratensis
	Setaria faberi
	Agrostis stolonifera

Tolerant weed species having $\geq$ 30 of I_{50} value (mg Cd kg^{-1}): the concentrations of Cd causing a 50% reduction in fresh weight of shoot to that of the untreated plant grown under sand culture conditions

Table 3. Tolerant weed species to Cd*

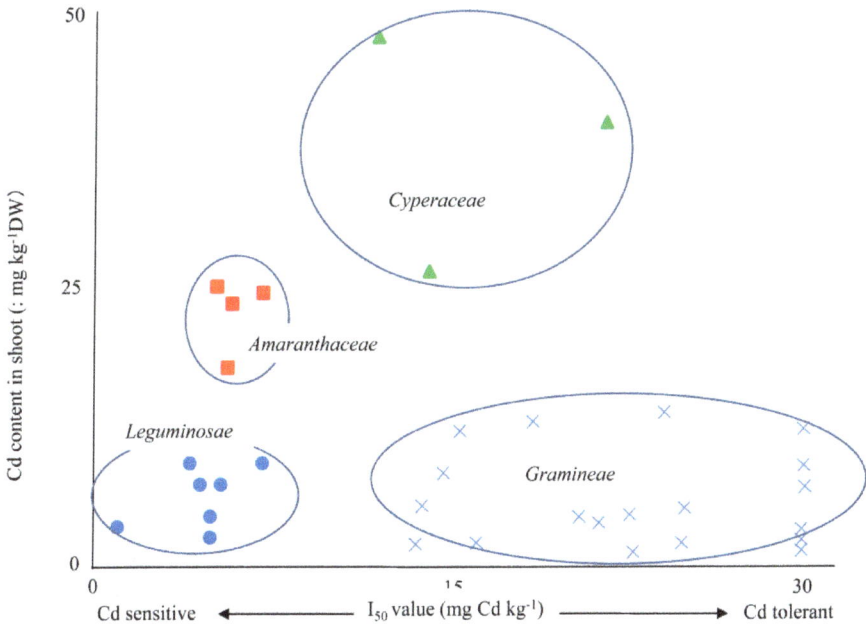

Fig. 3. Relation of susceptibility (I_{50} value) to Cd and Cd content in shoot of weeds belonged to 4 families

on about 200 weed species (Fig. 3 and Table-4). This result indicated that the Cd concentrations of soils in a certain areas could be predicted by analysis of weed vegetation. Among weed species belonging to *Gramineae* and *Leguminosae* species, appropriate Cd remediation plant and Cd indicator may be found.

Family name	Total number of plant species tested	Number of plant species (%)				Mean of I_{50} values (Cd mg kg^{-1})
		Group 1	Group 2	Group 3	Group 4	
Caryophyllaceae	17 (100%)	13 (76.4%)	2 (11.8%)	1 (5.9%)	1 (5.9%)	8.8
Cruciferae	15 (100%)	10 (66.7%)	5 (33.3%)	-	-	7.5
Leguminosae	15 (100%)	15 (100%)	-	-	-	4.6
Compositae	31 (100%)	13 (41.9%)	12 (38.7%)	4 (12.9%)	2 (6.5%)	13.2
Gramineae	45 (100%)	-	15 (33.3%)	18 (40.0%)	12 (26.7%)	22.6

Group 1, I_{50} value <10; Group 2, I_{50} value 10 ~ <20; Group 3, I_{50} value 20 ~ <30; Group 4, I_{50} value≧30

Table 4. Susceptibilities of *Caryophyllaceae*, *Cruciferae*, *Leguminosae*, *Compositae* and *Gramineae* weed species to Cd

3. Cd accumulation abilities of weeds

High Cd accumulation ability as well as Cd tolerance is required for the remediation plants; however, there is no positive correlation between Cd accumulation and Cd sensitivity. Weed species with highly tolerant to Cd can be divided into two groups: weeds that absorb hardly Cd (Cd exclusion type) and weeds that absorb and detoxify Cd (Cd detox-accumulation type). Abe *et al.* (2008a) revealed from a pot test that *Cichorium intybus* (77 mg kg^{-1} DW) and *Matricaria chamomilla* (64.4 mg kg^{-1} DW) can accumulate large amounts of Cd in their shoots; these accumulations are more than *Polygonum thunbergii* (56.2mg kg^{-1} DW) that is known as a hyper-Cd accumulator (Shinmachi *et al.* 2003) (Table-5). The mechanism involved in Cd accumulation by weeds are still not well understood; however, it can be predicted that the absorbed Cd is detoxified by forming chelates with phytochelatin, malic acid, and citric acid in plants. On the other hand, *Gramineae* plant, e. g., *Oryza sativa* (cv. Milyang) accumulate high concentrations of Cd (>100 mg kg^{-1} DW) in its roots but accumulate little Cd in their shoots, in contrary, *Compositae* plant, e. g., *Bidens frondosa* accumulate low concentration of Cd in the root but accumulate much Cd in its shoots (Table-6, Table-7, Table-8). The Cd content ratio between roots and shoots (SR ratio) of *C. intybus* (3.56; it means that shoots contain 3.56 times more Cd than roots) is higher than that of *O. sativa* (cv. Milyang; 0.02); therefore, the results indicate that the Cd is easily translocated from roots to shoots in dicoltyledons as compared to monocotyldons (*Gramineae*) (Table 6). Therefore, *Bidens frondosa*, *B. pilosa* and *Amaranthus viridis*, which accumulate more Cd in their shoots than in roots and possess a large biomass, may be useful for phyto-extraction.

Family name	Scientific name	Cd conc. (mg kg^{-1} dry weight)
	Plant species with relatively high Cd conc.	
Compositae	Cichorium intybus	77.0
Compositae	Matricaria chamomilla	64.4
Polygonaceae	Polygonum thunbergii	56.2
Cyperaceae	Cyperus brevifolius var. leiolepis	48.3
Cruciferae	Sisymbrium orientale	48.2
Cruciferae	Sisymbrium altissimum	48.2
Compositae	Picris echinoides	41.0
	Plant species with relatively low Cd conc.	
Gramineae	Oryza sativa (cv. Milyang 42)	0.8
Gramineae	Panicum dichotomiflorum	1.0
Gramineae	Dactylis glomerata	1.2
Plantagineaceae	Plantago virginica	1.4
Gramineae	Panicum bisulcatum	1.5
Onagraceae	Oenothera biennis	1.8
Gramineae	Digitaria ciliaris	1.9
Leguminosae	Sesbannia exaltata	2.1
Gramineae	Echinochloa crus-galli var. prachicola	2.1

Table 5. Weed species which accumulate relatively high (≥30) and low (≤3) concentrations (mg kg^{-1} DW) of Cd in their shoots

Family name	Scientific name	Cd conc. (mg kg^{-1} dry weight)
	Plant species with relatively high Cd conc.	
Onagraceae	Oenothera biennis	171.9
Convolvulaceae	Calystegia sepium var. 1 americana	122.6
Leguminosae	Cassia obtusifolia	122.2
Gramineae	Oryza sativa (cv. Milyang 25)	117.5
Lubiatae	Salvia plebeia	116.6
Gramineae	Festuca arundinaceae	115.4
Caryophyllaceae	Silene dioica	113.0
Malvaceae	Sida rhombifolia	111.9
	Plant species with relatively low Cd conc.	
Compositae	Bidens pilosa	7.7
Compositae	Bidens frondosa	9.4
Compositae	Lactuca indica	11.4
Polygonaceae	Rumex crispus subsp. japonicus	12.2
Compositae	Sochus oleraceus	14.2
Cruciferae	Camelina sativa	15.9

Table 6. Weed species which accumulate relatively high (≥110) and low (≤16) concentrations (mg kg^{-1} DW) of Cd in their roots

Family name	Scientific name	Shoot Cd content (µg plant^{-1})
Cyperaceae	Cyperus brevifolius var. leiolepis	50.0
Polygonaceae	Polygonum thunbergii	49.7
Compositae	Bidens frondosa	40.5
Compositae	Cichorium intybus	38.8
Cyperaceae	Cyperus globosus	30.3
Compositae	Bidens pilosa	24.1
Amaranthaceae	Amaranthus viridis	22.5
Malvaceae	Hibicus cannabinus	21.4
Cruciferae	Sinapis alba	21.4
Cruciferae	Sisymbrium altissimum	21.4
Caryophylaceae	Silene notiflora	21.1
Amaranthaceae	Amaranthus hybridus	20.7

Table 7. Weed species which accumulate relatively high (≥20 µg plant^{-1}) content of Cd in shoots

Family name	Scientific name	Cd conc. (mg kg^{-1} dry weight)
	Plant species with relatively high shoot-root ratio	
Compositae	Cichorium intybus	3.56
Compositae	Bidens frondosa	3.30
Compositae	Lactuca indica	2.27
Compositae	Bidens pilosa	2.21
	Plant species with relatively low shoot-root ratio	
Onagraceae	Oenothra biennis	0.01
Plantaginaceae	Plantago virginica	0.02
Gramineae	Oryza sativa (cv. Milyyang 42)	0.02
Gramineae	Dactylis glomerata	0.03

* shoot-root ratio : the ratio of Cd conc. in shoot and root in plant

Table 8. Weed species which have relatively high (≥2.0) and low (≤0.03) shoot-root ratios*

4. Utilization of PGRs to Cd remediation by use of weeds

So far, various cultivation methods and materials have been used to control the Cd elution into soils (Turgut et al. 2004; Hattori et al. 2006); e. g., treatment of Calcium materials and deep flooding (rice paddy fields) have been used to diminish Cd elution into the free water in soils; on the other hand, drying of soil and treatment of chlorides and EDTA have been used to enhance the Cd elution. In addition to these materials, it is reported that certain plant hormones directly affect the growth of remediation plants. Abscisic acid suppresses a decrease in chlorophyll content caused by Cd in *Brassica napus*, while *28-homobrassinolide* mitigates the growth inhibition caused by Cd in *Cicer arietium*. According to a hydroponic test using white Japanese millet (*Echinochloa frumentacea*) conducted by Abe et al. (2011), it was shown that HMI (3-hydroxy-5-methylisoxazole) at 2.5×10^3 mmol/l significantly reduced the growth inhibition (dry weight; % of control) of roots and shoots caused by Cd at 4.4×10 mmol/l from 4.2% to 23.9% and from 48.5% to 82.6%, respectively, while increasing Cd content (µg plant^{-1}) from

0.32 ± 0.08 to 0.53 ± 0.04 in roots. This result indicates that Cd tolerance of remediation plants can be further enhanced by HMI. (Fig. 4, Fig. 5, Fig. 6) HMI, plant growth regulator (PGR) that shows similar action to those of plant hormones, has been used as rooting agent for paddy rice (Ogawa and Ota 1976) and thus it is thought that HMI can be also easily applied for Cd remediation with weeds, particular in the mulching.

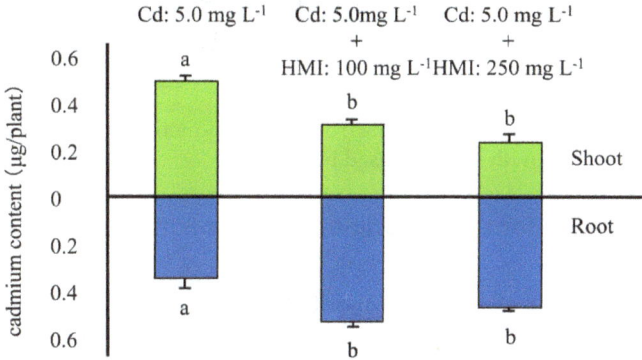

Fig. 4. Effect of HMI on Cd accumulation in shoot and root of *Echinochloa frumentacea*
The error bars represent the standard deviation of the mean (n=3).The means followed by the same letter within a column are not significantly different by Tukey's multiple range test (p<0.05)

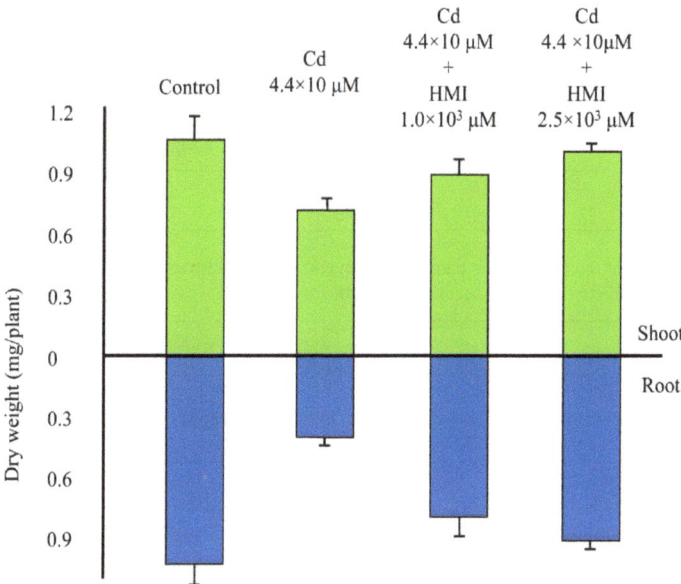

Fig. 5. Effect of HMI on *Echinochloa frumentacea* growth inhibited by Cd
The error bars represent the standard deviation of the mean (n=3)

| Control | Cd only
4.4 × 10 μmol L⁻¹Cd | Cd + HMl
4.4 × 10 μmol L⁻¹Cd + 1.0 × 10³ μmol L⁻¹HMI |

Fig. 6. Mitigation effect of hymexazole (HMI; 3-hydroxy-5-methylisoxazole) against *Echinochloa frumentacea* growth injury caused by Cd

5. Goal and approach for the restoration of Cd pollution soils by use of Weeds

In general, Cd pollution of soils are distributed over wide ranges, and recovery of vegetation at the pollution area and Cd extraction from the soils require a long time; thus it is important to make a grand design on goals and approach of the remediation of Cd pollution soils using weeds based on long term foresight. The remediation approach

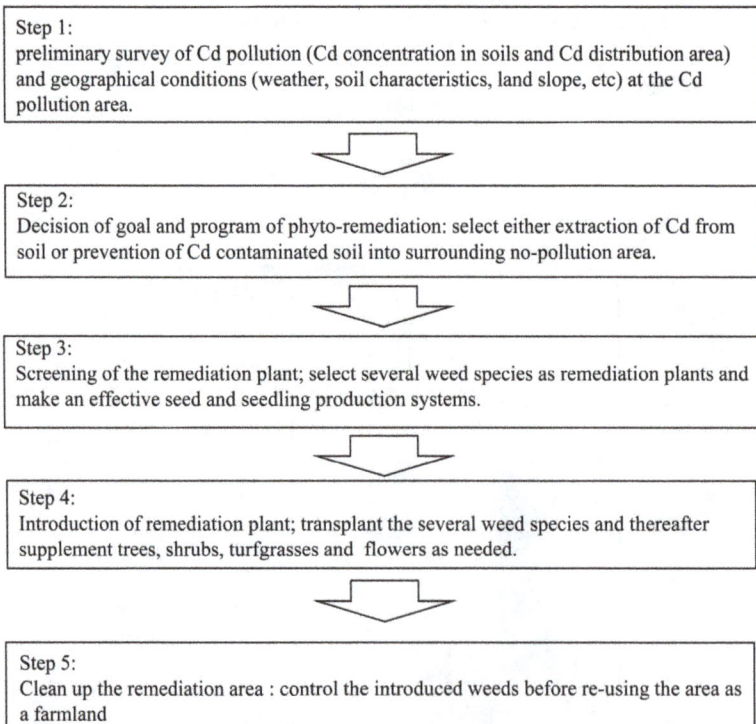

> **Step 1:**
> preliminary survey of Cd pollution (Cd concentration in soils and Cd distribution area) and geographical conditions (weather, soil characteristics, land slope, etc) at the Cd pollution area.

> **Step 2:**
> Decision of goal and program of phyto-remediation: select either extraction of Cd from soil or prevention of Cd contaminated soil into surrounding no-pollution area.

> **Step 3:**
> Screening of the remediation plant; select several weed species as remediation plants and make an effective seed and seedling production systems.

> **Step 4:**
> Introduction of remediation plant; transplant the several weed species and thereafter supplement trees, shrubs, turfgrasses and flowers as needed.

> **Step 5:**
> Clean up the remediation area : control the introduced weeds before re-using the area as a farmland

Fig. 7. Goal and approach of remediation of Cd pollution soils by use of weeds

depends on the not only Cd contamination levels in soil but also environmental and geographical conditions of the pollution areas, e. g., when the area is located in slopes, prevention of run-off of Cd contaminated soil by mulching should be preceded to Cd extraction. For the Cd remediation using weeds, several goals are suggested as follows; 1) prevention of Cd elution into the surrounding non-polluted areas; vegetation is restored by mulching with weeds and followed by introduction of grasses, turf grasses and shrubs as needed, but Cd extract is not involved in this case, 2) extraction of Cd from the pollution soils; Cd extraction is conducted by use of the remediation plants, and thereafter when post-mitigated area is reused as a farmland, introduced remediation plants (weeds) should be controlled completely by herbicides and practical methods before land reuses (Fig. 7). As mentioned above, Cd remediation approach varies with goals, however, preliminary survey of Cd concentrations in soil, weather conditions (temperature, sunlight and precipitation by month), physicochemical characteristics of soils, drainage, fertility and slope angle are needed regardless of the approach and the goal to decide the remediation plant (weed species) and the remediation methods (phyto-extraction, phyto-stabilization and/or malching). Moreover, in some cases, repeated introduction of the remediation plants into the Cd pollution areaa may be required, because seed production of the plant is possibly be inhibited by Cd even when Cd tolerant plants are used. For example, it is reported that the number of seeds / plant of *Portulaca oleracea* var. *sativa*, which is tolerant to Cd, is inhibited by 65% at 20 mgkg^{-1} of Cd (Abe *et al.* 2008b). Use of hormone and PGR may increase in the seed production and biomass of the remediation plants, and those seedlings habituated in the Cd contaminated soils before introducing them into the pollution areas, may be useful for the remediation.

6. Conclusion

Weeds are highly adaptive to adverse environmental conditions compared to crops, which is a crucial factor to consider weeds for developing soil remediation technologies. However, it might be considerably more difficult to ensure the necessary seed and seedling supply of weeds compare to crops, which can hamper the development of potential remediation technique. Thus when weeds are used for restoration of Cd pollution soils, to develop an effective seed and seeding production systems may become a crucial requirement. Besides, the proposed phyto-remediation with weeds might be envisioned as a long term approach due to the time requirements of the approach. In general, there is a great tendency that Cd extraction is much spotlighted than other goals, however, supplying the soils abundant in organic matter which absorb Cd and keep it in the soils by transplanting of herbaceous plants including weeds; resulting in the prevention of the run-off of Cd contaminated soil into surrounding non-polluted areas, may be efficient in case that Cd pollution is distributed over the extensive area.

7. References

Abe T., Fukami M., Ichizen N., and Ogasawara M., 2006: Susceptibility of weed species to cadmium evaluated in a sand culture. *Weed Biology and Management*, 6, 107-114.

Abe T., Fukami M., and Ogasawara M., 2008a: Cadmium accumulation in the shoots and roots of 93 weed species. *Soil Sci. Plant Nutr.*, 54, 566-573.

Abe T., SutoY., and Ogasawara M., 2008b: Effects of cadmium on the growth and the seed production of *Portulaca oleracea* L. var.*sativa* (Haw.) DC. *J. Weed Sci. Tech.*, 53, 1-7. (in Japanese with English summary)

Abe T., Fukami M., and Ogasawara M., 2011: Effect of hymexazole (3-hydroxy-5-methylisoxazole) on cadmium stress and accumulation in Japanese millet (*Echinochloa frumentacea* Link). *J. Pestic. Sci.*, 36, 48-52.

Asami T., 1972: The pollution of paddy soils by cadmium, zinc, lead, and copper in the dust, fume and waste water from Nisso Aizu Smelter. Jpn. J. Soil Sci. Plant Nutr., 43(9), 339-343 (In Japanese)

Brown S. L., Chaney R. L., Angel J. S., and Baker A. J. M., 1995: Zinc and cadmium uptake by hyperaccumulator *Thlaspi caerulescens* grown in nutrient solution. *Soil Sci. Soc. Am. J.*, 59, 125-133.

Bert V., Meerts P., Saumitou-Laprade P., and Salis P., 2003: Genetic basis of Cd tolerance and hyperaccumulation in *Arabidopsis halleri*. *Plant Soil*, 249, 9-18.

Elizabeth P. H., 2005: Phytoremediation. *Annu. Rev. Plant Biol.*, 56, 15-39.

Hattori H., Kuniyasu K., Chiba K., and Chino M., 2006: Effect of Chloride application and low soil pH on cadmium uptake from soil by plants. *Soil Sci. Plant Nutr.*, 52, 89-94.

Honma T., Ohba H., Kaneko A., and Hoshino T., 2009: Phytoremediation of cadmium by rice in low-level of Cd contaminated paddy field. *Jpn. J. Soil Sci. Plant Nutr.*, 80, 116-122. (in Japanese with English summary)

Hasan S. A., Hayat S., Ali B., and Ahmad A., 2008: 28-homobrassinolide protects chickpea (*Cicer arietinum*) from cadmium toxicity by stimulating antioxidants. *Environ.Pollut.*, 151, 60-66.

Keltjens W.G., and Beusichem M. L., 1998: Phytochelatins as biomarkers for heavy metal stress in maize (*Zea mays* L.) and wheat (*Triticum aestivum* L.): combined effects of copper and cadmium. *Plant Soil*, 203, 119-126.

Kaimi E., Mukaidani T., Miyoshi S., and Tamaki M., 2006: Ryegrass enhancement of biodegradation in diesel- contaminated soil. *Environ. Exp. Bot.*, 55, 110-119.

Lasat M. M., 2002: Phytoextraction of toxic metals: A review of biological mechanisms. *J. Environ. Qual.*, 31, 109-120.

Nishizono H., Suzuki S., and Ishii F., 1987: Accumulation of heavy metals in the metal-tolerant fern, *Athyrium yokoscense*, growing on various environments. *Plant Soil*, 102, 65-70.

Ogawa M., and Ota Y., 1976: 3-hydroxy-5-methylisoxazole as a plant growth stimulant. *Bull. Natl. Inst. Agric. Sci.* Series D 27, 103-137. (in Japanese with English summary).

Shinmachi F., Kumanda Y., Noguchi A., and Hasegawa I., 2003: Translocation and accumulation of cadmium in cadmium-tolerant *Polygonum thunbergii*. *Soil Sci. Plant Nutr.*, 49, 355-361.

Schat H., Llugany M., Vooijs R., and Hartley-Whitaker J., 2002: The role of phytochelatins in constitutive and adaptive heavy metal tolerance in hyperaccumulator and non-hyperaccumulator metallophytes. *J. Exp. Bot.*, 53, 2381-2392.

Takematsu T., and Ichizen N., 1987: Weeds of the world I. Zenkoku noson kyouiku kyoukai, Tokyo (in Japanese)

Takematsu T., and Ichizen N., 1993: Weeds of the world II. Zenkoku noson kyouiku kyoukai, Tokyo (in Japanese)

Takematsu T., Ichizen N., 1997: Weeds of the world III. Zenkoku noson kyouiku kyoukai, Tokyo (in Japanese)

Turgut C., Pepe M. K., and Cutright T. J., 2004: The effect of EDTA and citric acid on phytoremediation of Cd, Cr, and Ni from soil using *Helianthus annuus*. *Environ. Pollut.*, 131, 147-154.

12

Arsenic Behaviour in Polluted Soils After Remediation Activities

Francisco Martín[1], Mariano Simón[2], Elena Arco[1],
Ana Romero[1] and Carlos Dorronsoro[1]
*[1]Soil Science Department, Faculty of Science,
University of Granada, Campus Fuentenueva s/n, Granada,
[2]Soil Science Department, EPS CITE IIB,
University of Almería, Carretera Sacramento s/n, Almería
Spain*

1. Introduction

Arsenic (As) in soil is a serious environmental problem due to its potential high toxicity. Under field conditions As can accumulate in contaminated soils because it is only partially removed by leaching, methylation, and erosion or because it is only slightly taken up and accumulated by plants. Chemically, As exists as organic and inorganic species. It has two main oxidation states (+III and +V), depending on the type and amounts of sorbents, pH, redox potential (Eh), and microbial activity (Yong & Mulligan, 2004). Inorganic compounds are the most frequent in soil due to their water solubility. The most thermodynamically stable species within the pH range 4.0-8.0 include H_3AsO_3 of As^{III}, and $HAsO_4^{2-}$ and $H_2AsO_4^-$ of As^V (Smith et al., 1998). As^V species predominate in soil solutions under moderate reducing conditions, but As^{III} forms are more abundant when the redox potential is below 500 mV, according to Masscheleyn et al. (1991). These authors also indicate that a rise in pH, or a fall in As^V to As^{III}, boost the concentration of As in the solution, while its solubility under moderately reducing conditions is controlled by the dissolution of iron hydroxides (Marin et al. 1993). On the other hand, it is well known that the As concentration in a soil solution is governed by the physical and chemical properties of the soil, which influence adsorption-desorption processes. Arsenic has a high affinity for oxidic surfaces, and the reactivity of the oxides varies considerably with the pH, the charge density, and the composition of the soil solution. The soil texture and the nature of the mineral constituents also affect adsorption processes (Hiltbold, 1974). Pierce and Moore (1980) demonstrated the specificity of the surface of iron hydroxides and the influence of pH in As adsorption.

In soils, As has low mobility and under reducing conditions the concentration of dissolved As in soil solution declines. The availability of this element in soils can increase under acidic conditions (mainly pH below 5), due to the greater solubility of the iron and aluminium compounds, which augment As toxicity (O'Neill, 1995). In general, the mobility of this element is directly related to the total amount of As and inversely to time as well as to the iron and aluminium content; also, under oxidation conditions, its bioavailability is strongly limited (Kabata-Pendias & Pendias, 2001).

Compared to the abundant data on As adsorption, little information is available on As desorption in soils. In this sense, Carbonell et al. (1996) discovered that the adsorption of As^{III} forms is a reversible process, whereas the adsorption of As^V is a hysteresial process. Arsenic in soil is usually found in association with iron, aluminium, and manganese hydroxides, clays, and mineral oxyanions (sulphates, phosphates, and carbonates), which may serve as significant repositories of As due to their ubiquity in the environment (Foster, 2003). Iron hydroxides, such as goethite and ferrihydrite, are commonly found in soil and have been shown to be important in influencing the mobility behaviour of As (Foster, 2003; Jiang et al., 2005; Sun & Doner, 1996; Waychunas et al., 1993). Arsenic behaviour in soil is related to many factors. Microbial activity changes the oxidation state, and the formation of volatile As compounds by methylation leads to losses of this element in the superficial horizons (Dudas, 1987). In any case, these reactions depend both on the microorganism type as well as on the As compound (NRCC, 1978). The presence of organic matter has been studied as a key factor in the desorption of arsenic from iron oxides, in this way, Redman et al. (2002) found that the interaction between natural organic matter and hematite diminished the sorption of arsenate, promoting its mobility, and other authors considered that dissolved organic matter can mobilize arsenic from iron oxides, increasing the concentration of this element in the solution (Bauer & Blodau, 2006; Dobran & Zagury, 2006; Mladenov et al., 2010). Otherwise, significant desorption of As is observed with the rise in pH, in this case, the higher pH is related to the lower positive surface charge of the iron oxides, which facilitates the desorption of arsenate (Ghosh et al., 2006; Klitzke & Lang, 2009; Masscheleyn et al., 1991).

In this chapter, we present a general overview of the current stage of As content in the soils after one of the most important accidents involving soil pollution in Spain in recent decades: the Aznalcollar mine spill (Seville, SW Spain) in 1998. In this accident, the settling pond of the pyrite mine in Aznalcóllar broke open, spilling some 3.6×10^6 m^3 of water and 0.9×10^6 m^3 of toxic tailings into the Agrio and Guadiamar river basins (Aguilar et al., 2003; Simón et al., 1998). The toxic tailings spread approximately 40 km downstream, reaching the wetlands of the Doñana National Park (proclaimed world heritage by UNESCO in 1994). The total affected area was roughly 55 km^2. The disaster left sludge deposits between 1 cm to 1.5 m thick in different parts of the affected area (Simón et al., 1999; Lopez-Pamo et al., 1999). Arsenic was one of the major components of the toxic tailings, with a mean concentration of 4953 mg kg^{-1} (López-Pamo et al., 1999). The correlation between total As and sulphur strongly suggest that As was present in the tailings as arsenopyrite (FeAsS) and that the oxidation of the tailings would release iron and As (Simón et al., 2001). The remediation of soils was focused on the clean-up of the tailings and uppermost layer of the heavily polluted soils, together with the application of blocking agents to neutralize the acidity and to immobilize the highly soluble As concentrations (Aguilar et al., 2007b). Cleanup operations began almost immediately, so that by November 1998 the tailings were almost completely removed with heavy machinery and the acidic waters had been treated and discharged (Aguilar et al., 2003). To neutralize the acidity, liming material (sugar-refinery scum) was applied throughout the affected area at rates ranging from 20 to 150 Tm ha^{-1}. For arsenic immobilization, red soils rich in iron and located next to the affected area (Mudarra, 1988) were used; these soils had a concentration in free-iron oxides of between 2.26% and 6.31%, and the application rate ranged between 120 and 300 Tm ha^{-1}. Due to the climatic conditions (Mediterranean climate with ETP > precipitation), the soil solution tended to move upwards and concentrate the pollutants at the soil surface (Simón et al.,

2007). In the case of moderately polluted areas, when the aforementioned remediation actions were not feasible, the soils were tilled and homogenized to a depth of 25 cm, causing the As concentrations to decline in these soils to below the intervention level, although this action did not reduce the pollution.

Phytoremediation was also applied in some parts of the affected area (Clemente et al., 2006; Peñalosa et al., 2007), but the results at the plot scale were not effective enough to apply for the recovery of the whole area (Clemente et al., 2005; Madejón et al., 2003; Pérez de Mora et al., 2006). The final measurement was the stabilization of the soils by revegetation with native plants. A monitoring of the area in 2004 (6 years after the accident) revealed that although the remediation measurements lowered the As concentrations, the percentage of soils exceeding the maximum permitted level for agricultural soils was around 65% of the total affected soils, while around 30% of the soils had even doubled the maximum permitted of As (100 mg kg^{-1}; Simón et al., 2009).

In this chapter, we present a general overview of the time course of As content in soils during the remediation actions undertaken in the affected area. We present monitoring data 12 years after the accident and discuss the implications of the remediation measurements in relation to the mobility of As in soils over time.

2. Material and methods

To assess the arsenic contamination level in the basin, we made a systematic sampling in the affected area after the removal of the tailings covering the soils, using a network (400 x 400 m) and randomly selecting 100 sampling points. At each sampling point, samples were taken at the centre and four corners of a square (10 m side), at 0-10, 10-30, and 30-50 cm in depth. The samples for the same depth were mixed and homogenized, providing 3 samples per sampling point. Samples were also collected from uncontaminated soils in nearby areas unaffected by the spill.

After the study of the main soil properties between 0 and 50 cm in depth, samples were grouped into five different types using a cluster analysis via the k-mean method (Figure 1). Soil types 1 and 2 (no-carbonate sector), located in the upper part of the basin, closest to the tailing pond, were slightly acidic; type 1 had a loamy texture while type 2 was dominated by sand and gravel. Soil types 3, 4, and 5 (carbonate sector) were predominantly neutral or slightly alkaline, the main differences between them being texture (type 3, clay loam; type 4, loam; type 5, silty clay). The soils in the affected area were classified as Typic Xerofluvent (upper part of the basin) and Typic Xerorthent (middle and lower part of the basin) (Soil Survey Staff, 2003). Field descriptions of soils were based on procedures of the Soil Survey Staff (1951).

Soil samples were air-dried and sieved to 2 mm to estimate the gravel content. Soil analyses were made with the < 2 mm fraction. Sulphate was determined in the saturation extract (water-soluble sulphate) by ion chromatography in a Dionex-120 chromatograph. Particle-size distribution was determined by the pipette method after the elimination of organic matter with H_2O_2 and dispersion with sodium hexametaphosphate (Loveland & Whalley, 1991). The pH was measured potentiometrically in a 1:2.5 soil:water suspension. The $CaCO_3$ equivalent was determined by the method of Bascomb (1961). Total carbon was analysed by dry combustion with a LECO SC-144DR instrument. Organic carbon (OC) was determined by the difference between total carbon and inorganic carbon from $CaCO_3$. The cation-exchange capacity (CEC) was determined with 1N Na-acetate at pH 8.2, measuring the sodium in a

METEOR NAK-II flame-photometer. The total concentration of iron (Fe_T) was measured by X-ray fluorescence in a Philips PW-1404 instrument, from a disc of soil and lithium tetraborate in a ratio of 0.6:5.5. Poorly crystallized iron oxides (Fe_O) were extracted with ammonium oxalate (Schwertmann & Taylor, 1977), and measured by atomic absorption spectroscopy.

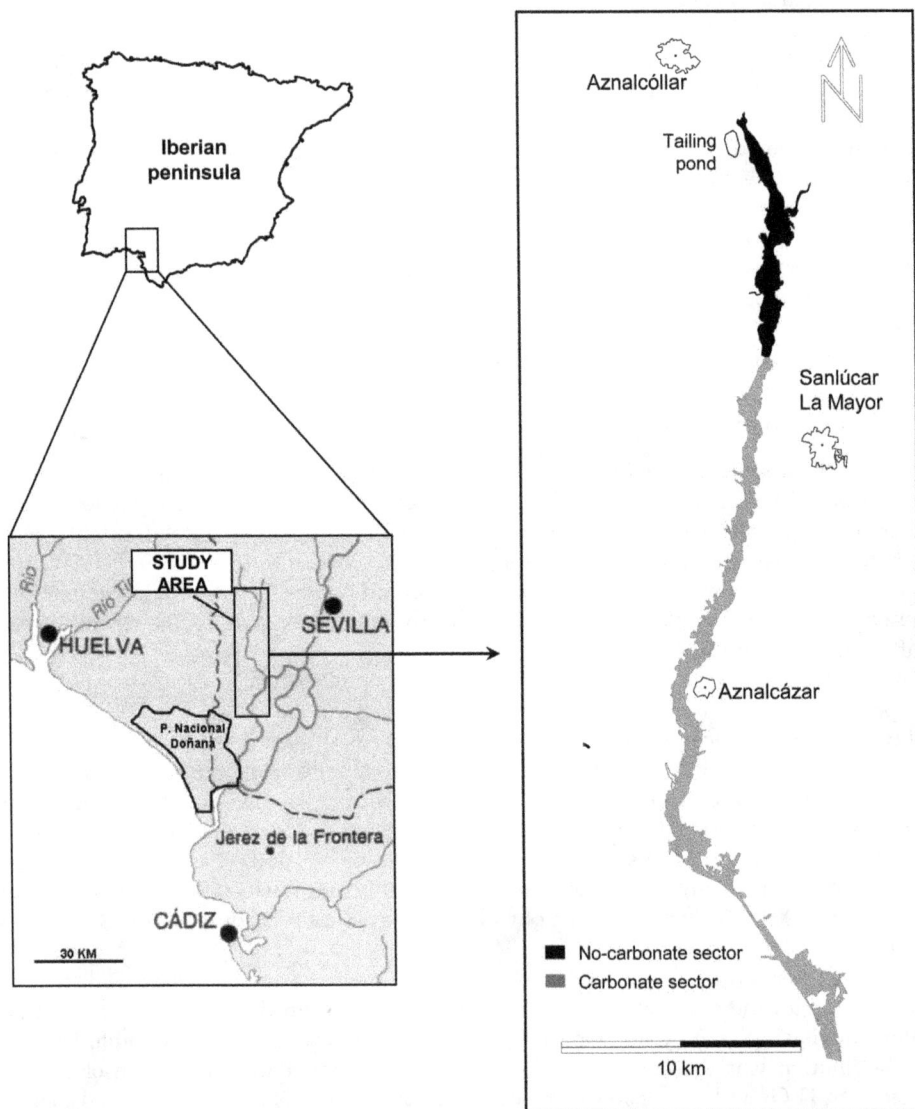

Fig. 1. Location of the area affected by the pyrite tailing spill

Soil samples were finely ground (<0.05 mm), and digested in strong acids (HNO_3 + HF); in these dissolutions, total As values were determined by ICP-MS in a Perkin Elmer Elan 5000 instrument. A Multi-element Calibration Standard 4 (Perkin-Elmer) was used with Rh as the internal standard. The detection limit for As in soils measured by this technique was 0.01 µg l[-1].The accuracy of the method was corroborated by analyses (six replicates) of a standard reference material: SRM 2711 (soil with moderately elevated trace-element concentrations). For As, the mean certified value was 105.0 mg kg[-1] with a standard deviation of 8.0; the mean experimental value was 102.4 mg kg[-1] with a standard deviation of 1.1. For the study of the mobility of As forms in soil samples, extractions were conducted with distilled water and EDTA (AsW and AsE, respectively; Quevauviller et al., 1998), and ammonium oxalate (AsO; Schwertmann & Taylor, 1977).

Toxicity bioassays were made with the water extract (1:5 soil:water ratio) of the affected soils to assess the potential risk of the soil-water solution. We used two types of bioassays: i) the response of bioluminescent bacteria *Vibrio fischeri* according to Microtox Basic Test for Aqueous Extracts Protocol (AZUR Environmental, 1998); and ii) seed-germination text with lettuce (*Lactuca sativa* L.) according to the US EPA (1996) procedure.

3. Results and discussion

3.1 Initial pollution (the year of the accident)

The accident (25th April, 1998) caused a spill that covered the soils with variable amounts of tailings and polluted waters. The mean concentration of As in the tailings was 4953 mg kg[-1], and in the waters 0.002 mg l[-1]. According to Simón et al. (2007) the As contamination entered the soil mainly from the solid phase, and estimation of the quantity of tailings that penetrated the soil ranged from 1.7 to 150.8 mg tailing per kg soil, depending on the affected sector considered. Nine days after the spill (May, 1998), with the soils still covered by the tailings, the mean As concentration in the uppermost 10 cm of the soils was 121.7 mg kg[-1], although the pollution was very heterogeneous, mainly due to the soil properties, essentially structure, which affected the penetration of the tailings into the soil (Simón et al., 1999).

A few weeks after the spill (June, 1998), as result of drying and aeration of the tailings, sulphides oxidized to sulphates (Nordstrom, 1982), the pH fell markedly due to the formation of sulphuric acid (Stumm & Morgan, 1981) and the formerly insoluble pollutants partly solubilized. During the weeks following the spill, this oxidation became evident, the sulphates increased rapidly in the tailings solution, this being accompanied by a sharp fall in the soil pH (with pH values up to 2.5 in the most polluted areas). The concentration of soluble As, measured in a water extract of tailings, was also found to vary over time. In this way, most of this solubilization occurred between 25 and 40 days from the spill (increasing more than 5-fold the water-soluble arsenic in soil in relation to the previous period), when the oxidation and solubilization of the sulphides bonding to arsenic in the tailings were highest and a rainfall period occurred. At 88 days from the spill (July, 1998), the oxidative pollution was negligible (Simón et al., 2007). This oxidative pollution sharply increased the arsenic contamination in the soils over time in the surface samplings (0-10 cm) but without increasing the pollution in the samples at 10-30 cm in depth (Aguilar et al., 2007a).

After the initial contamination (direct input of tailings into the soils) and the secondary contamination (infiltration of pollutant solutions coming from the oxidation of the tailings), the soils were considered strongly polluted. Because of the potential environmental risk of pollution, the first remediation measure was to clean-up the affected area; in this action, the tailings and the upper part of the soils (between 20 and 50 cm mean) were removed. The systematic sampling in the affected area after the clean-up (November, 1998) indicated that, after this measure the mean arsenic concentration in the uppermost 10 cm of the soils was

157.3 mg kg^{-1} (with maximum values up to 1,226.8 mg kg^{-1}). The soil pollution exceeded the permitted value for agricultural soils in Andalusia (50 mg kg^{-1}; Aguilar et al., 1999) for 82% of the overall surface area affected by the tailings, 96% in the no-carbonate sector (soil types 1 and 2), and 76.8 % in the carbonate sector (soil types 3, 4 and 5). Meanwhile, 53% of the affected soils surpassed the limit of 100 mg kg^{-1} maximum permitted for Natural Park in Andalusia (Aguilar et al., 1999).

3.2 Evolution of the soil pollution after the remediation actions

The rapid clean-up of the affected area (in eight months, 45 km^2 were cleaned) resulted in a deficient remediation action, as part of the tailings remained mixed with the soil, appearing residual tailings heterogeneously distributed throughout all the affected area. According to the high concentrations of As in soils after the clean-up conducted in 1998, secondary remediation actions were applied in the area between 1999 and 2001. These actions consisted in: i) repeat cleanup of the most polluted areas; ii) amendment applications (liming, organic matter, soils rich in iron oxides); and iii) tilling of the uppermost 25 – 30 cm to dilute the As concentrations in the upper 10 cm of the moderately contaminated soils. Phytoremediation was also applied in some parts of the affected area (Clemente et al., 2006; Peñalosa et al., 2007), but the results at the plot scale were not effective enough to apply for the recovery of the whole area (Clemente et al., 2005; Madejón et al., 2003; Pérez de Mora et al., 2006). The final measurement conducted in the area was the stabilization of the soils by revegetation with native plants.

After the end of all the remediation actions, the monitoring of the pollution made 6 years later (summer 2004) revealed that the initial pollution was reduced, but the reduction of the area exceeding the As concentration of 50 mg kg^{-1} was negligible, and around 30% of the area continued to exceed the intervention level of 100 mg kg^{-1} (Figure 2).

In all cases, the As pollution concentrated in the uppermost 10 cm of the soils and decreased sharply in depth, without significantly affecting the subsoil or groundwater (Simón et al., 2001, 2002, 2009; Dorronsoro et al., 2002).

To study the As retention in the soil, we made extractions with selective reagents (water, calcium chloride, acetic acid, EDTA, oxalic-oxalate). Arsenic was extracted mainly with oxalic-oxalate both in the non-carbonate sector (30% in relation to the total As) and in the carbonate sector (20% in relation to the total As). Considering oxalic-oxalate to be the reagent which extracts specifically the elements adsorbed onto oxides, we conclude that As is strongly retained by the iron oxides of the soil (Martin, 2002). The correlation between the main soil constituents and the total arsenic concentration (AsT) indicated that there was a significant correlation only with the total iron (FeT) concentration (Eq. 1), which indicates that As is likely to be absorbed as anionic forms by iron oxides precipitated in the soil (García et al., 2009).

$$AsT = 2 \cdot 10^{-5} \, FeT^{4.009} \quad (r^2 = 0.983) \tag{1}$$

The mineralogical study of the most polluted soils in the affected area (García et al., 2009; Martín et al., 2008), indicated that the arsenic retention in these media should be related to the neoformation of iron hydroxysulfate minerals (jarosite [$KFe_3SO_4(OH)_6$]; schwertmannite [$Fe_8O_8(OH)_6SO_4$]) and ferrihydrite [$5Fe_2O_3 \cdot 9H_2O$]), suggesting that the removal of As through co-precipitation and adsorption reactions is probably the dominant solid-phase control on the mobility of arsenic (Bigham et al., 1996; Dold, 2003; Sánchez España et al., 2005). In this way, the retention of As in these soils would be related to the precipitation of relatively stable forms of ferrihydrite and schwertmannite; this process is related to the reduction of soluble As concentrations in soils (Carlson et al., 2002; Courtin-Nomade et al., 2003).

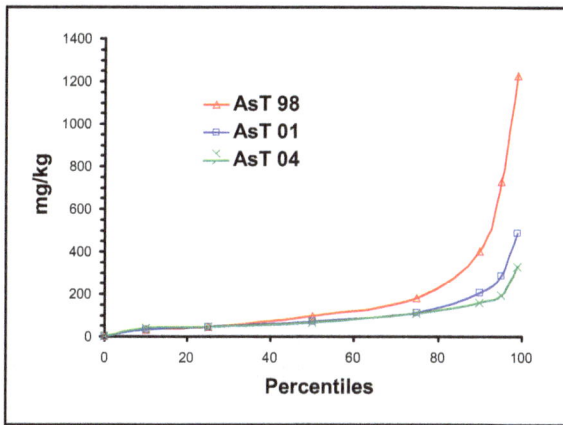

Fig. 2. Total arsenic concentrations (AsT) in the affected soils (0-10 cm) in 1998, 2001, and 2004

3.3 Mobility, bioavailability, and toxicity of arsenic in soil (evolution during 6 years)

In this type of pollution, toxicity and metal uptake are more accurately related to soluble and bioavalilable fractions than total metal concentrations in soil (Kumpiene et al. 2008). These forms are more related to the potential environmental risk of the pollution because they indicate the ability of the pollutant to be dispersed by the environment or transferred to organisms (animals and plants) living in the affected ecosystem. In this way, the soil is usually considered as a key factor to control and avoid environmental contamination. Soil acts as a buffering medium that receives contaminants and in many cases, due to the interaction with the soil constituents and properties, the pollution is minimized and the dispersion of the contaminants to other more sensitive media such as water or organisms is strongly reduced.

The easiest form of a pollutant to mobilize in the soil is the soluble-in-water form. Arsenic is a highly toxic element and therefore the toxic levels in soil solution are reported at very low concentrations (0.04 mg As kg^{-1} soil; Bohn et al., 1985). The soluble-in-water forms were extracted just after the first clean-up actions (at the end of 1998), three (in 2001) and six years (in 2004) later (Figure 3). In this period, the solubility of the As forms strongly reduced in the polluted soils. After the end of the first remediation actions in 1998, high amounts of As remained soluble in the affected soils, with more than 50% of the soils exceeding the toxic level. Three years later, there was a decline in the As solubility, although 19% of affected soils surpassed the toxic level, with differences depending on the soil properties (10% in the carbonate sector and 25% in the non-carbonate sector). Six years after the accident, there was a sharp reduction in the soluble-in-water forms of As, only the 5% of the affected soils exceeding the toxic level, with no significant differences being found between the carbonate and the no-carbonate sector.

The forms extracted with EDTA are considered to be bioavailable by many authors because they are related to the carbonates, inorganic precipitates, amorphous oxides or organic ligands that can be uptaken by most plants (Beckett, 1989; Quevauviller et al., 1998; Rendell et al., 1980; Sposito et al., 1982). In this way, arsenic extracted by EDTA significantly differed over time (Table 1).

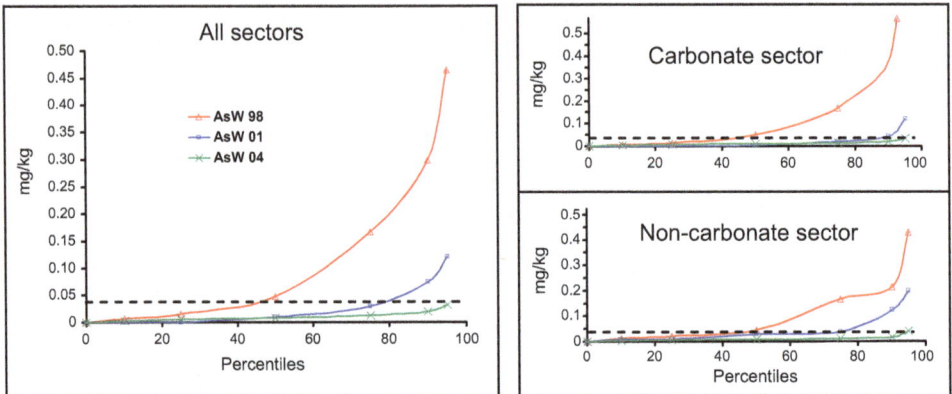

Fig. 3. Soluble-in-water arsenic concentrations (AsW) in all the affected soils and in the different sectors (carbonate and non-carbonate) in 1998, 2001, and 2004 (doted line: toxic level 0.04 mg kg-1; Bohn et al., 1985)

	1998	2001	2004
All soils	58	12	bl
Non-carbonate sector	69	25	bl
Carbonate sector	50	2	bl

Table 1. Percentage of soils exceeding the level of 2 mg kg-1 of As extracted with EDTA in all the affected soils and in the different sectors (carbonate and non-carbonate) in 1998, 2001, and 2004 (bl: below level)

After the end of the first remediation actions in 1998, high amounts of As remained bioavailabe in the affected soils of all sectors, with around 58% of the soils exceeding the level of 2 mg kg-1. Three years later, there was a reduction in the solubility of As and only 12% of soils had bioavailable concentrations in As above the reference level, although 25% of affected soils were located in the non-carbonate sector and only 2% in the carbonate sector. Six years after the accident, there was a notable reduction in the bioavailable forms of As, with no soil exceeding the reference level of 2 mg As kg-1 soil.

Bioassays were made to assess the toxicity of the affected soils 6 years after the accident. The toxicity was estimated from the soluble-in-water fraction using two types of bioassays: i) seed germination test (US EPA, 1996) with lettuce (*Lactuca sativa L.*); and ii) bioluminescent text (AZUR, 1998) with bacteria (*Vibrio fischeri*). These tests were applied because the toxicity related to the soluble-in-water fractions of contaminants reflects the behaviour of the most mobile fraction of pollutants and are strongly related to the high risk of dispersion, solubilization, and bioavailability of contaminants in the environment. Both assays used distilled water as a control.

This study was made in six georreferenced soils included in a heavily contaminated sector in 1998 (CS 98). The same study was repeated in the same six soils after the end of the remediation actions in 2004 (RS 04); six reference soils adjacent to the area but not affected by the spill were selected as uncontaminated soils (UCS). In all water extracts, pH, electric conductivity and the soluble concentration of arsenic and other elements were determined.

The results of the response of the bioluminescent bacteria after 15 min (I15) and the root elongation of lettuce seeds (RE) are presented in Table 2.

	pH	EC (dS m⁻¹)	CuW (mg l⁻¹)	ZnW (mg l⁻¹)	AsW (mg l⁻¹)	PbW (mg l⁻¹)	I15	RE
UCS	7.62	0.56	0.02	0.02	0.01	0.01	(+) 30.5	(+) 5.6
CS 98	3.67	2.78	135.30	632.09	0.18	0.31	(-) 73.9	(-) 66.8
RS 04	7.05	2.16	0.67	10.25	0.01	0.92	(+) 5.1	(+) 3.2

Table 2. Main properties of the water extracts [pH, electrical conductivity (EC) and soluble-in-water concentrations of pollutants] used in the bioassays with bioluminescent bacteria after 15 minutes (I15) and root elongation of lettuce seeds (RE)

The results indicate that the contaminated soils (CS 98) strongly inhibited bioluminescent bacteria (74% in relation to the control) and heavily reduced of the root elongation in lettuce (67% in relation to the control). The most significant variables related to the toxicity were pH, and the soluble concentrations in Cu and Zn. Therefore, in this type of multi-elemental pollution, the As concentration in the water extract was not the main element related to the toxic response in both bioassays. The comparison of these results with those registered six years later in the same area after the end of the remediation actions (RS 04) indicated a sharp decline in the soluble concentrations of the pollutants, and a positive response in both bioassays. This indicates that no toxicity occurred in the selected soils after remediation and that the organisms in contact with the soil solution increased in luminescence or in root elongation compared to the control (hormesis). This positive response was also found in the uncontaminated soils of the area (UCS), indicating that the remediation measurements successfully lowered potential toxicity related to the soluble fraction in the polluted soils.

3.4 Evolution of arsenic pollution twelve years after the accident

Due to the remediation actions, some of the soil properties changed significantly from 2004 to 2010 (twelve years after the accident). The influence of the liming applied over the soils to avoid the acidification caused by the sulphide oxidation and to promote the precipitation of the pollutants raised the pH over time (Figure 4). In the non-carbonate sector (NCS) the rise over time was notable (p<0.001), with mean values of 5.75 in 2004, rising to 7.63 in 2010. In the carbonate sector (CS), the rise was less remarkable but also significant (p=0.004), ranging from mean values of 7.64 in 2004 to 8.25 in 2010. The differences in pH between sectors decreased over time but remained significantly different in all years. The soil organic carbon (OC) also showed a sharp increase during the study period, and the differences between sectors were significantly different in 2010 (Figure 4), due to the addition of organic amendments to help stabilize the soils and regenerate the vegetation, being this recovery more rapid in the carbonate sector. The soil OC content in 2004 in the NCS was close to 1%, increasing significantly (p=0.03) to 1.3% in the last years sampled; meanwhile this increment was higher and statistically significant (p<0.001) in the CS, where the OC reached mean values of close to 1.7%.

After the potential contamination and the mobility of pollutants in the environment were minimized, a protected landscape configuration was established by the regional government of Andalusia (Spain), called the "Green Corridor of the Guadiamar River" (CMA, 2003). The use of all the agricultural soils within the affected area was prohibited to avoid the potential risk of high levels of total concentration in arsenic and other heavy metals in soils. Therefore, the remediation measurements during the study period 2004 – 2010 were focused on stabilizing the vegetation and protecting the area, causing the total concentration of arsenic to remain constant in the affected soils.

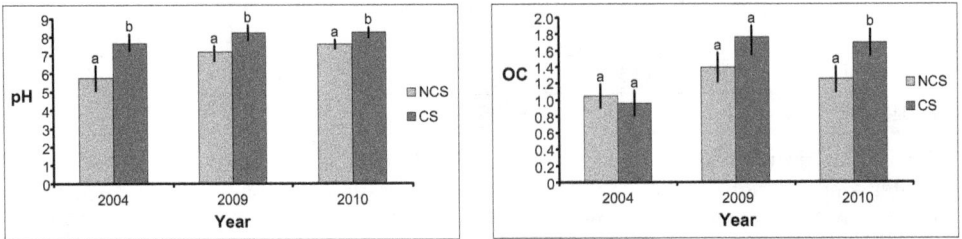

Fig. 4. Time course of pH and soil organic carbon (OC) in the non-carbonate sector (NCS) and carbonate sector (CS) in 2004, 2009, and 2010 (different letters indicate significant differences between sectors according Tukey test p<0.05)

To study the medium-term development of the pollution, we repeated the same systematic sampling of the affected area in 2009 and 2010, eleven and twelve years, respectively, after the accident. The total As concentration in the affected soils, had not significantly changed over time. Six years after the spill (in 2004), the soil exceeding the intervention level of 100 mg As kg^{-1} soil was around 30% (40% of the non-carbonate soils and 25% of the carbonate soils), and the percentage of soils exceeding this value in 2009 and 2010 remained approximately the same (Figure 5), with a similar ratio between the carbonate and non-carbonate sectors.

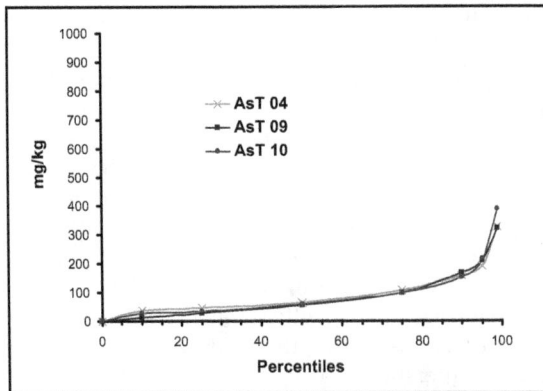

Fig. 5. Total arsenic concentrations (AsT) in the affected soils in 2004, 2009, and 2010

The arsenic mobility over time was also studied in 2009 and 2010 after analysing the soluble-in-water and the EDTA extracted forms. In relation to the previous sampling (in 2004) there was a sharp increase in the solubility of arsenic forms in remediated soils (Figure 6).
This increase occurred both in the non-carbonate as well as in the carbonate sector, indicating that the solubility of arsenic in soils has changed independently of the soil type. In relation to the EDTA extracted forms, considered to be bioavailable, the increase was also pronounced throughout the affected area. The concentration of As extracted with EDTA had

a mean value of 0.49 and 0.41 mg kg^{-1} in 2004 in the non-carbonate and carbonate sectors, respectively. Six years later (in 2010) these concentration increased to 0.97 and 0.58 mg kg^{-1} in both sectors, respectively.

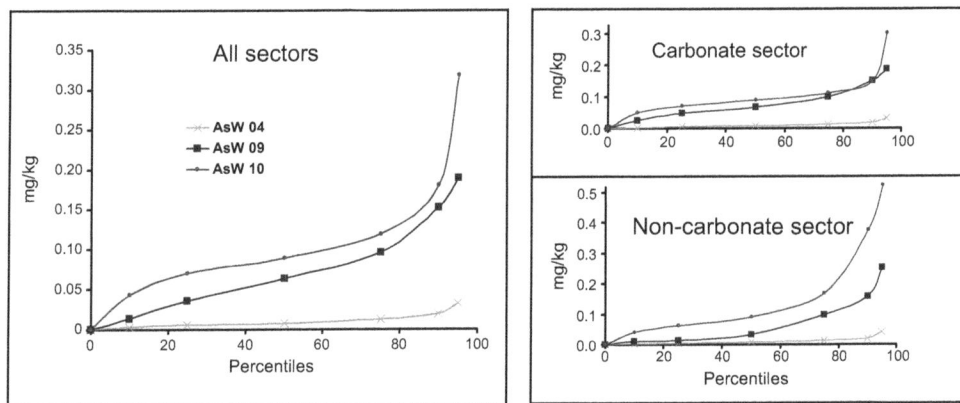

Fig. 6. Soluble-in-water arsenic concentrations (AsW) in all the affected soils and in the different sectors (carbonate and non-carbonate) in 2004, 2009, and 2010

The reduction of soluble As concentrations in these soils has been related, in the previous section, to As retention by iron oxi-hydroxides precipitated in the soils mainly as low crystallization forms (amorphous). In any case, some authors have described such adsorption as a reversible process (Carbonell et al., 1996). A correlation analysis was made between the mobile fractions of As and the main soil properties and constituents between the period 2004 to 2010. This analysis gave a significant (p<0.001) and negative correlation between the soluble-in-water As forms (AsW) and the amorphous iron oxides (Feo) in soil, indicating that the reduction of these amorphous phases should be related to the increase in the arsenic concentration in the soil solution.

The quantification of the reduction in the amorphous iron oxides in the soil is not enough to explain the strong increase in the solubility of arsenic in the affected area, so we studied the main constituents that may influence As mobilization in soils. According to the different soil types described in Section 2, we made a correlation analysis between the mean values of the main soil properties and the arsenic forms; the main coefficients are presented in Table 3. Significant correlations were found between the soluble-in-water arsenic (AsW) with the pH and the organic carbon (OC) of the soil; also, As extracted with EDTA (AsE) significantly correlated with organic carbon. However, we found no correlation between the mobile As forms and the total concentration of As in the soil (AsT), indicating that after the remediation measurements the initial contamination has little influence on the As mobility in soils.

According to these results, a rise in pH can intensify the As concentration in the solution. Arsenic in soils has a high affinity for oxidic surfaces, and the reactivity of the oxides varies considerably with the pH, the charge density, and the composition of the soil solution. Pierce and Moore (1980) demonstrated the specificity of the surface of iron hydroxides and the influence of pH in the As adsorption. In this way, significant desorption of As is observed with the rise of pH; in this case, the rise in pH is related to the decrease of the positive surface charge of the iron oxides, which facilitates the desorption of arsenate (Ghosh et al., 2006; Klitzke & Lang, 2009; Masscheleyn et al., 1991).

	AsW	AsE	AsT	pH
AsE	0.535 (*)			
AsT	-0.188	0.082		
pH	0.639 (**)	0.346	-0.634 (**)	
CO	0.704 (**)	0.531 (*)	-0.384	0.691 (**)

Table 3. Correlation matrix between the soluble-in-water (AsW), EDTA extracted (AsE), total arsenic (AsT) forms, pH, and organic carbon (OC), in the different soil types in the period 2004 – 2010 (p<0.05 = (*); p<0.01 = (**))

Another key factor influencing As mobility in soils over time is the total soil organic carbon (OC). According to our data, a significant and positive correlation was found between the AsE and AsW with the total soil organic carbon (Table 3). The regression plot between the AsW and the OC is presented in the Figure 7 by a potential equation.

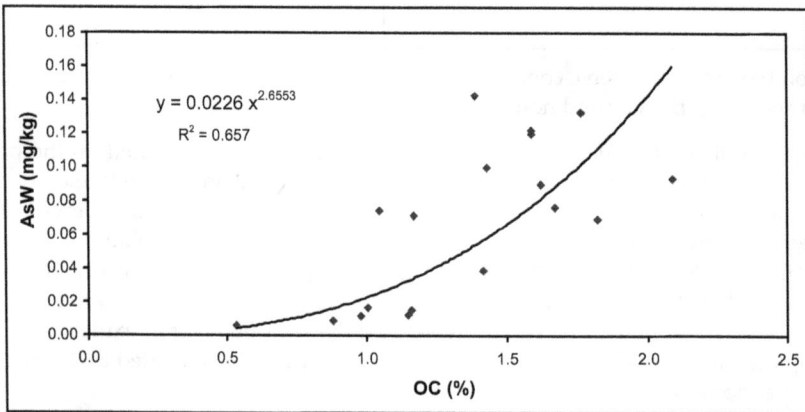

Fig. 7. Regression plot between soluble-in-water arsenic concentrations (AsW) and total soil organic carbon (OC) in all the different soil types between the period 2004 – 2010

The presence of organic matter has been studied as a key factor in the desorption of As from iron oxides. In this sense, Redman et al. (2002) reported that the interaction between natural organic matter and hematite diminished the sorption of arsenate, promoting its mobility, while other authors have observed that dissolved organic matter can mobilize As from iron oxides, augmenting the concentration of this element in the solution (Bauer & Blodau, 2006; Dobran & Zagury, 2006; Mladenov et al., 2010). Therefore, the competing effect of organic matter with arsenate for surface sites should also be related to the mobilization of As retained in soils after the remediation actions.

4. Conclusion

Twelve years after the Aznalcóllar pyrite tailing spill, and after the end of all remediation actions made in the affected area, we detected a notable increase in the soluble-in-water and bioavailable arsenic forms previously retained by the amorphous iron oxides of the soils. In the affected soils, two main parameters seems to be related to the increase of As solubility over time, soil pH and total soil organic carbon. The rise in pH values under alkaline conditions caused by the liming applications, can be related to the decrease of the positive surface charge of the iron oxides which can trigger the desorption of the retained arsenate. Otherwise, the increase in soil organic matter content related to the revegetation of the soils can be related to the greater mobility of the arsenic in soils due to the competing effect for the surface sites. Monitoring and assessment of soils polluted by arsenic and remediated with amendment applications (mainly liming and organic matter) are necessary over the short to medium terms in order to control the desorption or mobilization of arsenic forms over time and to avoid the environmental risk related to the potential toxicity of the soil solutions.

5. Acknowledgment

This study has been made possible by the research project RNM-3315 of the Regional Environmental Department of the Andalusian Government, and by the research project CGL2010-19902 of the Science and Innovation Ministry of Spain. Also thanks goes to Mr. David Nesbitt for correcting the English of the manuscript.

6. References

Aguilar, J.; Bellver, R.; Dorronsoro, C.; Fernández, E.; Fernández, J.; García I.; Martín, F.; Ortiz, I. & Simón, M. (2003). *Contaminación de suelos por el vertido tóxico de Aznalcóllar.* Junta de Andalucía, Consejería de Medio Ambiente, Sevilla, Spain.

Aguilar, J.; Dorronsoro, C.; Fernández, E.; Fernández, J.; García, I.; Martín, F.; Sierra, M. & Simón, M. (2007a). Arsenic contamination in soils affected by a pyrite-mine spill (Aznalcóllar, SW Spain). *Water, Air, and Soil Pollution*, Vol.180, pp. 271-281.

Aguilar, J.; Dorronsoro, C.; Fernández, E.; Fernández, J.; García, I.; Martín, F.; Sierra, M. & Simón, M. (2007b). Remediation of As-contaminated soils in the Guadiamar river basin (SW, Spain). *Water, Air, and Soil Pollution*, Vol.180, pp. 109-118.

Aguilar, J.; Dorronsoro, C.; Galán, E. & Gómez, J.L. (1999). Criterios y estándares para declarar un suelo como contaminado en Andalucía, In: *Investigación y Desarrollo Medioambiental en Andalucía,* Univ. Sevilla, (Ed.), 45-59, Sevilla, España.

AZUR Environmental. 1998. The Microtox ® Acute Basic, DIN, ISO and Wet Test Procedure. Carlsbad, Calif, USA.

Bascomb, C.L. (1961). A calcimeter for routine use on soil samples. *Chem. Ind.,* Vol.45, pp. 1826–1827.

Bauer, M. & Blodau, C. (2006). Mobilization of arsenic by dissolved organic matter from iron oxides, soils and sediments. *The Science of the Total Enviromment*, Vol.354, pp. 179-190.

Bigham, J.M.; Schwertmann, U.; Traina, S.J.; Winland, R.L. & Wolf, M. (1996). Schwertmannite and the chemical modelling of iron in acid sulphate waters. *Geochim. Cosmochim. Ac.,* Vol.60, No.12, pp. 2111–2121.

Bohn, H.L.; McNeal. B.L. & O'Connor, G.A. (1985). *Soil Chemistry.* Wiley Interscience, New York, USA.

Carbonell, A.; Burló, F. & Mataix, J. (1996). Kinetics of arsenite desorption in Spanish soils. *Communications in Soil Science and Plant Analysis,* Vol.27, pp. 3101–3117.

Carlson, L.; Bigham, J.M.; Schwertmann, U.; Kyek, A. & Wagner, F. (2002). Scavenging of As from acid mine drainage by schwertmannite and ferrihydrite: A comparison with synthetic analogues. *Environ. Sci. Technol.*, Vol.36, pp. 1712–1719.

Clemente, R.; Almela, C. & Bernal, M.P. (2006). A remediation strategy based on active phytoremediation followed by natural attenuation in a soil contaminated by pyrite waste. *Water, Air, and Soil Pollution*, Vol.177, pp. 349-365.

Clemente, R.; Walker, D.J. & Bernal, M.P. (2005). Uptake of heavy metals and As by *Brassica juncea* grown in a contaminated soil in Aznalcóllar (Spain): The effect of soil amendments. *Environmental Pollution*, Vol.138, No.1, pp. 46-58.

CMA. (2003) Decreto 112/2003 de 22 de Abril to declare the protected landscape of the Green Corridor of Guadiamar. Consejería de Medio Ambiente, Junta de Andalucía, Spain.

Courtin-Nomade, A.; Bril, H.; Neel, C. & Lenain, J. (2003). Arsenic in iron cements developed within tailings of a former metalliferous mine—Enguiales, Aveyron, France. *Appl. Geochem.*, Vol.18, pp, 395–408.

Dobran, S. & Zagury, G.J. (2006). Arsenic speciation and mobilization in CCA- contaminated soils: Influence of organic matter content. *The Science of the Total Environment*, Vol.364, pp. 239-250.

Dold, B. (2003). Dissolution kinetics of schwertmannite and ferrihydrite in oxidized mine samples and their detection by differential X-ray diffraction (DXRD). *Appl. Geochem.*, Vol.18, pp. 1531–1540.

Dorronsoro, C.; Martín, F.; Ortiz, I.; García, I.; Simón, M.; Fernández, E.; Fernández, J. & Aguilar, J. (2002). Migration of trace elements from pyrite tailings in carbonate soils. *J. Environ. Qual.*, Vol.31, pp. 829–835.

Dudas, M. J. (1987). Accumulation of native arsenic in acid sulphate soils in Alberta. *Canadian Journal of Soil Science*, Vol.67, pp. 317–321.

Foster, A. L. (2003). Spectroscopic investigation of arsenic species in solid phases, In: *Arsenic in ground water: geochemistry and occurrence*, A. H. Welch & K. G. Stollenwerk (Eds.), 27–65. Boston, USA.

García, I.; Diez, M.; Martín, F.; Simón, M. & Dorronsoro, C. (2009). Mobility of arsenic and heavy metals in a sandy-loam textured and carbonated soil. *Pedosphere*, Vol.19, No.2, pp. 166–175.

Ghosh, A.; Sáez, A.E. & Ela, W. (2006). Effect of pH, competitive anions and NOM on the leaching of arsenic from solid residuals. *Science of the Total Environment*, Vol.363, pp. 46– 59.

Hiltbold, A.; Hajek, B.F. & Buchanan, G.A. (1974). Distribution of arsenic in soil profiles after leaching. *Soil Science Society of America Proceedings*, Vol.38, pp. 647–652.

Jiang, W.; Zhang, S.; Shan, X.; Feng, M.; Zhu, Y. & McLaren, R. G. (2005). Adsorption of arsenate on soils. Part 2: Modeling the relationship between adsorption capacity and soil physiochemical properties using 16 Chinese soils. *Environmental Pollution*, Vol.138, pp. 285–289.

Kabata-Pendias, A. & Pendias, H. (2001). *Trace elements in soils and plants* (3rd ed.), CRC Press, Boca Raton, Florida, USA.

Klitzke, S. & Lang, F. (2009). Mobilization of Soluble and Dispersible Lead, Arsenic, and Antimony in a Polluted, Organic-rich Soil - Eff ects of pH Increase and Counterion Valency. *J. Environ. Qual*, Vol.38, pp. 933–939.

Kumpiene, J.; Lagerkvist, A. & Maurice, C. (2008). Stabilization of As, Cr, Cu, Pb and Zn in soils using amendments – A review. *Waste Management,* Vol.28, pp. 215-225.

López-Pamo, E.; Barettino, D.; Antón-Pacheco, C.; Ortiz, G.; Arránz, J.C.; Gumiel, J.C.; Martínez-Pledel, B.; Aparicio, M. & Montouto, O. (1999). The extent of the Aznalcóllar pyritic sludge spill and its effects on soils. *The Science of the Total Environment,* Vol.242, pp. 57–88.

Loveland, P.J. & Whalley, W.R. (1991). Particle size analysis. In: *Soil analysis: Physical methods,* K.A. Smith & C.E. Mullis (Eds.), Marcel Dekker, 271–328, New York, USA.

Madejón, P.; Murillo, J.M.; Marañón, T.; Cabrera, F. & Soriano, M.A. (2003). Trace element and nutrient accumulation in sunflower plants two years after the Aznalcóllar mine spill. *The Science of the Total Environment,* Vol.307, pp. 239-257.

Marin, A.; Masscheleyn, P.J. & Patrick, W.H. (1993). Soil redox-pH stability of arsenic speciation. *Environmental Science & Technology,* Vol.33, pp. 773–781.

Martín, F. (2002). *Pollution of soils by the spill of a pyrite mine (Aznalcóllar, Spain),* PhD Thesis, University of Granada, Spain.

Martín, F.; García, I.; Diez, M.; Sierra, M.; Simón, M. & Dorronsoro, C. (2008). Soil alteration by continued oxidation of pyrite tailings. *Applied Geochemistry,* Vol.23, pp. 1152–1165.

Masscheleyn, P.H.; Delaune, R.D. & Patrick, W.H. (1991). Effect of redox potential and pH on arsenic speciation and solubility in a contaminated soil. *Environmetal Science & Technology,* Vol.25, pp. 1414-1419.

Mladenov, N.; Zheng, Y.; Miller, M.P.; Nemergut, D.R.; Legg, T.; Simone, B.; Hageman, C.; Rahman, M.M.; Ahmed, K.M. & Mcknight, D.M. (2010). Dissolved organic matter sources and consequences for iron and arsenic mobilization in Bangladesh aquifers. *Environmental Science and Technology,* Vol.44, No.1, pp. 123-128.

Mudarra, J. L. (1988). Study of soils in the Aljarafe region. Regional Education and Science Department of the Andalusia Government, Seville, Spain.

N.R.C.C. (1978). Effects of arsenic in the Canadian environment. National Research Council of Canada, no. 15391, Ottawa, Canada.

Nordstrom, D.K. (1982). Aqueous pyrite oxidation and the consequent formation of secondary iron minerals. In: *Acid sulfate weathering,* J. A. Kitrick, D. S. Fanning, & L. R. Hossner (Eds.), Soil Science Society of America, 37–56, Madison, USA.

O'Neill, P. (1995). Arsenic. In: *Heavy metals in soils* (2nd ed.) B.J. Alloway (Ed.), 105–121. Glasgow, UK.

Peñalosa, J.M.; Carpena, R.O.; Vázquez, S.; Agha, R.; Granado, A.; Sarro, M.J. & Esteban, E. (2007). Chelate-assisted phytoextraction of heavy metals in a soil contaminated with a pyritic sludge. *The Science of the Total Environment,* Vol.378, pp. 199-204.

Pérez-de-Mora, A.; Madejón, E.; Burgos, P. & Cabrera, F. (2006). Trace elements availability and plant growth in a mine-spill-contaminated soil under assisted natural remediation I. Soils. *The Science of the Total Environment,* Vol.363, pp. 28–37.

Pierce, N.L. & Moore, C.B. (1980). Adsorption of arsenite and arsenate on amorphous iron hydroxide from dilute aqueous solutions. *Environmental Science & Technology,* Vol.14, pp. 214–216.

Quevauviller, Ph.; Lachica, M.; Barahona, E.; Gómez, A.; Rauret, G.; Ure, A. & Muntau, H. (1998). Certified reference material for the quality control of EDTA and DTPA extractable trace metal contents in calcareous soils (CRM 6000). *Presenius J. Anal. Chem.,* Vol.360, pp. 505–511.

Redman, A.D.; Macalady, D.L. & Ahmann, D. (2002). Natural organic matter affects arsenic speciation and sorption onto hematite. *Environmental Science & Technology*, Vol.36, pp. 2889-2896.

Sánchez España, J.; López Pamo, E.; Santofimia, E.; Aduvire, O.; Reyes Andrés, J. & Barettino, D. (2005). Acid mine drainage in the Iberian Pyrite Belt (Odiel river watershed, Huelva, SW Spain): Geochemistry, mineralogy and environmental implications. *Appl. Geochem.*, Vol.20, pp 1320-1356.

Schwertmann, U. & Taylor, R.M. (1977). Iron oxides. In: *Minerals in soil environments*, J. B. Dixon & S. B. Webb (Eds.), Soil Science Society of America, 148-180. Madison, USA.

Simón, M.; Díez, M.; García, I. & Martín, F. (2009). Distribution of As and Zn in Soils Affected by the Spill of a Pyrite Mine and Effectiveness of the Remediation Measures. *Water, Air, and Soil Pollution*, Vol.198, pp. 77-85.

Simón, M.; Dorronsoro, C.; Ortiz, I.; Martín, F. & Aguilar, J. (2002). Pollution of carbonate soils in a Mediterranean climate due to a mailing spill. *Eur. J. Soil Sci.*, Vol.53, pp. 321-330.

Simón, M.; García, I.; Martín, F.; Díez, M.; del Moral, F. & Sánchez, J.A. (2008). Remediation measures and displacement of pollutants in soils affected by the spill of a pyrite mine. *The Science of the Total Environment*, Vol.407, pp. 23-39.

Simón, M.; Martín, F.; García, I.; Dorronsoro, C. & Aguilar, J. (2007). Steps in the Pollution of Soils after a Pyrite Mine Accident in a Mediterranean Environment, In: *Environmental Pollution: New Research*, Nova Publishers, (Ed.), 185-224, ISBN 1-60021-285-9, New York, USA.

Simón, M.; Martín, F.; Ortiz, I.; García, I.; Fernández, J.; Fernández, E.; Dorronsoro, C. & Aguilar, J. (2001). Soil pollution by oxidation of tailings from toxic spill of a pyrite mine. *The Science of the Total Environment*, Vol.279, pp. 63-74.

Simón, M.; Ortiz, I.; García, I.; Fernández, E.; Fernández, J.; Dorronsoro, C. & Aguilar, J. (1999). Pollution of soils by the toxic spill of a pyrite mine (Aznalcóllar, Spain). *The Science of the Total Environment*, Vol.242, pp. 105-115.

Simón, M.; Ortiz, I.; García, I.; Fernández, E.; Fernández, J.; Dorronsoro, C. & Aguilar, J. (1998). El desastre Ecológico de Doñana. *Edafología*, Vol.5, pp. 153-161.

Smith, I. C.; Naidu, R. & Alston, A. M. (1998). Arsenic in the soil environment. A review. *Advances in Agronomy*, Vol.64, pp. 150-195.

Soil Survey Staff. (1951). *Soil survey manual. Handbook 18*. USDA, Washington DC, USA.

Soil Survey Staff. (2003). *Keys to soil taxonomy (9th ed.)*, Blacksburg, Virginia, USA.

Stumm, W. & Morgan, J.J. (1981). *Aquatic chemistry: An introduction emphasizing chemical equilibria in natural waters*. John Wiley and Sons, New York, USA.

Sun, X. & Doner, H. E. (1996). An investigation of arsenate and arsenite bonding structures on goethite by FTIR. *Soil Science*, Vol.161, pp. 865-872.

US EPA (US Environmental Protection Agency). 1996. Ecological effects test guidelines. Seed germination/root elongation toxicity test. OPPTS 850.4200.

Waychunas, G. A.; Rea, B. A.; Fuller, C. C. & Davis, J.A. (1993).Surface chemistry of ferrihydrite: Part 1. EXAFS studies of the geometry of coprecipitated and absorbed arsenate. *Geochimica et Cosmochimica Acta*, Vol.57, pp. 2251-2269.

Young, R.N. & Mulligan, C.N. (2004). Natural attenuation of contaminants in soils. CRC Press, Boca Raton, Florida, USA.

13

Herbicide Off-Site Transport

Timothy J. Gish[1], John H. Prueger[2], William P. Kustas[1],
Jerry L. Hatfield[2], Lynn G. McKee[1] and Andrew Russ[1]
[1]*USDA-ARS, Hydrology and Remote Sensing Laboratory, Beltsville, Maryland,*
[2]*USDA-ARS, National Laboratory for Agriculture and the Environment, Ames, Iowa*
USA

1. Introduction

Herbicides are an important part of modern agriculture as they control weeds that would otherwise reduce yields by competing for water and nutrients. The U.S. Environmental Protection Agency estimated that 226,000 metric tons of herbicides were used in the U.S, alone during 2007, which accounts for 25% of the globally usage in 2007 (USEPA, 2011). Additionally, during 2007, 89% of the herbicide usage in the U.S. was for agriculture (USEPA, 2011). Since it has been proposed that increasing agricultural food and fiber production will be necessary to maintain political and social stability in developing countries (Tilman et al., 2002), herbicide use will become increasingly important to meet these global needs, especially as marginal lands are converted to agriculture (Helling, 1993). Although critical to production, herbicides can be toxic to humans and other organisms, even at low concentrations (Jin-Clark et al., 2002; USEPA, 2008). To maintain productive and sustainable agricultural systems there is an immediate need to understand field-scale processes governing herbicide use and off-site transport.

During the past three decades several national surveys in the U.S. have shed light on the prevalence of herbicides in the environment. One of the first national surveys was conducted by the U.S. Environmental Protection Agency (USEPA, 1990) which determined that about 10% of community water system wells contained detectable amounts of at least one herbicide. From 1993 to 1995, the National Water-Quality Assessment program monitored 20 major basins in the U.S. and found herbicides in over 50% of the sites sampled (Koplin et al., 1998). Furthermore, the U.S. Geological Survey observed that 97 percent of all streams sampled from agricultural and urban areas contain detectable concentrations of at least one herbicide, while 65 percent of the streams in undeveloped areas contained observable levels of herbicide (Gillion et al., 2006). Clearly, herbicide occurrence in streams, groundwater aquifers, and community wells are well documented, but determining the relative importance of major off-site transports processes at the field and watershed scales is still in its infancy.

As summarized in Figure 1, herbicide off-site transport occurs primarily through surface runoff (Wauchope, 1978; Shipitalo and Owens, 2006), groundwater leaching (Isensee et al., 1990; Gaynor et al., 2001; Hansen et al., 2001), and/or volatilization (Taylor and Spencer, 1990; Gish et al., 2011). Precipitation events are especially crucial for determining which loss pathways are most critical in governing herbicide off-site transport. For example, if a precipitation event does not occur within a few weeks of application runoff and groundwater leaching losses will be negligible – however, herbicide volatilization can still be substantial (Prueger et al., 2005; Gish et al., 2011). To a large extent, how a given herbicide is partitioned between runoff, leaching and volatilization is a function of how the herbicide is distributed between three phases: 1) herbicide adsorbed to soil particles; 2) herbicide in the soil solution (liquid phase); and 3) the mass of the herbicide in the vapor phase (in soil pores and above the soil surface). Herbicides can also move from one phase to another, depending on a number of chemical and environmental factors. For example, although a particular herbicide may have a high affinity for the soil matrix, increases in soil water content move more herbicide from the adsorbed and liquid phases into the vapor phase (Prueger et al., 2005; Gish et al., 2009). Additionally, in a 2-year study (Weber et al., 2006) observed as much as 21 % of the applied metolachlor leached from field lysimeters when 316 mm of precipitation occurred during the first month after application, relative to only 2.8 % of the applied metolachlor when only 106 mm of precipitation occurred during the first month after application. Relative to a dry field conditions, a rainfall event shortly after application can enhance herbicide surface runoff (Pantone et al., 1992; Zhang et al., 1997), leaching (Gaynor et al., 2002; Weber et al., 2006), and volatilization (Prueger et al., 2005; Gish et al., 2011). As a result, if future off-site transport of herbicides is to be accurately quantified, field investigations where metrological conditions are also monitored will be essential.

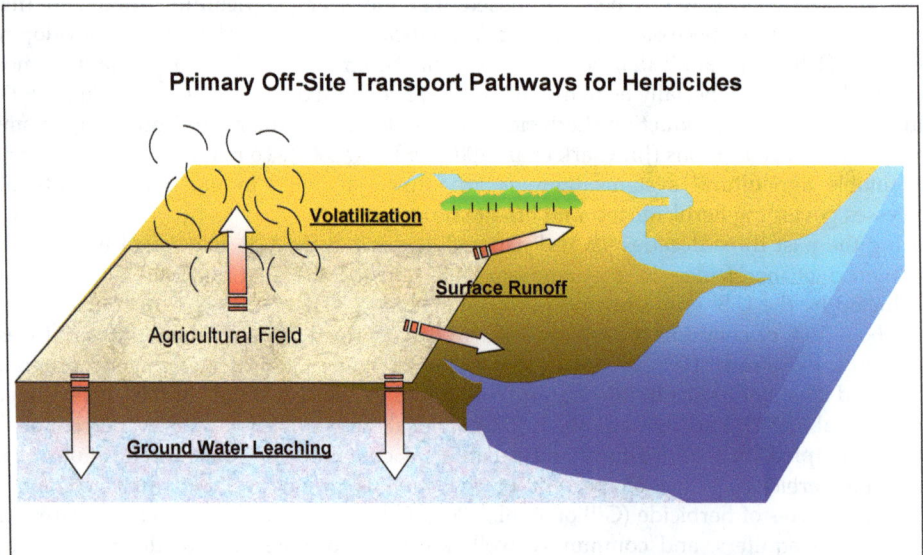

Fig. 1. Schematic of loss pathways critical to off-site transport of herbicides

As summarized in Figure 2, methods for evaluating off-site herbicide transport are expensive and difficult to interpret due to process interactions that vary both spatially and temporally and are a function of scale. The rate at which herbicides are lost from the three major loss pathways is influenced by a number of small scale factors which include soil water content, organic matter content, soil hydraulic properties, as well as larger scale influences such as wind speed profiles, agricultural management practices, timing of rainfall events relative to application, and field slope (Wauchope, 1978; Mojašević et al., 1996; Gaynor et al., 2001; Shipitalo and Owens, 2006). Thus, although the impact of various soil and environmental factors on herbicide behavior have been quantified in controlled laboratory and greenhouse environments, monitoring and interpreting field- scale herbicide behavior is more ambiguous (Helling and Gish, 1986). Furthermore, in addition to the three major loss pathways, herbicide emissions can occur as spray drift in concentrated droplets or as herbicide attached to dust particles (Symons, 1977; Majewski and Capel, 1995) or from herbicides deposited directly into streams via tree wash-off (personal communication, Dr. Clifford Rice, USDA-ARS Beltsville, MD). To reduce risks associated with herbicide use, the three major loss pathways (runoff, leaching and volatilization) must be simultaneously evaluated to avoid developing herbicide formulations or practices that simply shift herbicides loss from one off-site transport pathway to another.

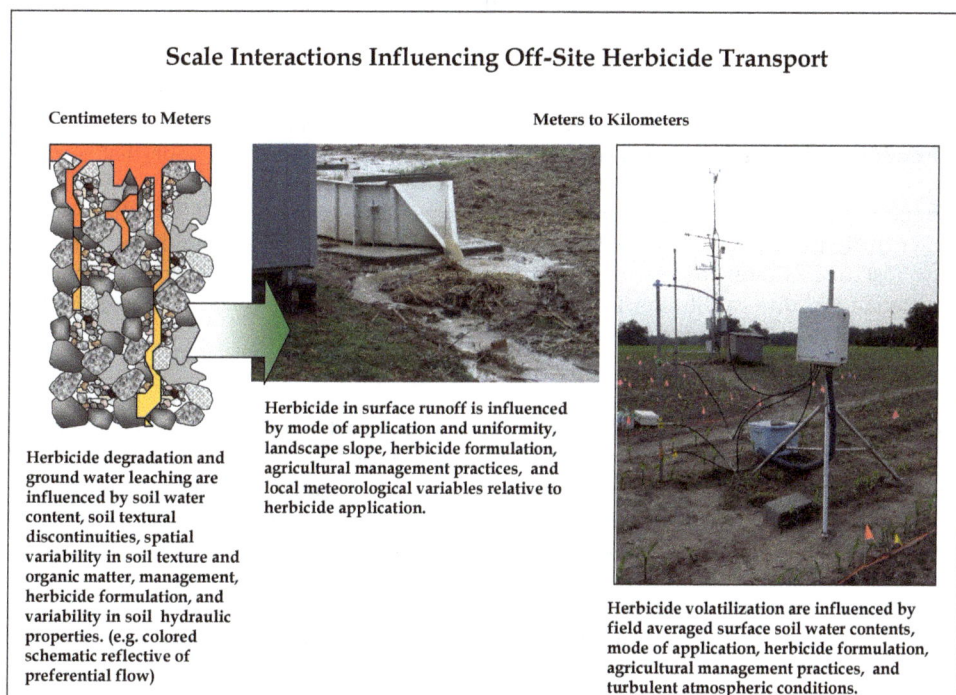

Scale Interactions Influencing Off-Site Herbicide Transport

Centimeters to Meters

Meters to Kilometers

Herbicide degradation and ground water leaching are influenced by soil water content, soil textural discontinuities, spatial variability in soil texture and organic matter, management, herbicide formulation, and variability in soil hydraulic properties. (e.g. colored schematic reflective of preferential flow)

Herbicide in surface runoff is influenced by mode of application and uniformity, landscape slope, herbicide formulation, agricultural management practices, and local meteorological variables relative to herbicide application.

Herbicide volatilization are influenced by field averaged surface soil water contents, mode of application, herbicide formulation, agricultural management practices, and turbulent atmospheric conditions.

Fig. 2. Scale of properties and processes that interact to influence herbicide off-site transport

This chapter will discuss the relative importance of herbicide loss through surface runoff, groundwater leaching, and volatilization including the impact of soil properties, agricultural management, and meteorological conditions. Since local climatic and surface soil conditions influence herbicide behavior, emphasis will be given to field-scale, long-term investigations. Additionally, methods for reducing herbicide off-site transport will be briefly discussed for each loss pathway.

2. Surface runoff

The highest concentrations of herbicides in surface streams are typically associated with this first significant runoff event after application to agricultural fields (Thurman et al., 1991; Battaglin and Goolsby, 1999; Gentry et al., 2000; Scribner et al., 2000). As a result, herbicide surface runoff is a concern in many watersheds where intensive agriculture may be adjacent to sensitive ecosystems (Capel et al., 2008). Herbicide in surface runoff occurs through two mechanisms: 1) erosion of herbicide adsorbed to soil sediment; and 2) dissolution of the herbicide into the surface runoff water. Although herbicide concentrations in runoff sediment can be several times higher than those observed in the water phase, most of the herbicide lost in runoff is from the water phase since runoff water volumes are typically much greater than sediment losses (Wauchope, 1978). Within a single runoff event the bulk of the herbicide loss occurs early in the event and decreases exponentially with time (Buttle, 1990; Reddy et al., 1994; Shipital and Owens, 2006). Seasonal runoff losses are predominately an accumulation of single-event losses with minor losses in-between major storm or irrigation events (Haith and Ross, 2003). Although rainfall timing, intensity, and duration are the most critical factors governing herbicide runoff, the rate of application, formulation, management practice, and landscape features are also important (Caro, 1976; Baker and Johnson, 1979; Wauchope, 1978; Hall et al., 1983; Felsot, 1990; Domagalski et al., 2008). Typically, annual losses from a single rainfall event are small, < 1 % of that applied (Shipitalo and Owens, 2006; Gish et al., 2011). However, in situations, such as when a major rainfall event follows herbicide applications, herbicide losses can exceed 2 % of that applied (Baker, 1980; Haith and Rossi, 2003, Shipital and Owens, 2006). Regardless of the herbicide mass lost from runoff, detrimental impacts decrease with increasing distance from the application site due to dilution from other runoff sites, streams, rivers, and lakes (Baker, 1980; Capel et al., 2008; Domagalski et al., 2008).

2.1 Rainfall impact

The primary factors governing herbicide off-site transport via runoff are the intensity, duration, and timing of the rainfall events relative to application (Baker and Johnson, 1979; Baker et al., 1978). Figure 3 depicts typical herbicide surface runoff concentrations with time from a 7 ha research site in Beltsville, Maryland over 8 years (for general site description see Chinkuyu et al., 2004). The exponential decrease in runoff and lack of herbicides in runoff after 20 days may be due to the sandy textured soil which dominates this site. However, similar trends were also observed by Pantone et al, (1992) who observed higher herbicide runoff losses the first day after application than 30 days later. Additionally, Shipitalo and Owens (2006) in a 9 year study over several small watersheds demonstrated that herbicide runoff concentrations also decreased exponentially with time after application. To account for the interaction between rainfall intensity, duration, and timing, classifying three types of runoff events have been proposed: minor, critical, and catastrophic (Wauchope, 1978). Minor runoff events are a product of rain events which produce small amounts of runoff shortly after herbicide application, generally within 1-2 days. These minor events typically have high concentrations of herbicide in a relatively small amount of surface runoff and account for herbicide losses < 1 % of that applied. However, the high concentration of herbicide in these minor runoff events may affect sensitive ecosystems adjacent to agricultural land. The second type, a critical runoff event occurs within two weeks of herbicide application and has about 50 % of the rainfall exiting the field through surface runoff. The amount of herbicide lost in a critical event is also dependent on herbicide soil persistence, adsorption affinity, and landscape features (Shipitalo et al., 2000; Ma et al., 2004). Herbicide runoff losses can be significant with a critical runoff event even if the herbicide has a high affinity for the soil matrix, since sediment loss is common with such events. As a result, critical runoff events typically produce the bulk of the herbicide runoff loss from agricultural fields and account for 1-4 % of the applied herbicide annually (Ma et al., 2004). The third type, catastrophic runoff events are rare and differ from a critical runoff event by the high intensity of the rainfall occurring shortly after herbicide application (Wauchope, 1978; Schulz et al., 1998; Shipitalo and Owens, 2006). For example, catastrophic runoff events are typically caused by severe thunderstorms that produce large amounts of rainfall within three days of the herbicide application and can account for > 4% of the herbicide applied. Although a higher percentage of the applied herbicide is lost in a catastrophic runoff event, the concentration of herbicide in the water phase removed from the field is relatively low due to dilution. Occurrences of catastrophic runoff events are rare, since timing of a major storm and application of herbicide must coincide and farmers typically attempt to avoid extreme weather situations.

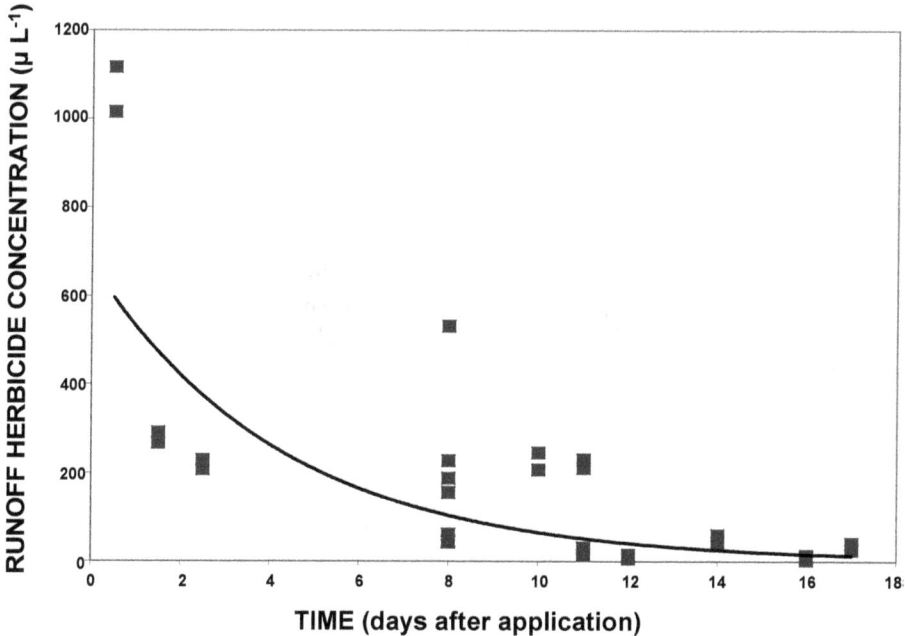

Fig. 3. Typical herbicide concentrations in surface runoff as a function of time. Data represents metolachlor and atrazine runoff over 8 years from a research site in Beltsville, Maryland (USA)

2.2 Management strategies
Since herbicide runoff is primarily a function of rainfall, best management practices for reducing herbicide runoff losses must consider reducing herbicide concentrations in both

the water phase and the soil sediment. Erosion control practices such as minimum tillage and grass buffers may be effective in reducing erosion and runoff water which greatly reduce pesticide loss via sediment erosion, while less ineffective in controlling herbicide losses in the water phase (Baker et al., 1978; Baker and Johnson, 1979; Cole et al., 1997; Shipitalo et al., 2000; Caron et al., 2010). If soils are frequently wet, installing tile-drains can reduce herbicide runoff. Southwick et al. (1990) reported that installing tile drains resulted in significantly reduced herbicide runoff due to a reduction in surface runoff volumes.

Application methods such as soil incorporation can reduce herbicide runoff, leaching and volatilization losses even though persistence may increase. Soil incorporation can reduce herbicide runoff losses from 1/4 to 1/20 of the surface applied herbicide loss (Baker and Laflen, 1979; Hansen et al 2001). Foliar applications of herbicide are to be avoided, if possible, since they tend to be easily washed off and transported off-site in runoff water before being adsorbed to the soil (Wauchope, 1978; Gevao et al., 2000).

An important management factor which influences herbicide runoff losses is the formulation used. Some herbicides are applied as a wettable powder and have high runoff potentials, resulting in annual runoff losses generally ranging from 2 – 5% of that applied (Wauchope, 1978). Since wettable powder formulations subjected to critical and catastrophic runoff events can result in herbicide runoff losses exceeding 5% of that applied they should be avoided if at all possible. Herbicides applied as an emulsion have the next highest potential for loss in runoff, and losses are typically about 1% of that applied. Water soluble herbicides have a much greater potential of being lost in water runoff, whereas, non-water soluble herbicides often have an affinity for soil particles and will most likely be lost in the sediment portion of runoff. An example of formulation influences on pesticide runoff is observed with ester and amine salt based herbicides (Barnett et al., 1967). The amine salt, which is water soluble, rapidly dissolves in water and can be leached into the soil or moved by the water phase of a runoff event. Ester herbicides are relatively insoluble, but are readily adsorbed to soil particles and are primarily lost through the erosion of sediment. For all other herbicide formulations (e.g., pelleted and micro-encapsulated) annual runoff losses are typically < 0.5% of that applied, except when a critical or catastrophic event occurs (Wauchope, 1978; Shipitalo and Owens, 2006).

2.3 Landscape and soil properties

Landscape attributes and soil properties interact with the type of runoff event, herbicide chemistry, and formulation to influence runoff losses. All other conditions equal, herbicide runoff losses increase with increasing slope (Hall et al., 1983; Felsot et al., 1990; Celis et al., 2007). For example, herbicide runoff from a 3% slope can be as high as 2% of that applied, while slopes of 10-15% may result in herbicide runoff losses exceeding 5% of that applied (Wauchope, 1978). Soil properties which influence herbicide runoff losses include soil organic matter content (Jenks et al., 1998; Spark and Swift, 2002), pH (Weber et al., 1972; Jenks et al., 1998), soil compaction (Baker and Laflen, 1979), soil moisture content (Spark and Swift, 2002), cation exchange capacity (Wauchope and Meyers, 1985), and clay mineral content (Baskaran et al., 1996). In general, soil properties influence herbicide runoff by affecting adsorption and desorption processes. High soil organic matter contents (> 5 %) will typically be the most important factor influencing herbicide absorption (Sparks and Swift, 2002). In soils with a low soil organic matter content (< 2 %) clay mineral content may be the dominate factor, because of the larger surface area of clay particles (Laird et al. 1992; Jenks et al., 1998). Soil pH has been shown to influence herbicide adsorption by altering the chemical composition of the herbicide resulting in a net positive charge (Jenks et al., 1998). For example, atrazine is more adsorptive in acidic soils because they react with H^+, making the herbicide cationic and chemically attracted to cation exchange sites (Bailey and White, 1964; Weber et al., 1972; Jenks et al., 1998). Although herbicides favoring adsorption are less susceptible to runoff from a minor runoff event, they are increasingly susceptible to critical or catastrophic runoff events where sediment erosion can be significant.

Soil moisture influences adsorption and desorption of herbicides due to water competition for adsorption sites on soil particles (Hamaker and Johnson, 1972; Cole et al., 1997). As soil moisture increases (through a rain or irrigation event) water is adsorbed to the soil matrix and herbicides desorb. Subsequently, the desorbed herbicide diffuses into the water phase where it can be more readily transported off-site. Studies examining the application of herbicides in wet and dry soils showed that runoff losses were significantly greater for the wet soil, because of the lower adsorption potential (Barnet et al., 1967; Baldwin et al., 1975; Asmussen et al., 1977).

Herbicides which persist in the soil for long periods of time pose an environmental threat to neighboring ecosystems simply because the window for a significant runoff event is larger. Herbicide persistence is typically quantified as a half-life ($T_{1/2}$) which represents the time taken for half of the herbicide to degrade. Major factors influencing herbicide soil persistence include application method (Hall et al., 1983), herbicide chemistry and soil affinity (Jenks et al., 1998; Spark and Swift, 2002), leaching potential (Webb et al., 2008), soil water content (Mojašević et al., 1996), formulation and volatilization potential (Gish et al., 1994); and degradation processes (Gan et al., 2005). Because properties like soil moisture and organic matter content influence persistence, herbicides half-lives typically exhibit considerable variability in the field (Mojašević et al., 1996; Sparks and Swift, 2002). Where herbicide persistence is high ($T_{1/2} > 2$ weeks) concentrations in the second and third runoff events may actually be higher than in the first runoff event (Gan et al., 2005). Furthermore, multiple applications of a particular herbicide may result in metabolic pathways being established which can increase biological degradation rates and reduce soil persistence (Kearney et al., 1969).

2.4 Strategies for reducing herbicide runoff
Practices for reducing herbicide losses in runoff include: 1) avoid pesticide application during adverse weather conditions such as when rain or high winds are anticipated within 48 hours of application; 2) use erosion control practices such as conservation tillage, contouring, and grass buffers around waterways to reduce runoff ; 3) determine appropriate herbicide type, rate, and persistence for weather and soil conditions; 4) incorporate the pesticide if possible; and 5) avoid wettable powder formulations.

3. Groundwater leaching

Herbicide leaching from agricultural land has long been considered a potentially serious problem. Accordingly, criteria have been introduced to regulate standards for potable water (USEPA, 2008). As with surface runoff, there is increasing evidence that the specific environmental and agricultural field conditions prevailing during herbicide application or shortly thereafter is critical in determining the extent of herbicide leaching. Additionally, since water is the driving force for herbicide leaching, there is an inextricable link between herbicide leaching and surface and subsurface hydrology. Unfortunately, most herbicide leaching studies have focused on concentrations in soil cores or soil solution as a function of depth and time– instead of monitoring herbicide fluxes. As a result, the lack of herbicide flux data through soil has fostered debates regarding the relevance of various flow processes on herbicide transport to groundwater. Although monitoring herbicide fluxes to groundwater is still in its infancy, tile-drain studies indicate that herbicide leaching is typically less than 2 % of that applied and so are less than runoff losses (Kladivko et al., 1991, Gaynor et al., 2002).

Monitoring herbicide transport through soil to groundwater is complex and early attempts that focused on analyzing herbicides in soil cores as a function of depth had limited success. For example, Wyman et al. (1985 and 1986) evaluated aldicarb transport through soil by collecting 48 soil cores, each 3.6 m long, four times during a growing season. They observed no aldicarb below 2.4 m over a three year period and so concluded that aldicarb had degraded and did not pose a threat to groundwater quality. In contrast, Brasino (1986) conducted a coincident experiment which found that peak aldicarb concentrations in three groundwater monitoring wells (6 m depth) ranged from 27 ppb to 76 ppb even though the same chemical application rate, irrigation scheme, and soil were used. Instead of analyzing soil cores Kladivko et al. (1991) and Gentry et al. (2002) monitored tile drains, which allowed herbicide fluxes to be calculated. Unfortunately, interpreting field-scale solute travel times from a tile drain involves considerable uncertainty since solutes applied immediately above a tile drain have a much shorter distance to travel (to exit the field) than solutes applied between the tile drains (Jury et al ., 1975a, 1975b).

The two major processes governing herbicide transport through soil to groundwater are matrix and preferential flow. In matrix flow, the movement of the herbicide is governed by adsorption to soil particles, hydrodynamic dispersion, and convection processes (Jury et al., 1984; Helling and Gish, 1986). Hydrodynamic dispersion accounts for both molecular diffusion and a spreading out of the herbicide as the moving soil solution interacts with soil pores of various sizes. Convection is the bulk transport of the herbicide with the moving soil solution. Typically, matrix flow processes are well described by the classical convective-dispersion equation, which is the back bone of many transport models (Pachepsky et al., 2006). On the other hand, preferential flow is poorly described by the classical convective-dispersion equations, since it assumes that a small fraction of the total soil pore space is responsible for rapidly conducting solutes and herbicides to groundwater (German and Beven, 1981; Gish and Jury, 1983; Isensee et al., 1990; Jarvis, 2002; Gish et al., 2004). Preferential flow typically occurs through structureless soils by means of flow instabilities (Glass et al., 1989; Ghodrati and Jury, 1992); flow through spatial voids resulting from decayed roots, shrinking clay minerals, sink holes, or created by soil fauna (Gish et al., 1983; Libra et al., 1984; Gish et al., 1998; Shipitalo et al., 1990; Ritsema and Dekker, 1995; Williams et al., 2000); and/or flow along subsurface restricting layers (Kung, 1990). Unfortunately, quantifying preferential flow at the field-scale is extremely difficult since there is no way of knowing where these flow pathways are or when samples should be collected, so a mass flux is typically calculated, not monitored. Without a mass flux procedure, it is nearly impossible to quantify the *relevance* of preferential flow at the field scale for various management, soil and climatic scenarios. However, due to the rapid transport of herbicides to groundwater, it appears that preferential and not matrix flow is likely the dominate flow mechanism governing herbicides leaching (Kladivko et al, 1991; Flury, 1996; Harris and Catt, 1999; Novak et al., 2001; Koplin et al., 1998; Jarvis, 2002; Vereecken, 2005).

While preferential flow has been observed on all soils regardless of texture, the impact of preferential flow as a function of pesticide chemistry and soil texture has not yet been fully quantified. Initially, preferential leaching of herbicides was thought to occur on heavy or clayey textured soils utilizing macropores or other spatial voids (Harris and Catt, 1999 Johnson et al., 1996). However, significant herbicide leaching has also been detected through loamy and silty textured soils (Kladivko et al., 1991; Brown et al., 1995; Zehe and Fluhler, 2001) as well as sandy soils (Ghodrati and Jury, 1992). On the other hand, there is some evidence that *total herbicide mass losses* in a clayey structured soil may be greater than from a sandy soil (Traub-Eberhand et al., 1995).

3.1 Rainfall impacts

Herbicide leaching is a function of local meteorology, management practice, herbicide formulation, soil type, soil hydraulic properties, and several environmental factors (Jury, 1986; Helling and Gish, 1986). Like surface runoff, the timing of a rainfall event relative to application greatly influences herbicide leaching to groundwater. Specifically, herbicide transport to groundwater is especially vulnerable to preferential transport shortly after application (Gaynor et al., 2002; Weber et al., 2006). However, variability in groundwater herbicide concentrations can be considerable as spatial variability of soil properties can influence herbicide transit times. As depicted in Figure 4a, surface applied atrazine and metolachlor were rarely detected in one observation well (3 m depth) at a sandy field site located in Beltsville, Maryland over an eight year period. However, in the same field, during the same sampling times herbicide concentrations were consistently larger in another 3 m observation well where preferential flow was more dominant (depicted in Figure 4 B). If rainfall occurs within a few days of application, herbicide well concentrations will typically display peaked concentrations shortly after application then decrease with time, until fall/winter recharge. When little or no rainfall occurred after application, herbicide groundwater concentrations were low throughout the year. Additionally, the low herbicide concentrations after a few months supports the hypothesis that once solutes or herbicides reside in the smaller pores of the soil matrix it will move predominantly by matrix rather than preferential flow (Kung et al., 2000b; Delphin and Chapot, 2006). Consequently, the first rainfall event after application has the highest risk of herbicide leaching to groundwater, but to account for low concentrations and fall/winter recharge new modeling approaches must include interaction between matrix and preferential flow processes.

3.2 Management impacts

With regard to agricultural management, there is some ambiguity regarding their benefits. For example, Gish et al. (1991a) reported that soil incorporated carbofuran leached less than surface broadcast applied atrazine despite the much larger inherent mobility of carbofuran. However, Jones et al. (1995) suggested that soil incorporation of herbicides after application had no impact on herbicide transport to tile-drains. In addition, conservation tillage practices may temporarily enhance preferential herbicide transport through void root channels and bio-pores. However, in time these same root channels and bio-pores will have clay and/or organic coatings where herbicides can readily adsorbed and subsequently broken down by microbial degradation. After only a few years Gish et al. (1998) observed higher herbicide metabolites under no-till relative to conventional tillage. This suggests that if the herbicide metabolites are less harmful than the parent compound, conservation tillage may be beneficial to groundwater quality.

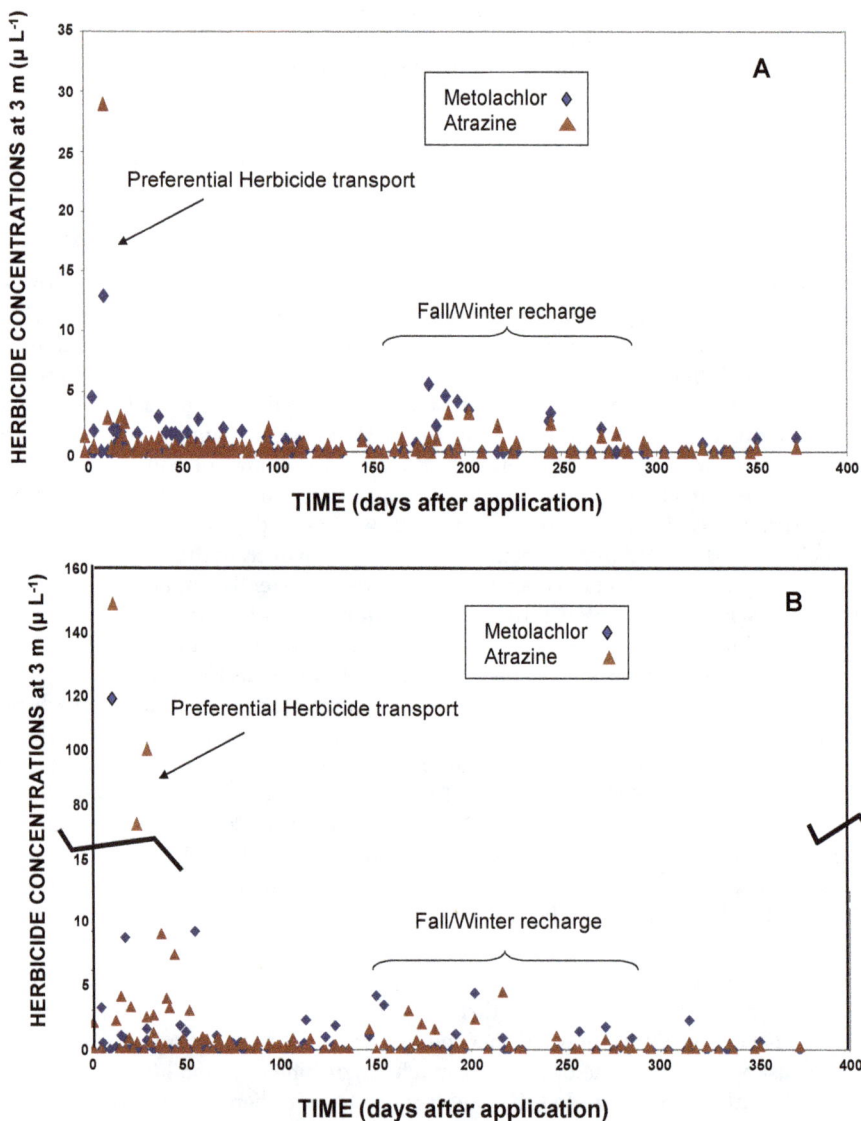

Fig. 4. Variability of atrazine and metolachlor well concentrations over eight years at a depth of 3 m. Figure 4A is representative of well locations where preferential has minimal impact, while Figure 4B denotes areas where preferential flow is more dominant. Notice scale change in Figure B

Controlled release formulations may reduce preferential transport of herbicides. In the laboratory and in field plots atrazine leachate concentrations were reduced by as much as 80% with starch encapsulation (Gish et al., 1991b; Schreiber et al., 1993). Surprisingly, Brown et al., (1995) reported unexpectedly high leaching losses with encapsulation. Later, field evaluations showed the starch encapsulated herbicides were more persistent in soil, which was attributed to being less available for leaching during rain events (Gish et al., 1994). Naturally, herbicide chemistry must be considered when optimizing an encapsulation matrix in order to be effective in controlling the targeted pest, as well as reducing environmental risks (Wienhold and Gish, 1994b).

3.3 Management strategies to reduce herbicide leaching

Herbicide transport through soil is primarily a function of preferential flow, so for herbicides with a low solubility encapsulation will reduce leaching by increasing diffusion into the smaller pores of the soil matrix. Reducing herbicide leaching through encapsulation may be countered by an enhanced runoff potential since encapsulation also increases soil persistence. Soil incorporation may also reduce leachability for some herbicides that are insoluble and have low soil persistence. Farm managers should avoid applying herbicide if rainfall is anticipated within 48 hours of application and should avoid using herbicides with a long half-life as this could increase susceptibility to groundwater leaching.

4. Herbicide volatilization

Volatilization is perhaps the principal loss pathway by which herbicides are transported off-site. Although herbicide volatilization can exceed 90% of that applied, typical losses for many herbicides range from 5 to 25% of that applied (Taylor and Spencer, 1990; Prueger et al., 1999; Glotfelty et al., 1989; Rice et al., 2002; Prueger, et al., 2005). The impact of herbicide volatilization generally decreases with time after application, but unlike surface runoff and groundwater leaching, volatilization occurs regardless of soil type, landscape features, or local meteorology – although these factors influence *how much* herbicide is volatilized. In an eight year field investigation in Beltsville, Maryland, volatilization losses of atrazine and metolachlor always exceeded runoff losses (details of the site set-up and for this multi-year investigation depicted in Figure 5 are given in Gish et al., 2011). Furthermore, once in the atmosphere, herbicides can be degraded or deposited in non-targeted areas via wet or dry deposition (Bidleman and Christensen, 1979; Bidleman, 1988; Burrows et al., 2002). Frequently, a portion of the applied herbicide volatilized into the atmosphere is transported and subsequently deposited in streams, rivers, and lakes (McConnell et al., 1998; Alegria and Shaw, 1999; Thurman and Cromwell, 2000; Kuang et al., 2003).

As depicted in Figure 6, herbicide volatilization occurs in two steps, evaporation of the herbicide from soil and/or plant residues followed by dispersion into the atmosphere by diffusion and turbulent mixing (Taylor, 1995; Prueger et al., 2005). Several methods have been developed to obtain estimates of herbicide volatilization at the field-scale. Parmele et al. (1972) developed an aerodynamic method based on gradients of wind speed, temperature and herbicide concentrations collected over a uniform area. Demmead et al. (1977) developed an integrated horizontal flux approach which uses herbicide concentration and horizontal wind speeds profiles. For certain conditions a theoretical profile shape method (Wilson et al., 1982) may be useful which measures wind speed and herbicide

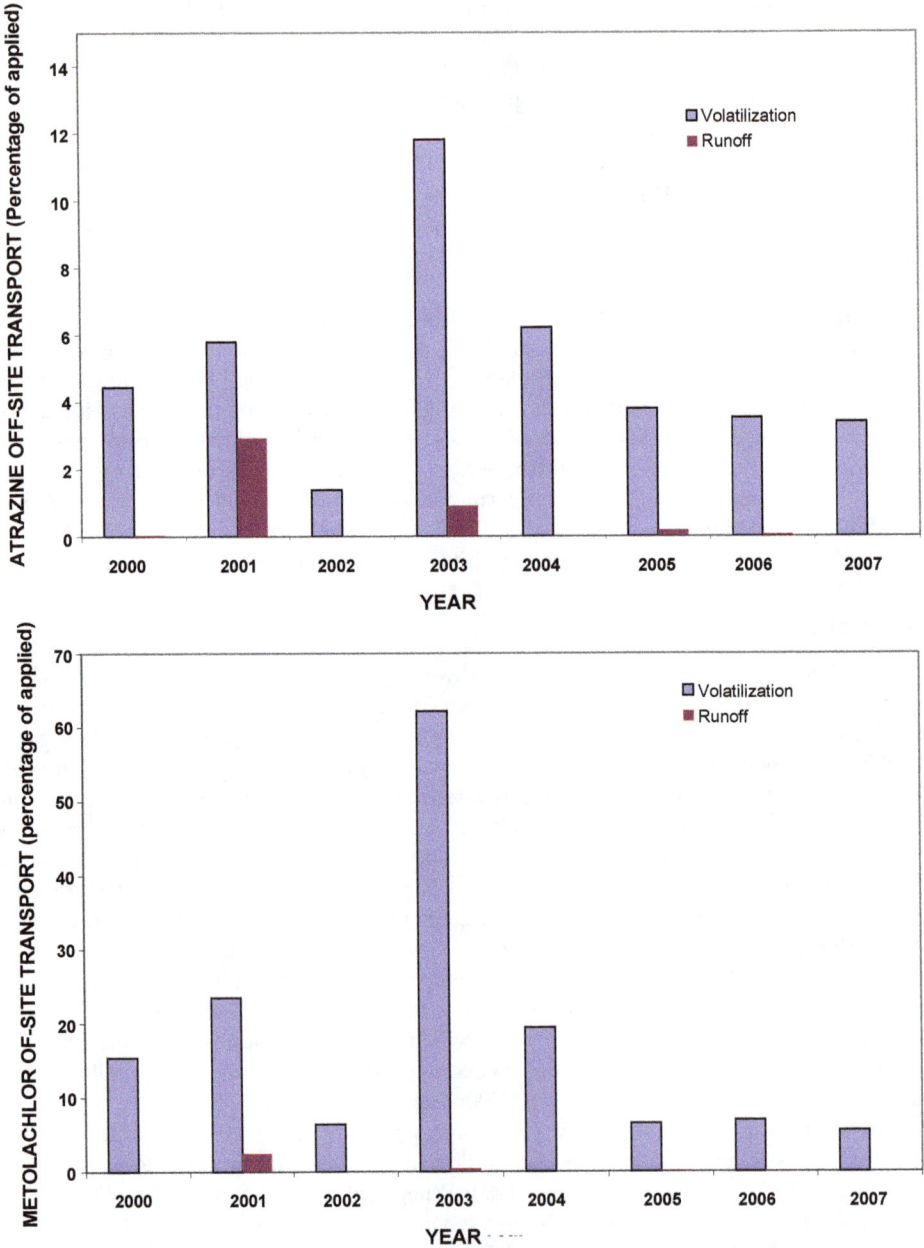

Fig. 5. Comparison of annual volatilization and runoff losses of atrazine and metolachlor on
the same site over eight years. Runoff was monitored from early May (before planting)
through November (well after harvest) while volatilization losses were monitored for only
the first 5 days after application

concentrations at a single height above the soil to calculate herbicide vapor losses. Recently, eddy covariance measurements of wind, temperature, and water vapor have been linked to and herbicide concentration profiles above the soil surface to quantify herbicide volatilization fluxes where turbulent flow conditions in the field may exist (Prueger et al., 2005, Gish et al., 2009).

Since herbicides are generally applied around crop planting, mass applications of herbicide within a given agricultural region may contribute to a large nonpoint pulses of herbicide entering the atmosphere. A regional signaling of herbicide volatilization, along with wet and dry deposition has been observed by Kuang et al. (2003) in the Chesapeake Bay, by Goolsby et al. (1997) and Majewski et al. (1998) in the Midwestern United States, and by Nations and Halberg (1992) and Hatfield et al. (1996) in Iowa. Factors influencing herbicide volatilization include vapor pressure of the herbicide, meteorological conditions, soil properties, and agricultural management practices.

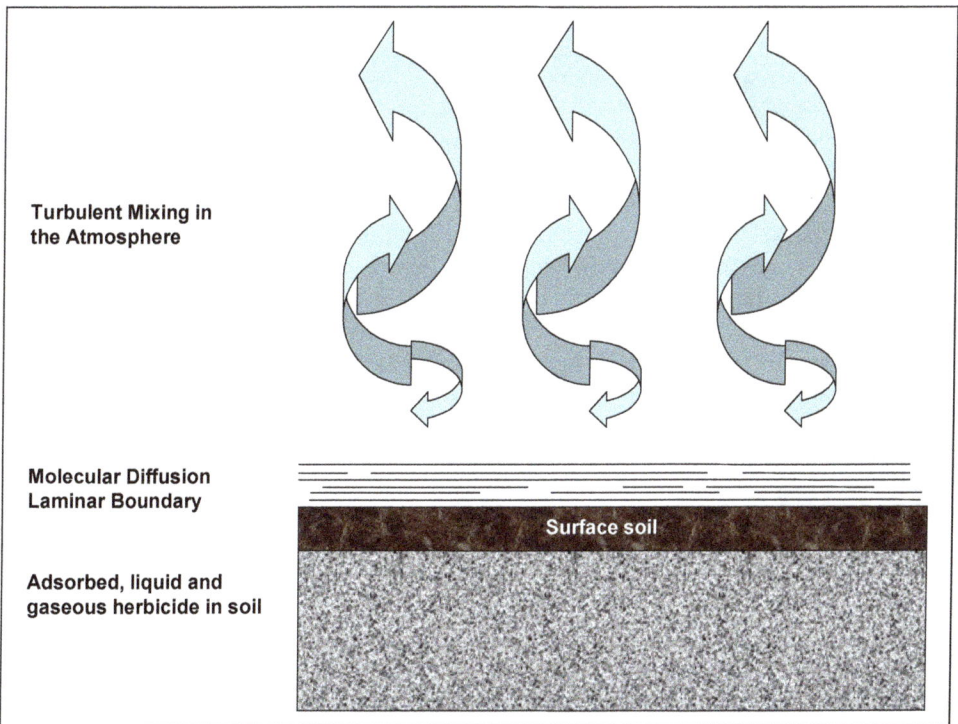

Turbulent Mixing in the Atmosphere

Molecular Diffusion Laminar Boundary

Surface soil

Adsorbed, liquid and gaseous herbicide in soil

Fig. 6. Schematic of herbicide volatilization processes

4.1 Vapor pressure

Perhaps the most crucial *herbicide property* influencing volatilization is its vapor pressure. As the vapor pressure increases, the herbicide increasingly favors the vapor phase and is more readily volatilized. In the field an "effective" herbicide vapor pressure is likely to be lower than the vapor pressure of the "pure" chemical due to interactions with the soil. For example, early studies detected a significant positive correlation between the pesticide vapor pressure and insecticide volatilization (Farmer et al., 1972; Glotfelty et al., 1984). Later it was observed that dry soil conditions favored soil adsorption which reduced the vapor pressure of the herbicide and lowered herbicide volatilization (Spencer and Cliath, 1974; Taylor and Spencer, 1990). By contrast, plant surfaces have a lower affinity for herbicides and as such exhibit vapor pressures that are closer to that of the "pure" herbicide (Taylor & Spencer, 1990). For most herbicides, soil adsorption is primarily governed by the soil organic fraction (Rao and Davidson, 1980; Karickhoff, 1981). Thus, soil properties like organic matter and to a lesser extent, texture (clay content), and soil pH affect herbicide volatilization by increasing adsorption, thus reducing the liquid phase concentration and vapor pressure in the soil.

4.2 Soil moisture and meteorological impacts

At the field-scale, the surface soil water content is perhaps the most critical *soil property* influencing herbicide volatilization. As the herbicide will be distributed among adsorbed, liquid, and vapor phases the amount of airspace within a soil volume and the thickness of the water molecule layer adsorbed onto the soil particles will influence herbicide volatilization. Spencer and Cliath (1974) measured the herbicide vapor pressures in soil at various soil water contents and demonstrated greater volatilization losses from wet than dry soils. Glotfelty et al. (1984) demonstrated that herbicide vapor losses increased more with soil water content than organic matter content or soil temperature. Furthermore, in a 5-year field investigation, Prueger et al. (2005) demonstrated that at high soil water contents, as much as 25% of surface applied herbicide metolachlor could be lost through volatilization, compared to as little as 5% when soils were dry. In a three year field investigation, Gish et al. (2009) compared metolachlor volatilization from two locations where soil texture, climatic inputs, formulation, and management practices were identical, but where soil moisture and surface organic matter content differed. They observed that when the "wet" location had surface soil water contents nearly twice that of the "dry" location that metolachlor volatilization losses were also doubled relative to the "dry"location. These results were surprising since the "wet" location had significantly higher organic matter content than the "dry location". Furthermore, when there was no significant difference in surface soil water contents between the two locations, both locations generated nearly identical metolachlor volatilization losses. As a result, it appears that surface soil moisture is more important than organic matter content (Gish et al., 2009). Several additional studies have shown the importance of soil moisture as peaks in early morning herbicide volatilization losses have been attributed to dew formation on the soil surface (Glotfelty et al., 1989; Taylor, 1995; Prueger et al., 2005). Lastly, increases in herbicide volatilization following a rain event are common and these spikes can be relatively large if the soil was dry prior to the rainfall event (Prueger et al., 2005). Although herbicide volatilization is influenced by soil bulk density, pH, soil mineralogy, and soil organic matter content, the dominant soil property appears to be soil moisture.

The impacts of meteorological conditions on herbicide volatilization were initially thought to be important only as they influence soil properties, such as surface soil water content. For example, relative humidity affects surface soil water content and can lead to enhanced herbicide volatilization if the fields are dry (Glotfelty et al., 1984; Taylor, 1995; Prueger et al.,

2005). Soil temperature was thought to increase herbicide volatilization by increasing diffusion rates and vapor pressure. However, increased soil temperatures can lead to drying of the soil surface, resulting in increased adsorption and a decrease in herbicide volatilization. Increased solar radiation can increase air and soil temperatures which influence herbicide volatilization, but the impact is dependent on the surface soil water content. Gish et al., (2009) demonstrated that when the surface soil was moist, increases in temperature led to sharp increases in metolachlor volatilization. However, when soils were dry increases in soil temperature had no discernible impact on metolachlor volatilization. Rainfall increases soil water contents, which in turn may increase pesticide volatilization, but depending upon rainfall intensity and duration it may actually move the herbicide deeper into the soil profile where it is less likely to volatilize. Increasing wind velocity could increase herbicide volatilization if soils are wet, but if the soil dries out the soil surface volatilization could decrease.

4.3 Management impacts
Agricultural management influences herbicide volatilization on several levels. First, soil incorporation of the herbicide decreases herbicide volatilization. Prueger et al. (1999) demonstrated that incorporating metolachlor in a band relative to a surface broadcast spray reduced herbicide volatilization losses from 22% to 6% of that applied. Although gas-phase diffusion is much greater than liquid-phase diffusion, only a small fraction of air-voids is present in soil (Spencer, 1970). Thus, by incorporating the herbicide gas diffusion may be limited, and volatility reduced. Second, increasing amounts of plant residue on the surface can increase herbicide volatilization since plants generally have a much lower affinity for pesticides than soil (Taylor & Spencer, 1990). Third, herbicide formulations such as control release or micro-encapsulated formulations can reduce volatility (Jackson and Lewis, 1978; Wienhold and Gish, 1994a; Gish et al., 1995). The effectiveness of the formulations is strongly dependent upon herbicide chemistry and the matrix encapsulating the herbicide. As herbicide solubility increases, the impact of formulation on reducing volatilization decreases (Wienhold and Gish, 1994b).

4.4 Reducing herbicide volatilization
Herbicide volatilization is governed by how the herbicide vapor pressure is influenced by interactions with soil properties, agricultural management practices, and local meteorology. In general, herbicides with a high vapor pressure should be avoided as they are more susceptible to volatilization. Since soil water content influences adsorption, applying herbicides to a wet soil, or applying the pesticide when precipitation is anticipated (e.g. shortly after application) will be detrimental to the environment. On the other hand, rainfall or irrigation *after* fumigants have been injected into the soil will decrease fumigant volatilization. Best management practices for reducing herbicide volatilization also includes the use of encapsulated formulations, and where possible, soil incorporation.

5. Summary
Among the major loss pathways, herbicide runoff has been the most rigorously studied. In general, herbicide annual runoff losses are less than 1% of that applied, with the largest

portion of this loss occurring near the time of application; however as much as 5%annually is common for worst case runoff scenarios' (Haith and Rossi, 2003; Shipitalo and Owens, 2006). Enhanced herbicide runoff relative to leaching from tile-drained fields supports the hypothesis that herbicide runoff is more detrimental to the environment than herbicide leaching (Lafleur et al., 1975; Muir and Baker, 1976; Ng et al., 1995). Unfortunately, field-scale herbicide leaching losses in non-tile-drained fields is difficult to quantify due to soil heterogeneity. Estimates of herbicide leaching are generally <1% of that applied and in a worst case scenario herbicide leaching losses are probably << 5% of that applied. The third loss pathway, volatilization, is a major environmental concern with herbicide losses commonly exceeding 15 % of that applied (Taylor and Spencer, 1990, Prueger et al. 1999; Prueger et al., 2005; Gish et al., 2009; Gish et al., 2011). Herbicide vapor pressure, soil moisture and meteorological interactions dominate herbicide volatility. Determining the impact of field scale off-site transport where all three major loss pathways are being monitored is rare and requires additional research on this critical topic.

6. References

Alegria, H. A., and T.J. Shaw. (1999). Rain deposition of atrazine and trifluralin in coastal waters of the South Atlantic Bight. *Environ Sci. Technol.* 33: 850-856.

Asmussen, L.E., A.W. White, E.W. Hauser, and J.A. Sheridan. (1977). Movement of 2,4-D in a vegetated waterway. *J. Environ. Qual.* 6:159-162.

Bailey, G.W., and J.L. White. (1964). Review of adsorption and desorption of organic pesticides by soil colloids, with implications concerning pesticide bioactivity. *J. Agr. Food Chem.* 12(4):324-332.

Baldwin, F.L., P.W. Santelmann, and J.M. Davidson. (1975). Movement of fluometuron across and trough the soil. *J. Environ. Qual.* 4:191-194.

Baker, J.L., and H.P. Johnson. (1979). The effect of tillage systems on pesticides in runoff from small watersheds. *Trans. Am .Soc. Agr. Eng.* 22:554-559.

Baker, J.L., J.M. Laflen, and H.P. Johnson. (1978). Effect of tillage systems on runoff losses of pesticides, a rainfall simulation study. *Trans Am Soc. Ag. Eng.* 21:886-892.

Baker, J.L., and J.M. Laflen. (1979). Runoff losses of surface applied herbicides as affected by wheel tracks and incorporation. *J. Environ. Qual.* 8:602-607.

Baker, J.L. (1980). Agricultural areas as nonpoint sources of pollution. In, *Environmental Impact of nonpoint Source Pollution.* Eds. M.R. Overcash and J.M. Davidson. pp 275-310. Ann Arbor Science Publication, Ann Arbor, Michigan.

Barnett, A.P., E.W. Hauser, A.W. White, and J.H. Holladay. (1967). Loss of 2,4-D in washoff of cultivated fallow lands. *Weeds* 15(2):133-137.

Baskaran, S., N.S. Bolan, A. Rahman, and R.W. Tillman. (1996). Pesticide sorption by allophonic and non-allophanic soils in New Zealand. *New Zealand J. Agric. Res.,* 39:297-310.

Battaglin W.A., and D.A. Goolsby. (1999). Are shifts in herbicide use reflected in concentration changes in Midwestern rivers? *Environ. Sci. Technol.* 33:2917-2925.

Bidleman, T.F. (1988). Atmospheric process – wet and dry deposition of organic compounds are controlled by their vapor-particle partitioning. *Environ. Sci. and Technol.* 22:361-367.

Bidleman, T.F., and E.J. Christensen. (1979). Atmospheric removal processes for high molecular weight organochlorines. *J. Geophys. Res.* 84:7857-7862.

Brasino, J. S. (1986). A simple stochastic model predicting conservative mass transport through the unsaturated zone into groundwater. *PhD thesis, Univ. Wisconsin-Madison, WI.*

Brown, C.D., R.A. Hodgkinson, D.A. Rose, J.K. Syers, and S.J. Wilcockson. (1995). Movement of pesticides to surface waters from a heavy clay soil. *Pestic. Sci.* 43:131-140.

Burrows, H.D., M. Canle, J.A. Santaballa, and S. Steenken. (2002). Reaction pathways and mechanisms of photodegradation of pesticides. *J. Photochem. Photobiol. B* 67:71-108.

Buttle, J.M. (1990). Metolachlor in surface runoff. *J. Environ. Qual.* 19:531-538.

Capel, P.D., K.A. McCarthy, and J.E. Barbash. (2008). National, holistic, watershed-scale approach to understand the sources, transport, and fate of agricultural chemicals. *J. Environ. Qual.* 37:983-993.

Caron, E., P. Lafrance, and J-C. Auclair. (2010). Impact of grass and grass popular buffer strips on atrazine and metolachlor losses in surface runoff and subsurface infiltration from agricultural plots. *J. Environ. Qual.* 39:617-629.

Caro, J. (1976). Pesticides in agricultural runoff. *In* Control of Water Pollution from Cropland, Vol. II. (Ed) B.A. Stewart. US *EPA-600/2-75-026b, pp 91-119.*

Celis R, C. Trigo, G. Facenda, M.C. Hermosin, and J. Cornejo. (2007). Selective modification of clay minerals for the adsorption of herbicides widely used in olive groves. *J. Agric. Food Chem* 55:6650-6658.

Chinkuyu, A.J., Meixner,T., Gish, T.J., Daughtry, C.S.T. (2004). The Importance of Seepage Zones in Predicting Soil Moisture Content and Surface Runoff from Watersheds with Gleams and RZWQM, *Trans. Am Soc. Ag. Eng.* 47:427-438. 2004.

Cole, J.T., J.H. Baird, N.T. Basta, R.L. Huhnke, D.E. Storm, G.V. Johnson, M.E. Payton, M.D. Smolen, D.L. Martin, and J.C. Cole. (1997). Influence of buffers on pesticides and nutrient runoff from bermudagrass turf. *J. Environ. Qual.* 26:1589-1598.

Denmead, O.T., J.R. Simpson, and J.R. Freney. (1977). A direct field measurement of ammonia emission after injection of anhydrous ammonia. *Soil Sci. Soc. Am. J.* 41:1001-1004.

Delphin, J.E., and J.Y. Chapot. (2006). Leaching of atrazine, metolachlor and diuron in the field in relation to their injection depth into a silt loam soil. *Chemosphere* 64:1862-1869.

Domagalski, J.L., S. Ator, R. Coupe, K McCarthy, D. Lampe, M. Sandstrom, and N. Baker. (2008). Comparative study of transport processes of nitrogen, phosphorus, and herbicides to streams in five agricultural basins, USA. *J. Environ. Qual.* 37:1158-1169.

Farmer, W.J., J.P. Martin, W.F. Spencer, and K. Igue. (1972). Volatility of organochlorine insecticides from soil: I. Effect of concentration, temperature, air flow rate, and vapor pressure. *Soil Sci. Soc. Am. Proc.* 36:443-447.

Fawcett, R.S., D.P. Tierney, and B.R. Christensen. (1994). The impact of soil conservation on pesticide runoff into surface water: A review and analysis. *J. Soil Water Conserv.* 49:126-135.

Felsot, A. S., J.K. Mitchell, and A.L. Kenimer. (1990). Assessment of management practices for reducing pesticide runoff from sloping cropland in Illinois. *J. Environ. Qual.* 19:539-545.

Flury, M. (1996). Experimental evidence of transport of pesticides through field soils: A review. *J. Environ. Qual.* 25:25-45.

Gan, J. S.J. Lee, W.P. Liu, D.L. Haver, and J.N. Kabashima. (2005). Distribution and persistence of pyrethroids in runoff sediments. *J. Environ. Qual.* 34:836-841.

Gaynor, J.D., C.S. Tan, C.F. Drury, H.Y.F. Ng, T.W. Welacky, and I.J. Wesenbeeck. (2001). Tillage, intercrop, and controlled drainage-subirrigation influence atrazine, metribuzin, and metolachlor loss. *J. Environ. Qual.* 30:561-572.

Gaynor, J.D., C.S. Tan, C.F. Dury, T.W. Welacky, H.Y.F. Ng, and W.D. Reynolds. (2002). Runoff and drainage losses of atrazine, metribuzin, and metolachlor in three water management systems. *J. Environ. Qual.* 31:300-308.

Gentry, L.E., M.B. David, K.M. Smith-Starks, and D.A. Kovacic. (2000). Nitrogen fertilizer and herbicide transport from tile drained fields. *J. Environ. Qual.* 29:232-240.

Gevao, B., K.T. Semple, and K.C. Jones. (2000). Bound pesticide residue in soils: a review. *Environ. Pollution* 108:3-14.

Ghodrati, M. and W.A. Jury. (1992). A field study of the effects of soil structure and irrigation method on preferential flow of pesticides in unsaturated soil. *J. Contam. Hydrol.* 11:101-125.

Germann, P.F. and K. Beven. (1981). water flow in soil macropores: I an experimental approach. *J. Soil Sci.* 18:363-368.

Gillion, R.J., J.E. Barbash, C.G. Crawford, P.A. Hamilton, J.D. Martin, N. Nakagaki, L.H. Nowell, J.C. Scott, P.E. Stackelberg, G.P. Thelin, and D.M. Wolock. (2006). *Pesticides in the nation's streams and groundwater, 1992-2001.* 172 pp. United States Geological Survey, Circular 1291.

Gish, T. J., D.Gimenez, and W.J. Rawls. (1998). Impact of roots on ground water quality. *Plant and Soil* 200:47-54.

Gish, T.J., A.R. Isensee, R.G. Nash, and C.S. Helling. (1991a). Impact of pesticides on shallow groundwater quality. *Trans. Am Soc. Agr. Eng.* 34:1745-1753.

Gish, T. J., and W. A. Jury. (1983). Effect of plant roots and root channels on solute transport. *Trans. Am Soc. Ag. Eng.* 26:440-444 & 451.

Gish, T.J., K.- J., S. Kung, D. Perry, J., Posner, G. Bubenzer, C.S. Helling, E.J., Kladivko, and T. S. Steenhuis. (2004). Impact of preferential flow at varying irrigation rates by quantifying mass fluxes. *J. Environ. Qual.* 33:1033-1040.

Gish, T.J., J.H. Prueger, C.S.T. Daughtry, W.P. Kustas, L.G. McKee, A.L. Russ, and J.L. Hatfield. (2011). Comparison of field-scale herbicide runoff and volatilization losses: An eight year field investigation. *J. Environ. Qual.* 40:1432-1442.

Gish, T.J., J.H. Prueger, W.P. Kustas, J.L. Hatfield, L.G. McKee, A.L. Russ. (2009). Soil moisture and metolachlor volatilization observations over three years. *J. Environ. Qual.* 38:1785-1795.

Gish, T. J., A. Sadeghi, B. J. Wienhold. (1995). Volatilization of alachlor and atrazine as influenced by surface litter. *Chemosphere* 31:2971-2982.

Gish, T. J., M. J. Schoppet, C. S. Helling, A. Shirmohammadi, M. M. Schreiber, and R. E. Wing. (1991b). Transport comparison of technical grade and starch-encapsulated atrazine. *Trans. Am. Soc. Ag. Eng.* 34:1738-1744.

Gish, T. J., A. Shirmohammadi, C. S. Helling. K.-J. S. Kung, B. J. Wienhold, and W. J. Rawls. (1998). Mechanisms of herbicide leaching and volatilization and innovative approaches for sampling, prediction and control. In *Integrated Weed and Soil Management.* p. 107-134. J. L. Hatfield, D.D. Buhler, and B.A. Stewart. (eds) Ann Arbor Press, Chelesa, Michigan.

Gish, T. J., A. Shirmohammadi, and B. J. Wienhold. (1994). Field-scale mobility and persistence of commercial and starch-encapsulated atrazine and alachlor. *J. Environ. Qual.* 23:355-359. 1994.

Glass, R. J., J. Y. Parlange, and T.S. Steenhuis. (1989). Wetting front instability. 1. Theoretical discussions and dimensional analysis. *Water Resour. Res.* 25:1187-1194.

Glotfelty, D.E., A.W.Taylor, B.J. Turner, and W.H. Zoller. (1984). Volatilization of surface applied pesticides from a fallow soil. *J. Agric. Food Chem.* 32:638-643.

Glotfelty, D.E., M.M. Leech, J. Jersey, and A.W. Taylor. (1989). Volatilization and wind erosion of soil-surface applied atrazine, simazine, alachlor, and toxaphene. *J. Agric. Food Chem.* 37:546-555.

Goolsby, D.A. E.M. Thurman, M.L. Pomes, M.T. Meyer, and W.A. Battaglin. (1997). Herbicides and their metabolites in rainfall: Origin transport, and deposition patterns across the Midwestern and Northeastern United States, 1990-1991. *Environ. Sci. Technol.* 31:1325-1333.

Hall, J.K., N.L. Hartwig, and L.D. Hoffman. (1983). Application mode and alternate cropping effects on atrazine losses from a hillside. *J. Environ. Qual.* 12:336-340.

Haith, D.A., and F.S. Rossi. (2003). Risk assessment of pesticides from turf. *J. Environ. Qual.* 32:447-455.

Hamaker, J.W., and J.M. Thompson. (1972). Adsorption. In *Organic chemicals in soil environment,* (ed.) C.A. I. Goring and J.W. Hamaker, eds. Vol. 1:49-143. Marcel Dekker, Inc. New York, NY.

Hansen,N.C., J.F. Moncrief, S.C. Gupta, P.D. Capel, and A.E. Olness. (2001). Herbicide banding and tillage system interactions on runoff losses of alachlor and cyanazine. *J. Environ. Qual.* 30:2120-2126.

Harris, G.L., and J.A. Catt. (1999). Overview of the studies on the cracking clay soil at Brimstone Farm, UK. *Soil Use Mange* 15:233-239.

Hatfield, J.L., C.K. Wesley, J.H. Prueger, and R.L. Pfeiffer. (1996). Herbicide and nitrate distribution in central Iowa rainfall. *J. Environ. Qual.* 25:259-264.

Helling, C.S. (1993). Pesticides, agriculture, and water quality. *Proc. Stockholm Water Symp..* Stockholm, 1992.

Helling, C. S., and T. J. Gish. (1986). Soil characteristics affecting pesticide movement into groundwaters. *In Evaluation of Pesticides in Ground Water.* W.Y. Garner, R.C. Honeycutt and H. N. Niggs (eds.). p 14-38.

Isensee, A.R., R.G. Nash, and C.S. Helling. (1990). Effects of no-tillage vs. conventional tillage corn production on the movement of several pesticides to ground water. *J. Environ. Qual.* 19:434-440.

Jackson, M.D. and R.G. Lewis. (1978). Volatilization of two methyl parathion formulations from treated fields. *Bull. Environ. Contam. Toxicol.* 20:793-796.

Jarvis, N.J. (2002). Macropore and preferential flow. In. *The encyclopedia of agrochemicals*. J. Plimmer (ed) Vol. 3. p 1005-1013. John Wiley & Sons, Chelsea, MI.

Jenks, B.M., F.W. Roeth, A.R. Martin, and D.L. McCallister. (1998). Influence of surface and subsurface soil properties on atrazine sorption and degradation. *Weed Sci.* 46:132-138.

Jin-Clark, Y., M.J. Lydy, and K.Y. Zhu. (2002). Effects of atrazine and cyanazine on chlorpyrifos toxicity in *Chironomus tentans* (Diptera: Chironomidae). *Environ. Toxicol. Chem.* 21: 598-603.

Johnson, A.C., A.H. Haria, C.L. Bhardwaj, R.J. Williams, and A. Walker. (1996). Preferential flow pathways and their capacity to transport isoproturon in a structured clay soil. *Pestic. Sci.* 48:225-237.

Jones, R.L., G.L. Harris, J.A. Catt, R.H. Bromilow; D.J. Mason, and D.J. Arnold. (1995). Management practices for reducing movement of pesticides to surface water in cracking clay soils. *Weeds* 2:489-98.

Jury , W.A. (1975a). Solute travel-time estimates for tile drained fields: I. Theory. *Soil Sci. Soc. Am. Proc.* 39:1020-1024.

Jury , W.A. (1975b). Solute travel-time estimates for tile drained fields: II. Application to experimental studies. *Soil Sci. Soc. Am. Proc.* 39:1024-1028.

Jury, W.A. (1986). Chemical movement through soil. In *Vadose zone modeling of organic pollutants*. P. 135-158. S.C. Hern and S. M. Melancon (ed.). Lewis Publi., Chlesa, MI.

Karickhoff, S.W. (1981). Semi-empirical estimation of sorption of hydrophobic pollutants on natural sediments. *Chemosphere* 10:833-846.

Kearney, P.C., R.G. Nash, and A.R. Isensee. (1969). Persistence of pesticide residues in soils. In *Current research on persistent pesticides*. P. 54-67. M. W. Miller and G.G. Berg eds (ed.). Chemical fallout. Charles C. Thomas Pub., Springfield, IL.

Kladivko, E.J., G.E. Van Scoyoc, E.J. Monke, K.M. Oates, and W. Pask. (1991). Pesticide and nutrient movement into subsurface tile drains on a silt loam soil in Indiana. *J. Environ. Qual.* 20:264-270..

Koplin, D.W., J.E. Barbash., and R.J. Gillion. (1998). Occurrence of pesticides in shallow groundwater of the United States: initial results from the National Water-Quality Assessment Program. *Environ. Sci. Technol.* 32:558-566.

Kuang, Z., L.L. McConnell, A. Torrents, D. Meritt, S. Tobash. (2003). Atmospheric deposition of pesticides to an agricultural watershed of the Chesapeake Bay. *J. Environ. Qual.* 32: 1611-1622.

Kung, K.-J.S. (1990). Preferential flow in a sandy vadose zone: 1 Field observations. *Geoderma* 46:51-71.

Kung, K-J.S., T.S. Steenhuis, E.J. Kladivko, T. Gish, G. Bubenzer, and C.S. Helling. (2000). Impact of Preferential Flow on the Transport of Adsorbing and non-Adsorbing Tracers. *Soil Sci. Soc. Am. J.* 64:1290-1296.

LaFleur, K.S., W.R. McCaskil, and D.S. Adams. (1975). Movement of prometryne through Congaree soil into groundwater. *J. Environ. Qual.* 4:132-133.

Laird, D.A., E. Barriuso, R.H. Dowdy, and W.C. Koskinen. (1992). Adsorption of atrazine on smectites. *Soil Sci. Soc. Am. J.* 56:62-67.

Ma, Q.L., R.D. Wauchope, L. Ma, K.W. Rojas, R.W. Malone, and L.R. Ahuja. (2004). Test of the root zone water quality model (RZWQM) for predicting runoff of atrazine, alachlor, and fenamiphos species from conventional-tillage corn mesoplots. *Pest Manag. Sci.* 60:267-276.

Majewski, M.S., and P.D. Capel. (1995). Pesticides in the atmosphere, distribution, trends, and governing factors. Ann Arbor Press, Inc., Chelsea, Michigan, USA.

Majewski, M.S., W.T. Forman, ED.A. Goolsby, and N. Nakagaki. (1998). Airborne pesticide residues along the Mississippi river. *Environ. Sci Technol.* 32:3689-3698.

McConnell, L.L., J.S. LeNoir, S. Datta, and J.N. Seiber. (1998). Wet deposition of current-use pesticides in the Sierra Nevada mountain range, California, USA. Environ. *Toxicol. Chem.* 17: 1908-1916.

Mojasevic, M., Helling, C.S., Gish, T.J., and Doherty, M.A. (1996). Persistence of seven pesticides as influenced by soil moisture. *J. Environ. Sci. Health B* 31(3) 469-476.

Muir, D.C., B.E. Baker. (1976). Detection of triazine herbicide and their degradation products in tile-drain water from fields under intensive corn (maize) production. *J. Agric. Food Chem.* 24:122-125.

Nations , B.K. and G.R. Hallberg. (1992). Pesticide in Iowa precipitation. *J. Environ. Qual.* 21:486-492.

Ng, H.Y.F., J.D. Gaynor, C.S. Tan, and C.F. Drury. (1995). Dissipation and loss of atrazine and metolachlor in surface and subsurface drain water: A case study. *Water Res.* 29:2309-2317.

Novak, S.M., J-M. Portal, and M. Schiavon. (2001). Effects of soil type on metolachlor losses in subsurface discharge. *Chemosphere* 42:235-244.

Pachepsky, Y.A., A.K. Guber, M.T. Van Genuchten, T.J. Nicholson, R.E. Cady, J.Simunek, M.G. Schapp. (2006). Model abstraction techniques for soil-water flow and transport. *NUREG/CR-6884. U.S. Nuclear Regulatory Commission, Washington DC.*

Parmele, L.H., E.R. Lemon, and A.W. Taylor. (1972). Micrometeorological measurement of pesticide vapor flux from bare soil and corn under field conditions. *Water, Air, Soil Pollut.* 1:433-451.

Pantone, D.J., R.A. Young, D.D. Buhler, C.V. Eberlein, W.C. Koskinen, and F. Forcella. (1992). Water quality impacts associated with pre- and post emergence applications of atrazine in maize. *J. Environ. Qual.* 21:567-573.

Prueger, J.H., J.L. Hatfield, and T.J. Sauer. (1999). Field-scale metolachlor volatilization flux estimates from broadcast and banded application methods in central Iowa. *J. Environ. Qual.* 1999: 28:75-81.

Prueger, J.H., T.J. Gish, L.L. McConnell, L.G. McKee, J.L. Hatfield, and W.P. Kustas. (2005). Solar radiation, relative humidity, and soil water effects on Metolachlor volatilization. *Environ. Sci. Technol.* 39:5219-5226.

Rao, P.S.C., and J.M. Davidson. (1980). Estimation of pesticide retention and transformation parameters required in nonpoint source pollution models. In *Environmental Impact of Nonpoint Source Pollution.* Pp 23-67. M.R. Overcash and J.M. Davidson (ed.). Ann Arbor Sci. Pub., Ann Arbor Michigan.

Reddy, K.N., M.A. Locke, and C.T. Bryson. (1994). Foliar washoff and runoff losses of lactofen, norflurazon, and fluometuron under simulated rainfall. *J. Agric. Food Chem.* 42:2338-2343.

Rice, C.P., C.B. Nochetto, and P. Zara. (2002). Volatilization of trifluralin, atrazine, metolachlor, chlorpyrifos, alpha—endosulfan, and beta-endosulfan from freshly tilled soil. *J. Agric Food Chem.* 50:4009-4017.

Ritsema, C.J. and L.W. Dekker. (1995). Distribution flow: A general process in the top layer of water repellent soils. *Water Resour. Res.* 31:1187-1200.

Schreiber, M.M., M.V. Hickman, and G.D. Vail. (1993). Starch-encapsulated atrazine – effects and transport. *J. Environ. Qual.* 22:443-453.

Scribner, E.A. W.A. Battaglin, D.A. Goolsby, and E.M. Thurman. (2000). Changes in herbicide concentrations in Midwestern streams in relationship to changes in use, 19891998. *Sci. Total Environ.* 248:255-263.

Shipitalo, M.J., W.A. Dick, W.M. Edwards. (2000). Conservation tillage macropore factors that affect water movement and the fate of chemicals. *Soil Till. Res.* 53:167-183.

Shipitalo, M.J., W.M. Edwards, W.A. Dick, and L.B. Owens. (1990). Initial storm effects on macropore transport surface-=applied chemicals in no-till soil. *Soil Sci. Soc. Am. J.* 54:1530-1536.

Shipitalo, Martin J. and Loyd B. Owens. (2006). Tillage system, application rate, and extreme event effects on herbicide losses in surface runoff. *J. Environ. Qual.* 35:2186-2194.

Shulz, R., M. Hauschild, M. Ebeling, J. Nankodrees, J. Wogram, and M. Liess. (1998). A qualitative field method for monitoring pesticides in the edge-of-field runoff. *Chemosphere* 36:3071-3082.

Southwick, L.M., G.H. Willis, R.L. Bengston, and T.J. Lormand. (1990). Effect of subsurface drainage on runoff losses of metolachlor and trifluralin from Mississippi River alluvial soil. *Environ. Contam. Toxicol.* 32:106-109.

Spark, K.M., and R.S. Swift. (2002). Effect of soil composition and dissolved organic matter on pesticide sorption. *Sci. Total Environ.* 298:147-161.

Spencer, W.F. and M.M. Cliath. (1970). Desorption of Lindane from soil as related to vapor density. *Soil Sci. Soc. Am. Proc.* 34:574-578.

Spencer, W.F. and M.M. Cliath. (1974). Factor ffecting vapor loss of trifuralin from soil. *J. Agric. Food Chem.* 20:987-991.

Stork, A., R. Witte, and F Fuhr. (1994). A wind tunnel for measuring the gaseous losses of environmental chemicals from the soil/plant system under field-like conditions. *Environ. Sci. Pollut. Res.* 1:234-245.

Symons, P.E.K. (1977). Dispersal and toxicology of the insecticide fenitrothian; predicting hazards of forest spraying. *Residue Rev.* 68:1-36.

Taylor, A.W. (1995). The volatilization of pesticide residues. In *Environmental Behavior of Agrochemicals*. Vol. 9, Chap.6, pp 257-306. T.R. Roberts and P.C. Kearney (ed.). Publisher John Wiley and Sons, New York, New York.

Taylor, A.W. and W.F. Spencer. (1990). Volatilization and vapor transport processes. p. 213-269. In *Pesticides in the soil environment: Processes, impacts, and modeling*, H.H. Cheng (ed.). SSSA, Madison, WI.

Thurman, E. M. and A.E. Cromwell. (2000). Atmospheric transport, deposition, and fate of triazine herbicides and their metabolites in pristine areas at Isle Royale National Park. *Environ. Sci. Technol.* 34: 3079-3085.

Tilman, D., K.G. Cassman, P.A. Matson, R. Naylor, and S. Polasky. (2002). Agricultural stability and intensive production practices. *Nature* 418:671-677.

Thurman E.M., D.A. Goolsby, M.T. Meyer, and D.W. Koplin. (1991). Herbicides in surface waters of the Midwestern United States. The effect of spring flush. *Environ. Sci. Technol.* 25:1794-1796.

Traub-Eberhard, U., K-P. Henschel, W. Kordel, and W. Klein. (1995). Influence of different field sites on pesticide movement into subsurface drains. *Pestic. Sci.* 43:121-129.

United States Environmental Protection Agency. (1990). *National Pesticide Survey: Project Summary.* EPA Report 570990NPSg, 12 pp.

United States Environmental Protection Agency. (2011). *Pesticide industry sales and usage 2006 and 2007 market estimates.* United States Protection Agency, Office of Pesticide Programs, Washington D.C. EPA733-R-11-001, 33p.

United States Environmental Protection Agency (USEPA). (2008). *Drinking Water Contaminants.* http://www.epa.gov/safewater/contaminants/index.html. EPA 816-F-03-016, June 2003. pp6.

Vereeckren. H. (2005). Mobility and leaching of glyphosate: A review. *Pest. Mang. Sci.* 61:1139-1151.

Wauchope, R.D. (1978). The pesticide content of surface water draining from agricultural fields. A review. *J. Environ. Qual.* 7:459-472.

Wauschope, R.D., and R.S. Meyers. (1985). Adsorption-desorption kinetics of atrazine and linuron in freshwater sediment aqueous slurries. *J. Environ. Qual,* 14:132-136.

Webb, R.M.T. M.E. Wieczorek, BT. Nolan, T.C. Hancock, M.W. Sandstrom, J.E. Barbash, E. R. Bayless, R.W. Healy, J. Linard. (2008). Variations in pesticide leaching related to land use, pesticide properties, and unsaturated zone thickness. *J. Environ. Qual.* 37:1145-1157.

Weber, J.B., K.A. Taylor, and G.G. Wilkerson. (2006). Soil and herbicide properties influenced mobility of atrazine metolachlor, and Primisulfuron-Methyl in field lysimeters. *Agron. J.* 98:8-18.

Weber, J.B., S.B. Weed, and T.J. Sheets. (1972). Pesticides-how they move and react in the soil. *Crops and Soils* 25:14-17.

Wienhold, B. J., and T. J. Gish. (1994a). Effect of Formulation and tillage practice on volatilization of atrazine and alachlor. *J. Environ. Qual.* 23:292-298.

Wienhold, B. J. and T. J. Gish. (1994b). Chemical properties influencing rate of release of starch encapsulated herbicides: implications for modifying environmental fate. *Chemosphere* 28:1035-1046.

Williams, A.G., D. Scholefield, J.F. Dowd, N. Holden, and L. Deeks. (2000). Investigating preferential flow in a large intact soil block under pasture. *Soil Use Manage.* 16:264-269.

Wilson, J.D., G.W. Thurtell, G. Kidd, and E. Beauchamp. (1982). Estimation of the rate of gaseous mass transfer from a surface source plot to the atmosphere. *Atmos. Environ.* 16:1861-1867.

Wyman, J. A, J. O. Jensen, D. Curwen, R. L. Jones, T. E., and Marquardt. (1985). Effects of application procedures and irrigation on degradation and movement of aldicarb residues in soil. *Environ. Toxicol. Chem.* 4:641-651.

Wyman, J. A., J. Medina, D. Curwen, J. L. Hansen, and R. L. Jones. (1986). Movement of aldicarb and aldoxycarb residues in soil. *Environ. Toxicol. Chem.* 5:545-555.

Zhang, X.C., L.D. Norton, and M. Hickman. (1997). Rain pattern and soil moisture content effects on atrazine and metolachlor losses in runoff. *J. Environ. Qual.* 26:1539-1547.

Zehe, E., H. Fluher. (2001). Slope scale variation of flow patterns in soil profiles. *J. Hydrol.* 247:116-132.

Part 7

Soil Pollution Assessment

14

Molecular Analyses of Soil Fungal Community – Methods and Applications

Yuko Takada Hoshino
National Institute for Agro-Environmental Sciences
Japan

1. Introduction

Fungi play important and diverse roles in soil ecosystems. They act as plant pathogens, mycorrhizal symbionts and most importantly, as the principal decomposers of organic materials (Christensen, 1989; Thorn, 1997). Fungi also represent a dominant component of the soil microflora in terms of biomass (Thorn, 1997). Compared with bacterial communities, however, knowledge regarding the diversity and functions of soil fungal communities remains limited.

Culture-independent molecular techniques, comprising of direct DNA extraction from soil followed by PCR and electrophoresis or cloning, have been introduced to investigate soil fungal communities (Anderson & Cairney, 2004). These techniques facilitate the detection of fungi, including fastidious or non-culturable strains, and an understanding of the fungal community structures and dynamics in soil (Hoshino & Matsumoto, 2007; Vandenkoornhuyse et al., 2002).

Molecular techniques have provided novel insights and significant advances in research on soil fungal ecology and have been applied to various soils in different ecosystems, such as forests (Perkiomaki et al., 2003), grasslands (Brodie et al., 2003), dunes (Kowalchuk et al., 1997), stream sediments (Nikolcheva et al., 2003) and agricultural fields (Gomes et al., 2003). For example, in agricultural soils, a fungal community is affected by plant growth (Gomes et al., 2003) and cultural practices, such as application of fertilizers and pesticides (Girvan et al., 2004).

With the development of new technologies, accumulating molecular data has contributed to the establishment of database combined with other environmental data and facilitated meta-analysis on a large scale. In agricultural soils, fungal communities are directly and indirectly related to crop production. Technological advances in molecular methods would help elucidate such a complicated relationship. Here I present molecular techniques applied to soil fungal community analyses, particularly in agricultural soils and discuss their limitations and future applications.

2. Molecular analysis techniques

In culture-independent molecular analysis of microbial community, DNA directly extracted from soil may be analyzed by PCR-based techniques targeting specific genes and by metagenomic approach using direct sequencing (Suenaga, 2011). For fungal community analyses, PCR-based techniques have been widely and generally used (Anderson & Cairney, 2004; Hoshino & Matsumoto, 2007). Fig. 1 showed the experimental scheme of PCR-based molecular analyses for soil fungal community, which consist of three steps: (i) direct extraction of DNA or RNA from soil, (ii) polymerase chain reaction (PCR) amplification of the 18S rRNA gene (rDNA) and internal transcribed spacer (ITS) region using fungal specific primers, and (iii) community profiling, including some electrophoresis techniques and sequence based techniques.

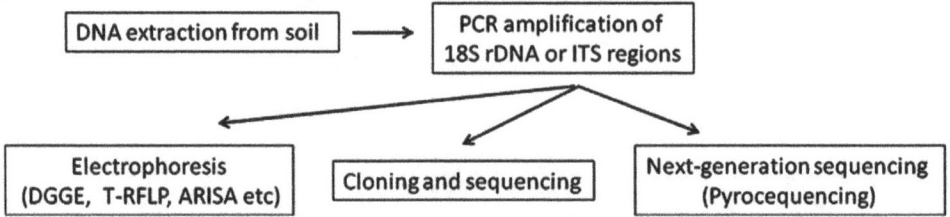

Fig. 1. Experimental scheme: molecular analyses of soil fungal community

2.1 DNA/RNA extraction

Many protocols for DNA/RNA extraction from soil have been developed (Robe et al., 2003) and used to extract fungal genomic DNA and fungal RNA. The majority of the direct extraction is the combination of chemical and/or enzymatic treatments and physical procedures. Bead-beating is most effective in cell disruption (Miller et al., 1999) and commercially available kits include this step (Borneman et al., 1996). In these procedures, soil samples are shaken vigorously with small glass beads in buffer including detergent. Microbial cells are disrupted within the soil matrix, and nucleic acids are released from lysed cells. DNA or RNA is, then, recovered and purified.

Because of soil diversity in terms of property and composition, extraction protocol of nucleic acids needs to be optimized for each soil type. For example, it was difficult to extract nucleic acids from Andisol, volcanic ash soils, which strongly adsorbed nucleic acids. The addition of adsorption competitors to the extraction buffer enabled to extract DNA and RNA, and increased the yield of DNA and RNA. From a variety of Andisols, we successfully extracted DNA and RNA for molecular analyses by using skim milk or RNA for DNA extraction (Fig.2A) (Hoshino & Matsumoto, 2004) and DNA for RNA extraction as adsorption competitors (Fig.2B) (Hoshino & Matsumoto, 2007).

The DNA extraction protocols, especially with different conditions of cell disruption, can affect the result: Martin-Laurent et al. (Martin-Laurent et al., 2001) showed that microbial community profiles were variable, according to DNA recovery methods used, both in terms of phylotype abundance and the composition of indigenous bacterial community. Soil sample size may also influence analysis results targeting fungal community. Ranjard et al. (Ranjard et al., 2003) reported that analytical results of fungal community structure varied among replicates from a single homogenized soil sample when sample size was less than 1 g, while sample range between 0.125 - 4 g had no effect on the assessment of bacterial community structure.

We also observed significant variation in fungal community profiles among replicates of the conventional sample size of 0.4 g soil from a single homogenized sample for DNA extraction, using commercially available kit (Fig. 3B), while such variation was not detected in bacterial community profiles (Fig. 3A), using denaturing gradient gel electrophoresis (DGGE), a community profiling method (see Section 2.3).

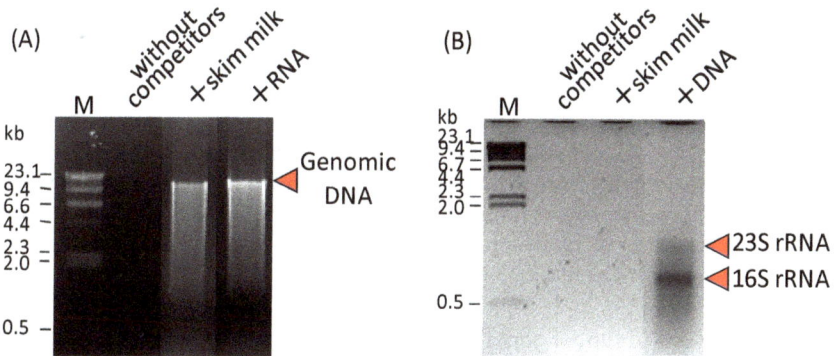

Fig. 2. Improvement of DNA and RNA extraction from Andisols, using adsorption competitors

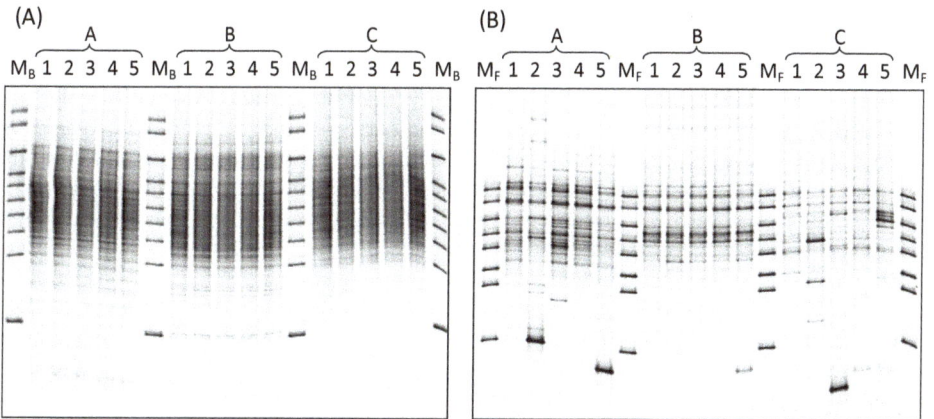

Fig. 3. Variation in bacterial (A) and fungal DGGE profiles (B) among replicates (1-5) from 0.4g of soil from a single homogenized sample. Soils A and B were taken from upland fields, and soil C from a paddy field. M_B: Marker for bacterial DGGE, M_F: Marker for fungal DGGE

Increasing sample size and pre-treatments for homogenization of soil samples decreased variation in fungal DGGE profiles among replicates (Fig. 4). Sample size increased by mixing DNA extracts from 0.4g soil to rule out the influence of sample size on the efficacy of DNA extraction (Fig. 4A). With regard to pre-treatment for soil homogenization, we found that grinding in liquid nitrogen was suitable for upland field soils while adding buffer to soil to obtain homogeneous soil suspension was suitable for paddy filed soils. These pre-treatments did not significantly affect fungal DGGE profiles under the experimental conditions (Fig. 4B).

(A)   (B)

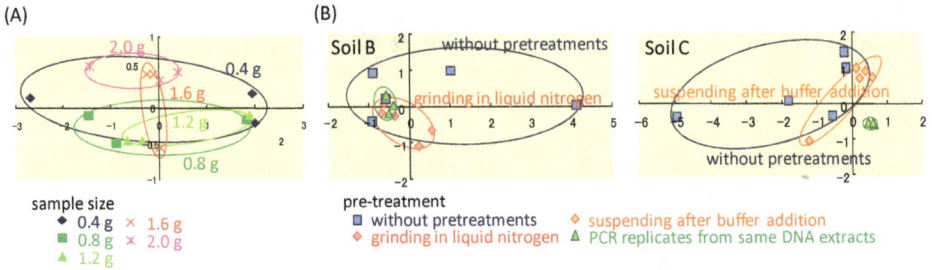

Fig. 4. Multi-dimensional scaling (MDS) map based on the similarities of fungal DGGE profile among replicates when increasing soil sample size from 0.4 to 2.0g (A) and when grinding in liquid nitrogen or suspending after buffer addition was done as a pre-treatment (B). In MDS map the closer the points to each other, the more similar the DGGE banding patterns represented by the points

2.2 PCR amplification of fungal genetic markers

Ribosomal RNA genes (rDNA), especially the small subunit ribosomal RNA genes, i.e., 18S rRNA genes (18S rDNA) in the case of eukaryotes, have been predominant target for the assessment of microbial community (Kowalchuk et al., 2006). The large subunit ribosomal RNA genes, 28S rDNAs, have been also targeted (Möhlenhoff et al., 2001) but been used less frequently than 18S rDNAs. The following properties of rDNAs are suitable for taxonomic identification: (i) ubiquitous presence in all known organisms; (ii) presence of both conserved and variable regions; (iii) the exponentially expanding database of their sequences available for comparison. In community analysis of environmental samples, the conserved regions serve as annealing sites for the corresponding universal PCR primers, whereas the variable regions can be used for phylogenetic differentiation. In addition, the high copy number of rDNA in the cells facilitates detection from environmental samples.

However, the lack of relative variation within 18S rDNA genes among closely related fungal species results in taxonomic identification commonly limited to genus or family level. For higher resolution in taxonomic identification, the internal transcribed spacer (ITS) region, which located between the 18S rDNA and 28S rDNA has been targeted (White et al. 1990; Gardes & Bruns, 1993). The ITSs, non-coding regions, have greater sequence variation among closely related species than the coding regions of rRNA genes because of their fast late of evolution. Protein-coding functional genes have been also employed as genetic markers to target a specific functional group (Ascomycetous laccase; Lyons et al., 2003) or to get higher resolution in specific fungal taxa (Fusarium elongation factors; Yergeau et al., 2005).

These genetic markers were amplified from soil DNA or RNA by PCR using fungal specific primers. Various PCR primer sets targeting fungal sequences are now available (Anderson & Cairney, 2004). Selection of PCR primer is one of the most important factors affecting outcome in fungal community analysis (Jumpponen, 2007; Hoshino & Morimoto, 2010). The properties of PCR primers will be described in Section 3.

2.3 Community profiling methods

PCR products are the mixtures of target genes, such as the rDNA and ITS region, derived from various kinds of fungi and are often of very similar size, differentiation must be achieved on the basis of the nucleotide composition. The compositions of these PCR products were analyzed by community profiling methods, including some electrophoresis techniques, such as denaturing gradient gel electrophoresis (DGGE), temperature gradient gel electrophoresis (TGGE), terminal restriction fragment length polymorphism analysis (T-RFLP), and automated ribosomal intergenic spacer analysis (ARISA), and sequence based techniques, such as cloning and sequencing and second-generation sequencing technologies (Nocker et al., 2007). Here we introduce the principles and the properties of the commonly used community profiling methods.

2.3.1 Electrophoresis techniques

Electrophoresis techniques are suitable for obtaining an overview of the total genetic diversity of a soil microbial community. PCR products are separated by electrophoresis based on the nucleotide composition. The data of electrophoretic profiles, i.e., the position and the relative intensity of different bands or peaks, could be transferred to numerical data which is applicable for calculation of diversity indices and several statistical analyses and enable comparison of numerous samples.

Currently, three electrophoresis methods have been mainly used: denaturing gradient gel electrophoresis (DGGE) (Muyzer et al., 1993), terminal restriction fragment length polymorphism analysis (T-RFLP) (Dunbar et al., 2000), and automated ribosomal intergenic spacer analysis (ARISA) (Ranjard et al., 2001). These fingerprinting approaches are based on different principles (Fig. 5).

DGGE separates DNA fragments of the same size but of different sequence based on the melting behaviour of DNA: double strands of the AT base pair more easily disassociate than those of the GC base pair. During electrophoresis in a denaturing gradient acrylamide gel, an increasing denaturing environment, partially dissociates DNA double strands, creating diverse, branched molecules. The partial melting of DNA strands reduces mobility. Because the melting concentration of the denaturant is sequence-specific, different sequences of DNA fragments have different mobility in denaturing gels, and each DNA fragment can be seen as a distinct band in the gel. T-RFLP is modified from the conventional RFLP approach using fluorescently labelled PCR primers before restriction digestion and size detection of fluorescently labelled terminal restriction fragments using a DNA sequencer. ARISA is simple and discriminates the length of whole PCR amplicons, generally targeting highly variable ITS regions. Automated DNA sequencer technology is applied for T-RFLP and ARISA.

The three fingerprinting techniques have both advantages and disadvantages. One of the main advantages of gel-based community profiling techniques like DGGE enables sequence analysis of each band in a gel, and therefore, facilitates more detailed phylogenetic analysis. In T-RFLP, each T-RFLP peak can be identified by using database of T-RF length in various microbial groups. However, the inability to get sequence data from T-RFLP peaks makes it difficult to identify unknown species.

The use of an automated DNA sequencer significantly increases throughput of T-RFLP and ARISA compared with gel-based techniques. It also improves the accuracy in sizing general fragments through the inclusion of an internal standard in each sample. On the other hand, reproducibility between gels has been highlighted as one of main pitfalls of DGGE (Fromin et al., 2002). As comparison between several different gels is required when dealing with large sample numbers, it is critical to standardize the resolution and quality of gels and to use suitable internal standards for the accuracy of analyses.

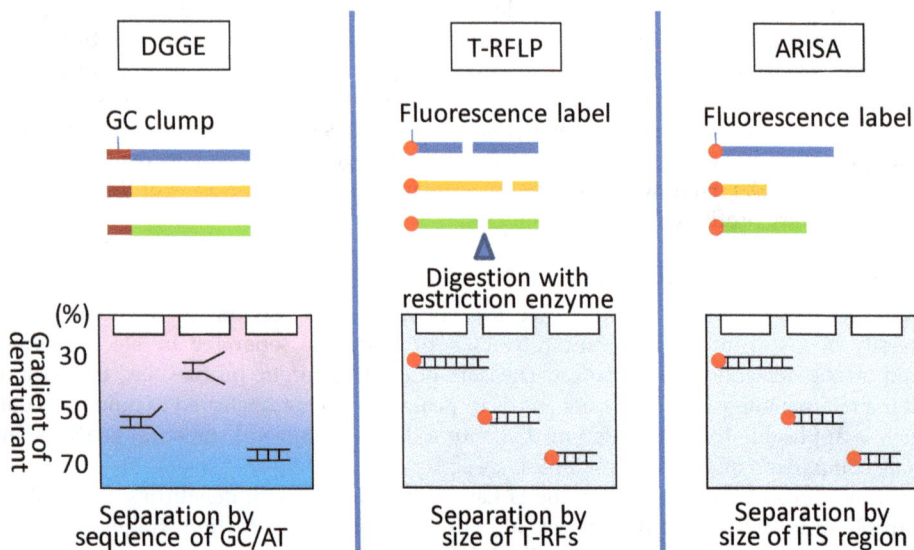

Fig. 5. Principles of three molecular fingerprinting methods

Okubo & Sugiyama (Okubo & Sugiyama, 2009) compared these fingerprinting methods, i.e., DGGE, T-RFLP and ARISA, by analyzing soil fungal communities. They reported that DGGE showed higher discrimination ability for soil fungal community rather than T-RFLP and ARISA, while ARISA exhibited the highest resolution ability.

From these properties, DGGE is suitable for analyses of highly heterogeneous communities including unknown members such as soil microbial community. T-RFLP is suitable for analyses of communities including known members and for specific taxonomic groups. ARISA appeared to be suitable for diversity analysis.

2.3.2 Cloning and high-throughput sequencing technology

Sequence-based community analyses can reveal fungal community structures with higher resolutions; fungal sequence data obtained can be identified or determined similarity to already known species through the use of extensive and rapidly growing sequence database (Nocker et al., 2007). Sequencing is the basis for construction of phylogenetic trees and for other comparative studies. Conversely, the sequence-based techniques are relatively time-consuming and costly. It depends on samples or target genes how many PCR amplicons were required to have fully analyzed the diversity contained within a single sample. Fierer et al. (2007) estimated that fungal 18S rDNA richness at the 97% sequence similarity level is likely to exceed 10^6 in 1.0 g of prairie soil, approximately 2 x 10^3 in rainforest soil and 2 x 10^4 in desert soil. Buée et al. (2009) reported that predicted richness of fungal ITS regions at the 97% sequence similarity level was approximately 2 x 10^3 in 4 g of forest soil.

Until recently, cloning and sequencing were primarily used to generate sequence data of environmental microbial community. PCR amplicons of rDNA and ITS region were cloned into an appropriate vector and clone libraries were sequenced by Sanger methods, also referred to as dideoxy chain termination sequencing (Sanger & Coulson, 1975). During the current decade, high-throughput second-generation sequencing technologies, such as pyrosequencing, have been developed and introduced to the research of microbial ecology (Petrosino et al., 2009; Roesch et al., 2007), including fungal community analyses (Buée et al., 2009; Lim et al., 2010; Lumini et al., 2010). Buée et al. assessed the fungal diversity in six different forest soils using 454 pyrosequencing (Buée et al., 2009). No less than 166350 reads were obtained from all samples. It enables reading of hundreds, thousands of PCR amplicons per 1 run.

DNA pyrosequencing, sequencing by synthesis, was developed in the mid 1990s as a fundamentally different approach to DNA sequencing (Ronaghi et al., 1996). Sequencing by synthesis occurs by a DNA polymerase-driven generation of inorganic pyrophosphate, with the formation of ATP and ATP-depending conversion of luciferin to oxyluciferin. The generation of oxyluciferin causes the emission of light pulses, and the amplitude of each signal is directly related to the presence of one or more nucleotides. Pyrosequencing can eliminate time and labours for cloning and has the 10-fold cost advantage per base pair over Sanger sequencing. The use of primer barcoding techniques enables to characterize many environmental samples in parallel on a single sequencing run. One important limitation of pyrosequencing is its relative inability to sequence longer stretches of DNA. With first- and second-generation pyrosequenceing chemistries, sequences rarely exceed 100-200 bases. Because of this limitation, cloning and Sanger sequencing are applied for the accurate recovery of longer sequence data at this stage.

3. Properties of PCR primers for fungal sequences

For fungal community analyses, PCR-based techniques are most powerful and generally used. The 18S rRNA gene (rDNA) and internal transcribed spacer (ITS) region are used widely as molecular markers for fungi, through the exploitation of both conserved and variable regions, and a large number of sequences are available in the data bank (Anderson & Cairney, 2004). Various PCR primer sets targeting 18S rDNA and ITS region are available for assessing fungal diversity in soil DNA samples (Table 1). Selection of PCR primer is one of the most important factors affecting outcome. Here, I will summarize their properties from our results and previous data.

3.1 PCR amplification and chimera formation

Although PCR-based strategies are the most powerful tools for the investigation of microbial diversity, they have a number of recognized limitations, perhaps the most insidious of which is the formation of recombinant or chimeric sequences during PCR amplification. Recombination can occur during PCR to jump from one template to another. Thus, whenever a heterogeneous pool of similar sequences, like rDNA and ITS regions, is amplified, chimera formation should be taken into account. The problem of chimeras in mixed DNAs from environmental samples has been highlighted several times in the literature (Kopczynski et al., 1994; Liesack et al., 1991; Wang & Wang, 1997). The existence of chimeras in PCR products may result in the overestimation of community diversity (Wintzingerode et al., 1997) and the occurrence of artificial novel taxa (Jumpponen, 2007). Chimeras seem to comprise a large proportion of the environmental sequence data in public databases (Ashelford et al., 2005; Hugenholtz & Huber, 2003). Jumppone (Jumpponen, 2007)

reported that a large proportion (40 or 31%) was chimeric in clone libraries obtained from
soil fungal analyses.

Genomic target	PCR primer	Primer sequence (5'-3')	reference
18S rDNA	NS1	GTAGTCATATGCTTGTCTC	(White et al., 1990)
	NS2	GGCTGCTGGCACCAGACTTGC	
	NS3	GCAAGTCTGGTGCCAGCAGCC	
	NS8	TCCGCAGGTTCACCTACGGA	
	EF3	TCCTCTAAATGACCAAGTTTG	(Smit et al., 1999)
	EF4	GGAAGGGRTGTATTTATTAG	
	Fung5	GTAAAAGTCCTGGTTCCCC	
	nu-SSU-0817	TTAGCATGGAATAATRRAATAGGA	(Borneman & Hartin, 2000)
	nu-SSU-1196	TCTGGACCTGGTGAGTTTCC	
	nu-SSU-1536	ATTGCAATGCYCTATCCCCA	
	FR1	AICCATTCAATCGGTAIT	(Vainio & Hantula, 2000)
	FF390	CGATAACGAACGAGACCT	
	Fun18S1	CCATGCATGTCTAAGTWTAA	(Lord et al., 2002)
	Fun18S2	GCTGGCACCAGACTTGCCCTCC	
	Fung	ATTCCCCGTTACCCGTTG	(May et al., 2001)
ITS	ITS1	TCCGTAGGTGAACCTGCGG	(White et al., 1990)
	ITS2	GCTGCGTTCTTCATCGATGC	
	ITS4	TCCTCCGCTTATTGATATGC	
	ITS1F	CTTGGTCATTTAGAGGAAGTAA	(Gardes & Bruns, 1993)
	ITS4B	CAGGAGACTTGTACACGGTCCAG	
	ITS4A	CGCCGTTACTGGGGCAATCCCTG	(Larena et al., 1999)
	2234C	GTTTCCGTAGGTGAACCTGC	(Sequerra et al., 1997)
	3126T	ATATGCTTAAGTTCAGCGGGT	
	PN3	CCGTTGGTGAACCAGCGGAGGGATC	(Viaud et al., 2000)
	PN34	TTGCCGCTTCACTCGCCGTT	

Table 1. Sequences of PCR primers used for assessing soil fungal diversity

PCR protocol was reported to affect the frequency of chimera formation (Qiu et al., 2001;
Wang & Wang, 1997; Wintzingerode et al., 1997). To evaluate the significance of primer
selection, we compared the compositions of 18S rDNA libraries amplified from upland and
paddy field soils using four primer sets: for single PCR, NS1/GCFung, FF390/FR1(N)-GC,
and NS1/FR1(N)-GC; and for nested PCR, NS1/EF3 for the first PCR and NS1/FR1(N)-GC
for the second PCR (Fig. 6) (Hoshino & Morimoto, 2010).

```
                        SSU-nu-0817 ►              ◄ SSU-nu-1196  ◄ SSU-nu-1536
      EF4 ►                ◄ fung5                              EF3 ◄
   [V1 | V2  |      | V3  |      V4      |   |V5|         |V7| V8 |   | V9 ]
  NS1 ►           ◄ Fung                          FF390 ►              ◄ FR1
```

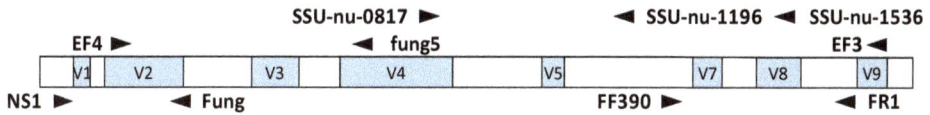

Fig. 6. Positions of PCR primers for fungal 18S rDNA. The variable regions (V) are highlighted in blue

The frequency of chimera sequences was related to the length of target regions of each primer set (Table 2). Long amplicons (targeted by primer sets NS1/FR1(N)-GC and NS1/EF3 & NS1/FR1(N)-GC) were more liable to produce chimeras than short amplicons (targeted by primer sets NS1/GCFung and FF390/FR1(N)-GC). Incomplete, prematurely terminated 18S rDNA sequences were also more frequent in the libraries obtained with primer sets NS1/FR1(N)-GC and NS1/EF3 & NS1/FR1(N)-GC. The concentration of amplified DNA initially increases exponentially and then gradually approaches a plateau when the depletion of reagents results in the generation of prematurely terminated strands. Such fragments seldom anneals with DNA strands of the same species among many homologous sequences of soil DNA (Torsvik et al., 1990), and the recombination events could be maximally expressed as chimeric molecules (Wang & Wang, 1996).

	Primer set	Expected products size (bp)	Number of PCR cycle	upland fiield soil				paddy field soil			
				mature 18S rDNA	immature 18S rDNA	chimeric 18S rDNA	non-18S rDNA	mature 18S rDNA	immature 18S rDNA	chimeric 18S rDNA	non-18S rDNA
1	NS1/GCFung	350	30	94.6	0.0	5.4	0.0	95.5	1.1	3.4	0.0
2	FF390/FR1-GC	390	40	89.0	0.0	9.9	1.1	89.9	2.2	7.9	0.0
3	NS1/FR1-GC	1650	40	62.5	11.4	18.2	8.0	79.1	4.4	8.8	7.7
4	NS1/EF3 NS1/FR1-GC	1650	25 30	29.5	17.9	48.4	4.2	41.1	22.1	34.7	2.1

Table 2. Composition of clone libraries obtained from upland and paddy field soils using primer sets 1. NS1/GCFung, 2. FF390/FR1(N)-GC, and 3. NS1/FR1(N)-GC; and for nested PCR, 4. NS1/EF3 for the first PCR and NS1/FR1(N)-GC for the second PCR. Figures indicated the percentage of the clones within each library (Hoshino & Morimoto, 2010)

Our results indicated that the numbers of PCR cycle was also related to chimera formation (Table 2). Because the efficacy of PCR amplification differed among primer sets, we used 30, 40, and 40 cycles for primer sets NS1/GCFung, FF390/FR1(N)-GC, and NS1/FR1(N)-GC, respectively, to produce a sufficient amount of PCR products for DGGE. In the case of nested PCR, 25 cycles were made for first PCR using NS1/EF3, and 20 cycles for second PCR using NS1/FR1(N)-GC. The higher the number of PCR cycle, the higher the frequency of chimeras. These results suggested that a smaller number of PCR cycles worked better. When using primer sets FF390/FR1(N)-GC and NS1/FR1(N)-GC; however, high PCR cycle numbers are needed to obtain enough product, because the efficacy of the amplification is low (Hoshino & Matsumoto, 2008).

Reducing chimera formation is required to provide a more accurate estimation of community diversity. Although the sequences with the potentiality of chimera could be identified and eliminated from data set, this procedure is often difficult and largely depends on personal judgement (Anderson & Cairney, 2004). The existence of chimera sequences also reduced available data in clone libraries (Table 2). Our results showed that properties of primer sets affected the frequency of chimera formation and that PCR protocol may be modified to decrease PCR cycles and to extend elongation time so that chimera contamination may minimized. Thus, PCR efficacy is an important factor, as well as the length of target fragment.

3.2 PCR primers specificity and bias in detection of fungal sequence from environmental samples

For accurate fungal community analyses, desirable PCR primer sets could exhaustively amplify fungal sequences without bias and strictly avoid the amplification non-fungal sequences from DNA pools extracted from environmental samples. However, primer sets targeting fungal 18S rDNA or ITS regions were designed for a broad range of fungi and consequently amplify genes of non-fungal organisms because of the high level of sequence similarity between 18S rDNAs of fungi and some closely related eukaryotes (Anderson & Cairney, 2004). Conversely, when increasing the specificity of primers for fungal genes, they may preferentially amplify a certain group of fungi, resulting in bias (Anderson & Cairney, 2004).

We evaluated single and nested PCR systems in terms of the frequency of non-fungal sequences and the diversity of fungal sequences in clone libraries (Hoshino & Morimoto, 2010). Four primer sets, i.e., for single PCR: NS1/GCFung, FF390/FR1(N)-GC, and NS1/FR1(N)-GC; and for nested PCR: NS1/EF3 for the first PCR and NS1/FR1(N)-GC for the second PCR, were compared using soil samples from upland and paddy fields. The rate of non-fungal eukaryotic 18S rDNAs amplified by single PCR ranged between 7 to 16 % for upland soil and between 20 to 31% for paddy field soil, whereas nested PCR produced a single eukaryotic clone in each library. The difference indicates that nested PCR increased the specificity to fungal sequences. Although the detection range of fungal taxa by 18S rDNA was generally similar among primer sets for single PCR, the fungal community detected by nested PCR was biased to specific sequences: diversity indices were significantly lower than those from single PCR in both libraries. These differences indicate that nested PCR system using primer set NS1/EF3 & NS1/FR1(N)-GC is not appropriate for diversity analysis on a wide range of taxonomic groups (i.e., total fungi).

The specificity of PCR primers varies depending on the composition of eukaryotic DNA contained in the extracted DNA pool. For example, although primer sets of EF4/EF3 and EF4/fung5 exclusively amplified fungal sequences from DNA extracted from wheat rhizosphere soil (Smit et al., 1999), they also amplified some non-fungal sequences from cultured organisms and avocado grove soil (Borneman & Hartin, 2000). Single PCR primer sets, NS1/GCFung, FF390/FR1(N)-GC and NS1/FR1(N)-GC, amplified more clones of non-fungal eukaryotic from the paddy soil than from the upland soil (Table 3)(Hoshino & Morimoto, 2010). These results may reflect the actual ratio of non-fungal related eukaryotic DNA to fungal DNA. In addition, phylogenetic groups of non-fungal eukaryotes detected were variable according to primer sets used (Table 3). The specificity of set NS1/FR1(N)-GC for fungi was higher than that of the other sets in the upland soil library but lower in the paddy soil library. The ratio of non-fungal gene preferentially detected by NS1/FR1(N)-GC was assumed to be higher in the DNA pool of paddy field soil than in that of upland field soil. Primer specificity is also affected by compositions of non-fungal eukaryotic sequences in samples. It showed that it is critical to check the specificity of primer sets for environmental samples to be studied.

Anderson et al. (Anderson et al., 2003) reported that the relative proportion of sequences representing the four main fungal phyla was similar in clone libraries from grassland soil with primer sets nu-SSU-817/nu-SSU-1196, nu-SSU-817/nu-SSU-1536, EF4/EF3 and ITS1F/ITS4. On the other hand, Jumpponen (Jumpponen, 2007) reported that EF4/EF3 biased toward Basidiomycota as predicted (Smit et al., 1999) and that nu-SSU-817/nu-SSU-1536 mainly amplified Ascomycota from soil samples of underneath willow canopies. We found that fungal 18S rDNA fragments showed a similar distribution at the phylum level in the upland and paddy soil libraries amplified with primer sets NS1/GCFung, FF390/FR1(N)-GC and NS1/FR1(N)-GC (Hoshino & Morimoto, 2010). The detection frequency of the Chytridiomycota, however, differed among these primer sets. NS1/GCFung failed to detect the Chytridiomycota, while FF390/FR1(N)-GC amplified it more efficiently (Table 3). At the class level, especially in the paddy soil libraries, the difference was evident in the distribution of fungal taxa inferred from 18S rDNA with these primer sets (Table 3). Primer selection has a pivotal importance on the community structure to be investigated although primer bias may not be as significant as previously thought, as Anderson et al. (Anderson et al., 2003) concluded.

kingdom	subkingdom	phylum	subphylum	class	upland field soil				paddy field soil			
					1	2	3	4	1	2	3	4
fungi	Dikarya	Ascomycota	Pezizomycotina	Dothideomycetes	19	11	9.1	11	9.5	10	4.2	5.1
				Eurotiomycetes	3.4	3.7	5.5				8.3	
				Leotiomycetes					25	14	4.2	2.6
				Orbiliomycetes		3.7				2.5		
				Pezizomycetes	4.5	3.7	1.8					
				Sordariomycetes	27	35	42	29	2.4	8.8	8.3	5.1
				unindentified						1.3		
		Basidiomycota	Basidiomycota	Agricomycets	6.8	9.9	7.3	3.6	21	16	26	67
				Tremellomycetes			1.8		2.4	2.5	1.4	
			Pucciniomycotina	Atractiellomycetes					1.2			
				Microbotryomycetes						2.5	1.4	
	incertae sedis	Chytridiomycota		Chytridiomycetes		8.6	7.3		1.2	6.3	1.4	
				Monoblepharidomycetes						1.4		
		Blastocladiomycota		Blastocladiomycetes							1.4	2.6
		incertae sedis	Mucoromycotina	Mucoromycetes	22	15	18	57	6	16	11	15
			Zoopagomycotina			1.2				1.4		
			Kickxellomycotina		1.1							
Amoebabiota		Amoebazoa							1.2			
Animalia	Amoebidiobiotina	Amoebidiozoa							1.2			
(=Metazoa)	Bilateralia	Annelida							25		13	
		Arthropoda			1.1							
Rhizaria		Cercozoa			6.8	2.5	3.6		1.2	19	9.7	2.6
Alveolata		Apicomplexa							3.6			
stramenopiles		Oomycota			8	1.2	1.8				1.4	
		Ochrophyta							1.2		4.2	
		incertae sedis					1.8				1.4	
Plantae		Viridiplantae							1.3			
incertae sedis		Apusozoa				1.2						
		Heliozoa				1.2						

Table 3. Distribution of 18S r RNA gene sequenses in clone libraries from upland and paddy field soils using different primer sets for single PCR (1-3) and nested PCR (4), i.e., 1, NS1/GCFung; 2, FF390/FR1-GC; 3, NS1/FR1-GC; and 4, NS1/EF3 for first PCR and NS1/FR1-GC for second PCR. Figures indicated the percentage of the clones within each library

These results indicate that appropriate primers should be selected according to the aims and the origin of samples and/or that more than two primer sets with different properties should be used to obtain a more comprehensive view of the fungal communities.

3.3 Applicability for DGGE

Community analysis by DGGE is sensitive to choice of primer sets because the separation of each DNA fragment in denetuaring gradient gels largely depends on the sequences of target regions. Okubo & Sugiyama (Okubo & Sugiyama, 2009) compared five fungal primer sets in terms of band separation of four fungal species in DGGE gels; when using EF4/GCFung, bands of the four species showed the same mobility in DGGE gels and were not separable, while they separated but smeared with EF4/Fung5 or ITS1F/ITS2-GC. On the other hands, NS1/GCFung and FF390/FR1-GC produced separate and single bands.

We evaluated primer sets for fungal 18S rDNA DGGE using agricultural soils in terms of the following features: detection and reproducibility of DGGE banding profiles, obtained diversity indices, and ability to discriminate fungal communities by DGGE (Hoshino & Matsumoto, 2008). Four primer sets, i.e., for single PCR, NS1/GCFung, FF390/FR1(N)-GC, and NS1/FR1(N)-GC; and for nested PCR, NS1/EF3 for the first PCR and NS1/FR1(N)-GC for the second PCR, were compared using six soil samples from upland (F1, F2, F3 and F4) and paddy fields (P1 and P2) in Japan (Fig. 6).

PCR products with different primer sets under the appropriate experimental regimes showed clear band separation in DGGE analysis, as reported previously (May et al., 2001; Oros-Sichler et al., 2006; Vainio & Hantula, 2000). In addition, repeated trials with the same samples produced virtually identical profiles in the same DGGE gels with primer sets NS1/GCFung and FF390/FR1(N)-GC. However, when primer set NS1/FR1(N)-GC was used, aggregates present in the middle of DGGE gel that sometimes interfered with the detection of target bands. We also detected smiling and distortion of banding patterns in DGGE with these primer sets. The presence of nonspecific aggregates and the distortion of banding pattern reduced reproducibility especially between different gels.

Although there was no significant difference in the number of bands that appeared in each sample among primer sets (Fig. 7A), the Shannon diversity indices, used to measure diversity in categorical data (Krebs, 1989), were lowest for primer set FF390/FR1(N)-GC, and tended to be higher for primer sets NS1/FR1(N)-GC and NS1/EF3 & NS1/FR1(N)-GC (Fig. 7B). Two main bands were highly dominant in DGGE profiles of these six samples with primer set FF390/FR1(N)-GC. However, sequence diversity in clone library with primer set FF390/FR1(N)-GC was the higher than libraries with other primer sets (Hoshino & Morimoto, 2010). The main reason for the difference may be ascribed to the low band separability in DGGE.

To evaluate the ability to discriminate fungal communities, multidimensional scaling (MDS) maps were generated from DGGE profiles for each primer set. Each MDS map showed a similar tendency (Fig. 8). Samples from upland and paddy field soils were positioned separately, with the exception of sample F4, which was always distant from other samples in the MDS maps. The MDS map with primer set NS1/GCFung showed the highest differentiation, with samples distantly located from one another, whereas with primer set FF390/FR1(N)-GC, except F4 differentiation among samples was lower in the MDS map. With primer set NS1/FR1(N)-GC, samples F1 and F3 and samples P1 and P2 were plotted close together.

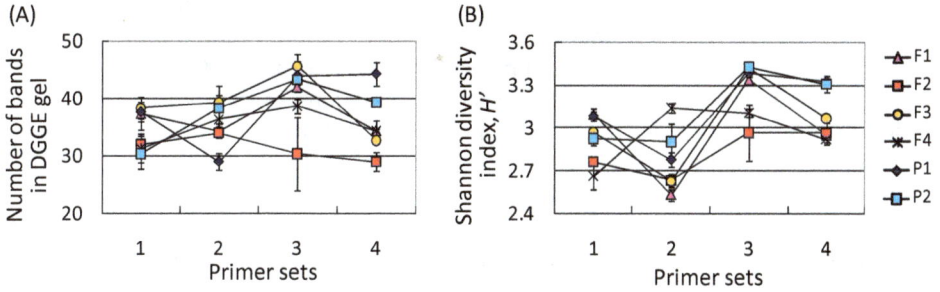

Fig. 7. Number of bands in DGGE gel (A), Shannon diversity index (B) of 18S rDNA DGGE profiles of upland field soils (F1, F2, F3, F4) and paddy field soils (P1, P2) using primer sets: 1. NS1/GCFung, 2. FF390/FR1(N)-GC, 3. NS1/FR1(N)-GC, and 4. NS1/EF3 for the first PCR and NS1/FR1(N)-GC for the second PCR (Modified from Fig. 4 in (Hoshino & Matsumoto, 2008))

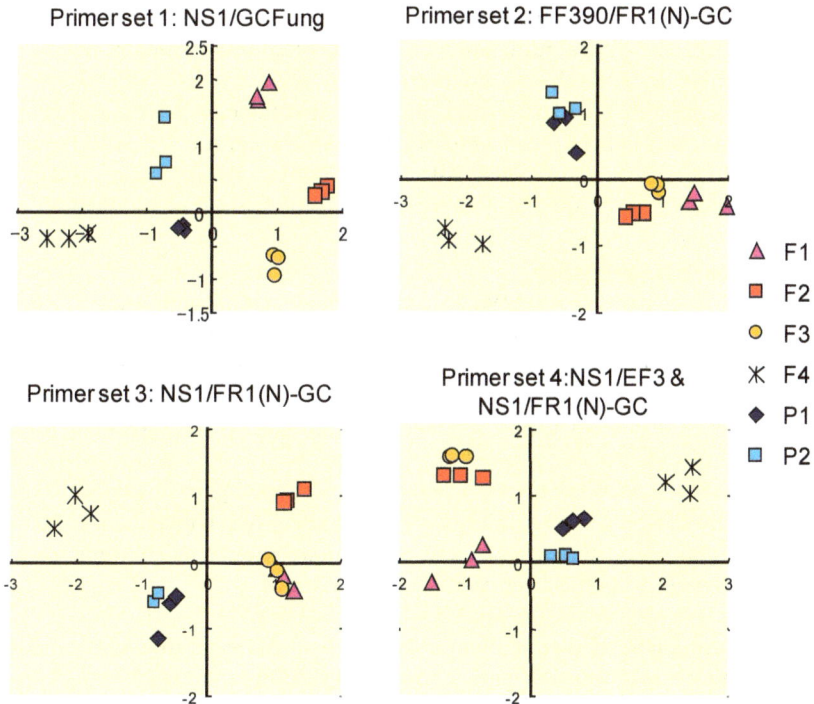

Fig. 8. Multidimensional scaling (MDS) map based on the squared distance of similarity of the 18S rDNA DGGE profiles of upland field soils (F1, F2, F3, F4) and paddy field soils (P1, P2) using primer sets: 1. NS1/GCFung, 2. FF390/FR1(N)-GC, 3. NS1/FR1(N)-GC, and 4. NS1/EF3 for the first PCR and NS1/FR1(N)-GC for the second PCR (Modified from Fig. 5 in (Hoshino & Matsumoto, 2008))

These data suggested that primer sets NS1/GCFung and FF390/FR1-GC were applicable for soil fungal community DGGE analysis and that primer set NS1/GCFung was the most suitable, considering all the various factors together. Comparison of DGGE profiles among each study is required to standardize experimental conditions, especially PCR primers. We selected primer set NS1/GCFung for this purpose and established DGGE experimental conditions using this primer set to prepare experimental protocols and technical reports of bacterial and fungal DGGE analyses (Morimoto & Hoshino, 2010).

4. Application examples of molecular techniques to fungal community analyses in agricultural field soils

Molecular analyses of fungal community have been reported for various soils in different ecosystems, such as forests (Perkiömäki et al., 2003), grasslands (Brodie et al., 2003), dunes (Kowalchuk et al., 1997), and stream sediments (Nikolcheva et al., 2003), as well as agricultural fields (Gomes et al., 2003). In agricultural soils, many field trials have shown the effect of plant cultivation (Gomes et al., 2003), fertilizer and pesticide application on fungal community (Girvan et al., 2004). Here, I will show two examples of the application of molecular techniques to fungal community analyses in Japanese agricultural soils. We analyzed the impact of chemical fumigation on fungal community structure of bulk soil and spinach rhizosphere in a field and monitored their recovery from the drastic change (Hoshino & Matsumoto, 2007). The results suggested that the effects were different among the chemicals and between bulk soil and rhizosphere. In addition, it was reported that fungal communities were most obviously affected by fertilizer treatment, i.e., changes in soil nutrient status, rather than edaphic factors such as soil type (Suzuki et al., 2009).

4.1 Effect of chemical fumigants

Pre-planting soil fumigation is used widely around the world in high-value crops and has been shown to be effective to control soil-borne pathogens, weeds, and plant-parasitic nematodes. In Japan, especially in areas that produce vegetables such as spinach, lettuce, and tomato, continuous monoculture is widely adopted to increase profit, often resulting in outbreak of pests. Many areas utilize soil chemical fumigation for consistent production. Most of chemical fumigants have a broad range of biocidal activity and can potentially harm beneficial organisms, in addition to target pests. Although methyl bromide (MeBr) had been widely used in the past, the use of MeBr in soil fumigation was banned since 2005 because of its environmental risk. Therefore, the use of alternatives, such as chloropicrin (CP) and 1,3-dichloropropene (1,3-D), has been increasing (Dungan et al., 2003). Their effect on non-target organisms was also of concern and should be evaluated.

However, there are relatively few studies on the effect of chemical fumigants on non-target soil fungal community as compared with a number of studies reporting the effect on specific plant pathogens (Browning et al., 2006; Hamm, 2003; Takehara et al., 2003) and on bacterial community (Dungan et al., 2003; Ibekwe et al., 2001). Itoh et al. (Itoh et al., 2000) and Tanaka et al. (Tanaka et al., 2003) reported that the count of viable fungi decreased after CP fumigation. De Cal et al. (De Cal et al., 2005) used a culture-dependent method with selective media to show that chemical fumigants reduced certain members of soil fungi, such as *Fusarium* spp., *Pythium* spp., and *Verticillium* spp. We aimed to analyze the effect of two chemical fumigants (CP and 1,3-D) and spinach growth on fungal community structure in a field using molecular techniques (Hoshino & Matsumoto, 2007).

Experiments were performed in an experimental field in Tsukuba, Japan. Annual cropping system consisted of soil fumigation in September followed by two consecutive spinach cultivations. Soil was treated with fumigants (CP at 20 ml m^{-2} or 1,3-D at 32 ml m^{-2}) and covered with polyethylene film for about two weeks. Bulk soil and rhizosphere samples were taken periodically during the three fumigation trials. DNA was extracted directly from soil samples and fungal 18S rRNA genes were amplified by nested PCR with primer pairs AU2/AU4 and GC-FR1/FF390 for DGGE analyses. Dezitized data of DGGE profiles were analyzed by (i) diversity indices and (ii) multivariate statistical technique. Dominant bands in DGGE gels were excised for sequencing. Sequences of DGGE bands were identified with the FASTA search from the database of the DNA Data Bank of Japan (DDBJ).

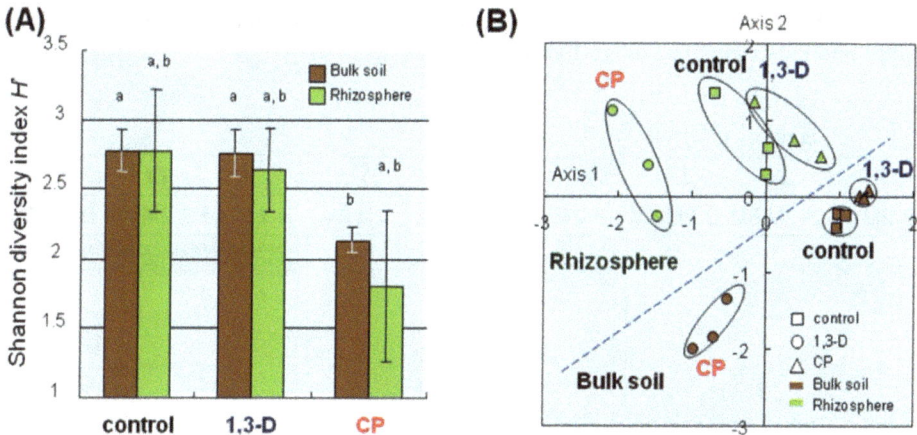

Fig. 9. Quantitative analysis of the 18S rDNA DGGE profiles 2 months after fumigation in the second trial (year 2). (A) Shannon's diversity index and (B) multi-dimensional scaling (MDS) map based on the squared distance of similarity. (Modified from Fig. 3 in (Hoshino & Matsumoto, 2007))

We compared the fungal 18S rDNA DGGE profiles among each treatment two month after fumigation both in bulk soil and rhizosphere. The Shannon diversity index H' (Fig. 9A) was calculated from these profiles. The index for bulk soil in the CP plots was significantly lower than that in the control plots ($P < 0.05$). The index for the rhizosphere soil in the CP plots also tended to be lower than that of control plots, but the difference was not significant. These values of the 1,3-D plots were almost equivalent to those of the control plots.

These DGGE profiles were also analyzed by multi-dimensional scaling (MDS) (Fig. 9B). The MDS map shows every band pattern on one plot, where relative changes in community structure can be visualized and interpreted as the distances between the points (Araya et al., 2003). The closer the points with each other, the more similar the DGGE banding patterns represented by the points are. In the MDS map, samples from the bulk soil and the rhizosphere soil were positioned separately, indicating that spinach cultivation affected soil fungal community structure, too (Fig. 9B). The MDS map also showed that the difference in DGGE profiles was greater between CP and control plots than between 1,3-D and control plots, both in bulk soil and rhizosphere soil (Fig. 9B). When the magnitude of the impact was compared between samples of bulk soil and rhizosphere soil, the differences in DGGE profiles between control and chloropicrin plots were smaller in rhizosphere soil than in bulk soil.

We monitored changes in fungal DGGE profiles in bulk soil after chemical fumigation of this field over three years. Fig. 10 shows the change in DGGE profiles of plots fumigated with chloropicrin or 1,3-D before fumigation, two months after fumigation, and six months after fumigation for each year. DGGE profiles drastically changed after CP treatment and did not recover completely 1 year after, e.g., before treatments in years 2 and 3 (Fig.10). In contrast, DGGE profiles of 1,3-D plots revealed a smaller change 2 months after fumigation but became indistinguishable from those of control plots after 6 months. These results indicated that the impact of fumigation on the soil fungal community was greater in the CP treatment than in the 1,3-D treatment both in terms of the magnitude of the effect after 2 months and the extent of recovery 1 year after.

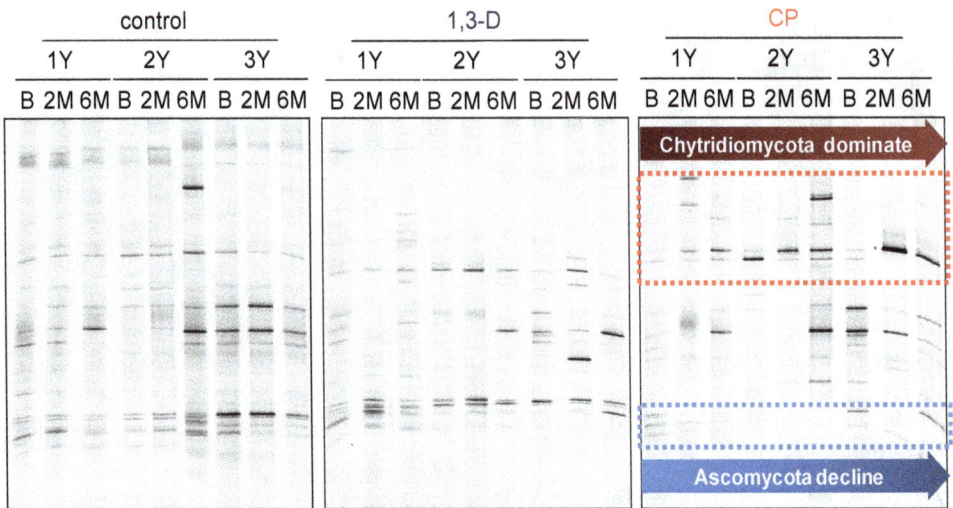

Fig. 10. Temporal change after fumigation in fungal 18S rDNA DGGE profiles from bulk soil samples in (A) untreated control plot, (B) CP plot, and (C) 1,3-D plot. Samples were taken before (B) and 2 and 6 months (2M and 6M, respectively) after fumigation for each year (1Y, 2Y and 3Y) in three-year trials

Between treatments of CP and 1,3-D, there are differences in fungal species affected. In CP-treated plots, bands with high sequence similarity to *Myrothecium cinctum* (100%, Ascomycota), *Bionectria ochroleuca* (99.7%, Ascomycota), *Metarhizium anisopliae* (100%, Ascomycota), *Dentipellisseparans* (96.7%, Basidiomycota), *Verticillium dahliae* (98.4%, Ascomycota) and *Exophiala dermatitidis* (100%, Ascomycota) decreased in band intensity. On the other hand, band intensity decreased in *Basidiobolus microsporus* (83.7%, Basidiomycota) and *Bensingtonia ciliate* (88.9%, Basidiomycota) in 1,3-D treated plots.

After CP treatment, bands inferred to represent chytridiomycota became dominant (Fig.10). Chytrids can rapidly reproduce and increase their populations in response to disturbance (Lozupone & Klein, 2002). These characteristics could allow them to quickly exploit nutrients released after soil disturbance, such as fumigation, increasing their overall population. Chytridiomycota cannot be detected by conventional dilution-plate counting, and are usually studied using culture and microscopic protocols based on baiting techniques, using a "bait" substrate to attract chytrids under flooded conditions (Lozupone & Klein, 2002). Our results indicated the advantages of molecular techniques to detect whole fungal community including such fungal groups.

4.2 Effect of soil types and fertilizers

Soil microbial communities are influenced by various factors such as cropping system (Kuske et al., 2002), tillage (Peixoto et al., 2006), fertilization (Marschner et al., 2003) and application of pesticide and herbicide (Yang et al., 2000). On the other hand, environmental factors, including soil characteristics, also affect microbial communities, e.g., soil type (Girvan et al., 2003), soil particle size (Sessitsch et al., 2001), soil air composition (Øvreås et al., 1998) and season (Girvan et al., 2004). Bacteria have been well documented for agricultural soils; many field trials have shown that the composition of the entire bacterial community is determined primarily by soil type (Girvan et al., 2003; Xu et al., 2009), emphasizing the effect of soil chemistry and structure, especially pH and soil texture (Fierer & Jackson, 2006; Lauber et al., 2008), rather than cultural practices. However, little has so far been known about factors affecting fungal community structure.

Suzuki et al. (Suzuki et al., 2009) studied the effect of soil type and fertilizer type on bacterial and fungal communities in a long-term experimental field in Tsukuba, Japan. Upland field plots containing four different soil types, i.e., Gleyic Mollic-Umbric Andosols (Cumulic Andosol), Gleyic Haplic Andosols (Low-humic Andosol), Gleyic Haplic Alisols (Yellow Soil), and Entric Fluvisols (Gray Lowland Soil) were maintained under three different fertilizer management systems (chemical fertilizer rice husks plus cow manure, and pig manure) for 5 years. Carrot and maize were annually cropped in the fields once every summer. Bulk soil samples were taken in May prior to fertilization and cultivation. From directly-extracted soil DNA, bacterial 16S rDNA and fungal 18S rDNA were amplified using primer pairs 968g-GC/1378r and NS1/GCFung, respectively and subjected to DGGE analyses.

Fungal DGGE profiles based on the 18S rRNA gene were analyzed by principal component analysis (PCA) to separate plots based on fertilization practices. This result showed that fungal community composition was more directly related to fertilization than soil type. On the other hand, PCA of bacterial DGGE profiles indicated that the plots were separated by soil type. Lauber et al. (Lauber et al., 2008) reported that fungal community composition was most closely associated with changes in soil nutrient status, i.e., concentration of total nitrogen and extractable phosphate, and the ratio of total carbon and nitrogen concentration. Suzuki et al. (Suzuki et al., 2009) described that fungi may be more suitable as microbial indicators of soil quality because the dynamics of fungal community were more reflected soil nutrient status than that of bacterial community.

5. Future perspectives

PCR-based techniques targeting 18S rDNA are powerful tools for fungal community analysis and have revealed phylogenetic compositions and dynamics of fungal communities in the environment. The accumulating molecular data has facilitated fungal community analyses on a large scale. Several previous reports, including examples shown in Section 4, have indicated that the fungal community could be substantially altered by cultural practices. However, most results were obtained from a few experimental fields. Large-scale and comprehensive analyses using enormous amounts of data on soils from various regions are required to determine whether the results presented in those reports are universally applicable or represent specific examples.

In Japan, the Environmental DNA database for agriculture soils (eDDASs) was established, which included not only DGGE profiles of bacteria, fungi and nematodes but also relevant information on soil, cultural practices, crop yield, etc. eDDASs facilitates large-scale analyses of the relationships between soil microbial communities and various environmental factors and may facilitate resolution of problems, such as disease forecasting, soil fertility evaluation, etc, in agricultural fields (Tsushima et al., 2011). The introduction of the next generation of sequencers combined with the development of bioinformatics tools will accelerate such large-scale analyses.

The abovementioned molecular techniques have limitations for the analyses of environmental fungal communities. The sequences of 18S rDNA or ITS regions only reflect the phylogenetic positions of target microbes but not necessarily their metabolic functions. The existence of DNA in soil, even functional genes, only demonstrates the potential of fungal activity not a confirmation of its actual presence. Analyses based on the utilisation of soil RNA and/or other genetic markers associated with metabolic function should fortify fungal community analyses. The next step should focus on the functional aspects of fungal communities.

PCR can cause biased detections that prevent the complete recognition of microbial diversity through primer specificity and simultaneous amplification of different targets. New approaches that do not depend on PCR, such as metagenomic or metatranscriptomic analysis, can provide less biased data on fungal community structures and functional aspects, although some problems remain, particularly in data analyses (Suenaga, 2001). Currently, it is difficult to directly assign individual sequences that were directly recovered from soils or to construct contigs from them because a single soil sample may contain several thousand microbial genotypes, whereas most of their genomic sequences are still unrevealed.

The development of new molecular technologies should alleviate the problems associated with rDNA-based methods and PCR amplification and promote the investigation of current topics, such as the effect of pollution and global warming on fungal communities and their functions and roles in soil ecosystems.

6. Conclusion

Culture-independent molecular techniques, such as direct DNA extraction from soil followed by PCR-based community analysis techniques, provide novel insights and significant research advances in soil microbial ecology. Compared with bacterial communities, however, the results of soil fungal community analyses using molecular techniques are limited. One reason for these limitations is the lack of sufficient information, in the case of fungi, about the influence of various experimental parameters, particularly PCR primer selection, on the results of diversity studies.

We evaluated various PCR primer sets targeting the 18S rRNA gene (rDNA), a widely used molecular marker for fungi, as well as other experimental parameters and established a standard DGGE protocol for soil fungal community analysis. Molecular methods revealed that soil fungal communities were affected by cultural practices, such as chemical fumigation and fertilization, in agricultural fields. These techniques undergo constant improvement and should continue to promote research based on fungal ecology in soil ecosystems.

7. Acknowledgment

We thank Dr. Naoyuki Matsumoto (HokkaidoUniversity) for useful suggestions regarding an early draft of this manuscript. This work was partially supported by a Grant-in-Aid (Soil eDNA) from the Ministry of Agriculture, Forestry and Fisheries of Japan (eDNA-07-101-3).

8. References

Anderson IC & Cairney JWG (2004). Diversity and ecology of soil fungal communities: increased understanding through the application of molecular techniques. *Environmental Microbiology*, Vol.6, No.8, (August 2004), pp. 769-779, ISSN 1462-2912

Anderson IC, Campbell CD & Prosser JI (2003). Potential bias of fungal 18S rDNA and internal transcribed spacer polymerase chain reaction primers for estimating fungal biodiversity in soil. *Environmental Microbiology*, Vol.5, No.1, (January 2003), pp. 36-47, ISSN 1462-2912

Araya R, Tani K, Takagi T, Yamaguchi N & Nasu M (2003). Bacterial activity and community composition in stream water and biofilm from an urban river determined by fluorescent in situ hybridization and DGGE analysis. *FEMS Microbiology Ecology*, Vol.43, No.1, (February 2003), pp. 111-119, ISSN 0168-6496

Ashelford KE, Chuzhanova NA, Fry JC, Jones AJ & Weightman AJ (2005). At least 1 in 20 16S rRNA sequence records currently held in public repositories is estimated to contain substantial anomalies. *Applied and Environmental Microbiology*, Vol.71, No.12, (December 2005), pp. 7724-7736, ISSN 0099-2240

Borneman J & Hartin RJ (2000). PCR primers that amplify fungal rRNA genes from environmental samples. *Applied and Environmental Microbiology*, Vol.66, No.10, (October 2000), pp. 4356-4360, ISSN 0099-2240

Borneman J, Skroch PW, O'Sullivan KM , Palus JA, Rumjanek NG, Jansen JL, Nienhuis J & Triplett EW (1996). Molecular microbial diversity of an agricultural soil in Wisconsin. *Applied and Environmental Microbiology*, Vol.62, No.6, (June 1996), pp. 1935-1943, ISSN 0099-2240

Brodie E, Edwards S & Clipson N (2003). Soil fungal community structure in a temperate upland grassland soil. *FEMS Microbiology Ecology*, Vol.45, No.2, (July 2003), pp. 105-114, ISSN 0168-6496

Browning M, Wallace DB, Dawson C, Alm SR & Amador JA (2006). Potential of butyric acid for control of soil-borne fungal pathogens and nematodes affecting strawberries.

Soil Biology and Biochemistry, Vol.38, No.2, (February 2006), pp. 401-404, ISSN 0038-0717

Buée M, Reich M, Murat C, Morin E, Nillson RH, Uroz S & Martin F (2009). 454 Pyrosequencing analyses of forest soils reveal an unexpectedly high fungal diversity. *New Phytologist,* Vol.184, No.2, (October 2009), pp. 449-456, ISSN 1469-8137

Christensen M (1989). A view of fungal ecology. *Mycologia,* Vol.81, No.1, (March 1989), pp. 1-19, ISSN 0027-5514

De Cal A, Martinez-Treceno A, Salto T, Lopez-Aranda JM & Melgarejo P (2005). Effect of chemical fumigation on soil fungal communities in Spanish strawberry nurseries. *Applied Soil Ecology,* Vol.28, No.1, (January 2005), pp. 47-56, ISSN 0929-1393

Dunbar J, Ticknor LO & Kuske CR (2000). Assessment of Microbial Diversity in Four Southwestern United States Soils by 16S rRNA Gene Terminal Restriction Fragment Analysis. *Applied and Environmental Microbiology,* Vol.66, No.7, (July 2000), pp. 2943-2950, ISSN 0099-2240

Dungan RS, Ibekwe AM & Yates SR (2003). Effect of propargyl bromide and 1,3-dichloropropene on microbial communities in an organically amended soil. *FEMS Microbiology Ecology,* Vol.43, No.1, (February 2003), pp. 75-87, ISSN 0168-6496

Fierer N & Jackson RB (2006). The diversity and biogeography of soil bacterial communities. *Proceedings of the National Academy of Sciences of the United States of America,* Vol.103, No.3, (January 2006), pp. 626-631, ISSN 0027 -8424

Fierer N, Breitbart M, Nulton J, Salamon P, Lozupone C, Jones R, Robeson M, Edwards RA, Felts B, Rayhawk S, Knight R, Rohwer F & Jackson RB (2007). Metagenomic and Small-Subunit rRNA Analyses Reveal the Genetic Diversity of Bacteria, Archaea, Fungi, and Viruses in Soil. *Applied and Environmental Microbiology,* Vol.73, No.21, (November 2007), pp. 7059-7066, ISSN 0099-2240

Fromin N, Hamelin J, Tarnawski S, Roesti D, Jourdain-Miserez K, Forestier N, Teyssier-Cuvelle S, Gillet F, Aragno M & Rossi P (2002). Statistical analysis of denaturing gel electrophoresis (DGE) fingerprinting patterns. *Environmental Microbiology,* Vol.4, No.11, (November 2002), pp. 634-643, ISSN 1462-2912

Gardes M & Bruns TD (1993). ITS primers with enhanced specificity for basidiomycetes - application to the identification of mycorrhizae and rusts. *Molecular Ecology,* Vol.2, No.2, (April 1993), pp. 113-118, ISSN 0962-1083

Girvan MS, Bullimore J, Ball AS, Pretty JN & Osborn AM (2004). Responses of Active Bacterial and Fungal Communities in Soils under Winter Wheat to Different Fertilizer and Pesticide Regimens. *Applied and Environmental Microbiology,* Vol.70, No.5, (May 2004), pp. 2692-2701, ISSN 0099-2240

Girvan MS, Bullimore J, Pretty JN, Osborn AM & Ball AS (2003). Soil type Is the primary determinant of the composition of the total and active bacterial communities in arable soils. *Applied and Environmental Microbiology,* Vol.69, No.3, (March 2003), pp. 1800-1809, ISSN 0099-2240

Gomes NCM, Fagbola O, Costa R, Rumjanek NG, Buchner A, Mendona-Hagler L & Smalla K (2003). Dynamics of fungal communities in bulk and maize rhizosphere soil in the tropics. *Applied and Environmental Microbiology,* Vol.69, No.7, (July 2003), pp. 3758-3766, ISSN 0099-2240

Hamm PB, Ingham RE, Jaeger JR, Swanson WH & Volker KC (2003). Soil fumigant effects on three genera of potential soilborne pathogenic fungi and their effect on potato yield in the columbia basin of oregon. *Plant Disease,* Vol.87, No.12, (December 2003), pp. 1449-1456, ISSN 0191-2917

Hoshino YT & Matsumoto N (2004). An Improved DNA Extraction Method Using Skim Milk from Soils That Strongly Adsorb DNA. *Microbes and Environments,* Vol.19, No.1, (March 2004), pp. 13-19, ISSN 1342-6311

Hoshino YT & Matsumoto N (2007). Changes in fungal community structure in bulk soil and spinach rhizosphere soil after chemical fumigation as revealed by 18S rDNA PCR-DGGE. *Soil Science and Plant Nutrition,* Vol.53, No.1, (February 2007), pp. 40-55, ISSN 0038-0768

Hoshino YT & Matsumoto N (2008). Comparison of 18S rDNA primers for estimating fungal diversity in agricultural soils using polymerase chain reaction-denaturing gradient gel electrophoresis *Soil Science and Plant Nutrition,* Vol.54, No.5, (October 2008), pp. 701-710, ISSN 0038-0768

Hoshino YT & Morimoto S (2010). Soil clone library analyses to evaluate specificity and selectivity of PCR primers targeting fungal 18S rDNA for denaturing-gradient gel electrophoresis (DGGE). *Microbes and Environments,* Vol.25, No.4, (December 2010), pp. 281-287, ISSN 1342-6311

Hugenholtz P & Huber T (2003). Chimeric 16S rDNA sequences of diverse origin are accumulating in the public databases. *International Journal of Systematic and Evolutionary Microbiology,* Vol.53, No.1, (January 2003), pp. 289-293, ISSN 1466-5026

Ibekwe AM, Papiernik SK, Gan J, Yates SR, Yang C-H & Crowley DE (2001). Impact of fumigants on soil microbial communities. *Applied and Environmental Microbiology,* Vol.67, No.7, (July 2001), pp. 3245-3257, ISSN 0099-2240

Itoh K, Takahashi M, Tanaka R, Suyama K & Yamamoto H (2000). Effect of fumigants on soil microbial population and proliferation of *Fusarium oxysporum* inoculated into fumigated soil. *Journal of Pesticide Science,* Vol.25, No.2, (August 2000), pp. 147-149, ISSN 1348-589X

Jumpponen A (2007). Soil fungal communities underneath willow canopies on a primary successional glacier forefront: rDNA sequence results can be affected by primer selection and chimeric data. *Microbial Ecology,* Vol.53, No.2, (February 2007), pp. 233-246, ISSN 0095-3628

Kopczynski ED, Bateson MM & Ward DM (1994). Recognition of chimeric small-subunit ribosomal DNAs composed of genes from uncultivated microorganisms. *Applied and Environmental Microbiology,* Vol.60, No.2, (February 1994), pp. 746-748, ISSN 0099-2240

Kowalchuk GA, Gerards S & Woldendorp JW (1997). Detection and characterization of fungal infections of *Ammophila arenaria* (marram grass) roots by denaturing gradient gel electrophoresis of specifically amplified 18S rDNA. *Applied and Environmental Microbiology,* Vol.63, No.10, (October 1997), pp. 3858-3865, ISSN 0099-2240

Kowalchuk GA, Drigo B, Yergeau E & van Veen JA (2006): Assessing bacterial and fungal community structure in soil using ribosomal RNA and other structural gene markers. In: *Nucleic Acids and Proteins in Soil,* Nannipieri P & Smalla K (Eds), 159-188. Springer Berlin Heidelberg, ISBN 978-3-540-29448-1, Germany

Krebs CJ (1989). (January 1989). *Ecological Methodology*, Harpercollins, ISBN 978-0060437848, New York

Kuske CR, Ticknor LO, Miller ME, Dunbar JM, Davis JA, Barns SM & Belnap J (2002). Comparison of Soil Bacterial Communities in Rhizospheres of Three Plant Species and the Interspaces in an Arid Grassland. *Applied and Environmental Microbiology*, Vol.68, No.4, (April 2002), pp. 1854-1863, ISSN 0099-2240

Larena I, Salazar O, González V, Julián MC & Rubio V (1999). Design of a primer for ribosomal DNA internal transcribed spacer with enhanced specificity for ascomycetes. *Journal of Biotechnology*, Vol.75, No.2-3, (October 1999), pp. 187-194, ISSN 0168-1656

Lauber CL, Strickland MS, Bradford MA & Fierer N (2008). The influence of soil properties on the structure of bacterial and fungal communities across land-use types. *Soil Biology and Biochemistry*, Vol.40, No.9, (September 2008), pp. 2407-2415, ISSN 0038-0717

Liesack W, Weyland H & Stackebrandt E (1991). Potential risks of gene amplification by PCR as determined by 16S rDNA analysis of a mixed-culture of strict barophilic bacteria. *Microbial Ecology*, Vol.21, No.1, (December 1991), pp. 191-198, ISSN 0095-3628

Lim Y, Kim B, Kim C, Jung HS, Kim BS, Lee JH & Chun J (2010). Assessment of soil fungal communities using pyrosequencing. *The Journal of Microbiology*, Vol.48, No.3, (June 2010), pp. 284-289, ISSN 1225-8873

Lord NS, Kaplan CW, Shank P, Kitts CL & Elrod SL (2002). Assessment of fungal diversity using terminal restriction fragment (TRF) pattern analysis: comparison of 18S and ITS ribosomal regions. *FEMS Microbiology Ecology*, Vol.42, No.3, (December 2002), pp. 327-337, ISSN 0168-6496

Lozupone CA & Klein DA (2002). Molecular and cultural assessment of chytrid and Spizellomyces populations in grassland soils. *Mycologia*, Vol.94, No.3, (May – June 2002), pp. 411-420, ISSN 0027-5514

Lumini E, Orgiazzi A, Borriello R, Bonfante P & Bianciotto V (2010). Disclosing arbuscular mycorrhizal fungal biodiversity in soil through a land-use gradient using a pyrosequencing approach. *Environmental Microbiology*, Vol.12, No.8, (August 2010), pp. 2165-2179, ISSN 1462-2912

Lyons JI, Newell SY, Buchan A & Moran MA (2003). Diversity of Ascomycete laccase gene sequences in a southeastern US salt marsh. *Microbial Ecology*, Vol.45, No.3, (April 2003), pp. 270-281, ISSN 0095-3628

Marschner P, Kandeler E & Marschner B (2003). Structure and function of the soil microbial community in a long-term fertilizer experiment. *Soil Biology and Biochemistry*, Vol.35, No.3, (March 2003), pp. 453-461, ISSN 0038-0717

Martin-Laurent F, Philippot L, Hallet S, Chaussod R, Germon JC, Soulas G & Catroux G (2001). DNA extraction from soils: old bias for new microbial diversity analysis methods. *Applied and Environmental Microbiology*, Vol.67, No.5, (May 2001), pp. 2354-2359, ISSN 0099-2240

May LA, Smiley B & Schmidt MG (2001). Comparative denaturing gradient gel electrophoresis analysis of fungal communities associated with whole plant corn silage. *Canadian Journal of Microbiology*, Vol.47(September 2001), pp. 829-841, ISSN 0008-4166

Miller DN, Bryant JE, Madsen EL & Ghiorse WC (1999). Evaluation and optimization of DNA extraction and purification procedures for soil and sediment samples. *Applied and Environmental Microbiology*, Vol.65, No.11, (November 1999), pp. 4715-4724, ISSN 0099-2240

Möhlenhoff P, Müller L, Gorbushina AA & Petersen K (2001). Molecular approach to the characterisation of fungal communities: methods for DNA extraction, PCR amplification and DGGE analysis of painted art objects. *FEMS Microbiology Letters*, Vol.195, No.2, (February 2001), pp. 169-173, ISSN 0378-1097

Morimoto S & Hoshino YT (November 2010). Technical Report on the PCR-DGGE Analysis of Bacterial and Fungal Soil Communities, In: *National Institute for Agro-Environmental Sciences*, 02.09.2011, Available from http://www.niaes.affrc.go.jp/project/edna/edna_jp/manual_bacterium_e.pdf

Muyzer G, de Waal EC & Uitterlinden AG (1993). Profiling of complex microbial populations by denaturing gradient gel electrophoresis analysis of polymerase chain reaction-amplified genes coding for 16S rRNA. *Applied and Environmental Microbiology*, Vol.59, No.3, (March 1993), pp. 695-700, ISSN 0099-2240

Nikolcheva LG, Cockshutt AM & Bärlocher F (2003). Determining diversity of freshwater fungi on decaying leaves: comparison of traditional and molecular approaches. *Applied and Environmental Microbiology*, Vol.69, No.5, (May 2003), pp. 2548-2554, ISSN 0099-2240

Nocker A, Burr M & Camper A (2007). Genotypic Microbial Community Profiling: A Critical Technical Review. *Microbial Ecology*, Vol.54, No.2, (August 2007), pp. 276-289, ISSN 0095-3628

Okubo A & Sugiyama S (2009). Comparison of molecular fingerprinting methods for analysis of soil microbial community structure. *Ecological Research*, Vol.24, No.6, (November 2009), pp. 1399-1405, ISSN 0912-3814

Oros-Sichler M, Gomes NCM, Neuber G & Smalla K (2006). A new semi-nested PCR protocol to amplify large 18S rRNA gene fragments for PCR-DGGE analysis of soil fungal communities. *Journal of Microbiological Methods*, Vol.65, No.1, (April 2006), pp. 63-75, ISSN 0167-7012

Øvreås L, Jensen S, Daae FL & Torsvik V (1998). Microbial Community Changes in a Perturbed Agricultural Soil Investigated by Molecular and Physiological Approaches. *Applied and Environmental Microbiology*, Vol.64, No.7, (July 1998), pp. 2739-2742, ISSN 0099-2240

Peixoto RS, Coutinho HLC, Madari B, Machado PLOA, Rumjanek NG, Van Elsas JD, Seldin L & Rosado AS (2006). Soil aggregation and bacterial community structure as affected by tillage and cover cropping in the Brazilian Cerrados. *Soil and Tillage Research*, Vol.90, No.1-2, pp. 16-28, ISSN 0167-1987

Perkiömäki J, Tom-Petersen A, Nybroe O & Fritze H (2003). Boreal forest microbial community after long-term field exposure to acid and metal pollution and its potential remediation by using wood ash. *Soil Biology and Biochemistry*, Vol.35, No.11, (November 2003), pp. 1517-1526, ISSN 0038-0717

Petrosino JF, Highlander S, Luna RA, Gibbs RA & Versalovic J (2009). Metagenomic Pyrosequencing and Microbial Identification. *Clinical Chemistry*, Vol.55, No.5, (May 2009), pp. 856-866, ISSN 0009-9147

Qiu X, Wu L, Huang H, McDonel PE, Palumbo AV, Tiedje JM & Zhou J (2001). Evaluation of PCR-generated chimeras, mutations, and heteroduplexes with 16S rRNA gene-based cloning. *Applied and Environmental Microbiology*, Vol.67, No.2, (February 2001), pp. 880-887, ISSN 0099-2240

Ranjard L, Lejon DPH, Mougel C, Schehrer L, Merdinoglu D & Chaussod R (2003). Sampling strategy in molecular microbial ecology: influence of soil sample size on DNA fingerprinting analysis of fungal and bacterial communities. *Environmental Microbiology*, Vol.5, No.11, (November 2003), pp. 1111-1120, ISSN 1462-2912

Ranjard L, Poly F, Lata JC, Mougel C, Thioulouse J & Nazaret S (2001). Characterization of Bacterial and Fungal Soil Communities by Automated Ribosomal Intergenic Spacer Analysis Fingerprints: Biological and Methodological Variability. *Applied and Environmental Microbiology*, Vol.67, No.10, (October 2001), pp. 4479-4487, ISSN 0099-2240

Robe P, Nalin R, Capellano C, Vogel TM & Simonet P (2003). Extraction of DNA from soil. *European Journal of Soil Biology*, Vol.39, No.4, (October - December 2003), pp. 183-190, ISSN 1164-5563

Roesch LFW, Fulthorpe RR, Riva A, Casella G, Hadwin AKM, Kent AD, Daroub SH, Camargo FAO, Farmerie WG & Triplett EW (2007). Pyrosequencing enumerates and contrasts soil microbial diversity. *The ISME Journal*, Vol.1, No.4, (August 2007), pp. 283-290, ISSN 1751-7362

Ronaghi M, Karamohamed S, Pettersson B, Uhlén M & Nyrén P (1996). Real-Time DNA Sequencing Using Detection of Pyrophosphate Release. *Analytical Biochemistry*, Vol.242, No1, (November 1996), pp. 84-89, ISSN 0003-2697

Sanger F & Coulson AR (1975). A rapid method for determining sequences in DNA by primed synthesis with DNA polymerase. *Journal of Molecular Biology*, Vol.94, No.3, (May 1975), pp. 441-446, ISSN 0022-2836

Sequerra J, Marmeisse R, Valla G, Normand P, Capellano A & Moiroud A (1997). Taxonomic position and intraspecific variability of the nodule forming Penicillium nodositatum inferred from RFLP analysis of the ribosomal intergenic spacer and Random Amplified Polymorphic DNA. *Mycological Research*, Vol.101, No.4, (April 1997), pp. 465-472, ISSN 0953-7562

Sessitsch A, Weilharter A, Gerzabek MH, Kirchmann H & Kandeler E (2001). Microbial Population Structures in Soil Particle Size Fractions of a Long-Term Fertilizer Field Experiment. *Applied and Environmental Microbiology*, Vol.67, No.9, (September 2001), pp. 4215-4224, ISSN 0099-2240

Smit E, Leeflang P, Glandorf B, Dirk van Elsas J & Wernars K (1999). Analysis of fungal diversity in the wheat rhizosphere by sequencing of cloned PCR-amplified genes encoding 18S rRNA and temperature gradient gel electrophoresis. *Applied and Environmental Microbiology*, Vol.65, No.6, (June 1999), pp. 2614-2621, ISSN 0099-2240

Suenaga H (2011). Targeted metagenomics: a high-resolution metagenomics approach for specific gene clusters in complex microbial communities. *Environmental Microbiology*, (March 2011), pp. 1-10, ISSN 1462-2920

Suzuki C, Nagaoka K, Shimada A & Takenaka M (2009). Bacterial communities are more dependent on soil type than fertilizer type, but the reverse is true for fungal communities. *Soil Science and Plant Nutrition*, Vol.55, No1, (February 2009), pp. 80-90, ISSN 1747-0765

Takehara T, Kuniyasu K, Mori M & Hagiwara H (2003). Use of a nitrate-nonutilizing mutant and selective media to examine population dynamics of *Fusarium oxysporum* f. sp. spinaciae in soil. *Phytopathology*, Vol.93, No.9, (September 2003), pp. 1173-1181, ISSN 0031-949X

Tanaka S, Kobayashi T, Iwasaki K, Yamane S, Maeda K & Sakurai K (2003). Properties and metabolic diversity of microbial communities in soils treated with steam sterilization compared with methyl bromide and chloropicrin fumigations. *Soil Science and Plant Nutrition*, Vol.49, No.4, (August 2003), pp. 603-610, ISSN 0038-0768

Thorn G (1997). The fungi in soil. In: *Modern Soil Microbiology*, van Elsas JD, Trevors JT & Wellington EMH (Eds), 63-127. Marcel Decker, ISBN 978-0824794361, New York, USA

Tsushima S (March 2011). eDNA Database for Agricultural Soils, In: *National Institute for Agro-Environmental Sciences*, 02.09.2011, Available from http://eddass.niaes3.affrc.go.jp/hp/index.html

Torsvik V, Goksøyr J & Daae FL (1990). High diversity in DNA of soil bacteria. *Applied and Environmental Microbiology*, Vol.56, No.3, (March 1990), pp. 782-787, ISSN 0099-2240

Vainio EJ & Hantula J (2000). Direct analysis of wood-inhabiting fungi using denaturing gradient gel electrophoresis of amplified ribosomal DNA. *Mycological Research*, Vol.104, No.8, (August 2000), pp. 927-936, ISSN 0953-7562

Vandenkoornhuyse P, Baldauf SL, Leyval C, Straczek J & Yong JPW (2002). Extensive fungal diversity in plant roots. *Science*, Vol.295, No.5562, (March 2002), pp. 2051, ISSN 0036-8075

Viaud M, Pasquier A & Brygoo Y (2000). Diversity of soil fungi studied by PCR-RFLP of ITS. *Mycological Research*, Vol.104, No.9, (September 2000), pp. 1027-1032, ISSN 0953-7562

Wang GC & Wang Y (1996). The frequency of chimeric molecules as a consequence of PCR co-amplification of 16S rRNA genes from different bacterial species. *Microbiology*, Vol.142, No.5, (May 1996), pp. 1107-1114, ISSN 1350-0872

Wang GC & Wang Y (1997). Frequency of formation of chimeric molecules as a consequence of PCR coamplification of 16S rRNA genes from mixed bacterial genomes. *Applied and Environmental Microbiology*, Vol.63, No.12, (December 1997), pp. 4645-4650, ISSN 0099-2240

White TJ, Bruns T, Lee S & Taylor J (1990) Amplification and direct sequencing of fungal ribosomal RNA genes for phylogenetic In: *PCR Protocols: A Guide to Methods and Applications*, Innis MA, Gelfand DH, Sninsky JJ & White TJ (Eds), 315-322. Academic Press, ISBN 978-0123721808, New York, USA

Wintzingerode FV, Göbel UB & Stackebrandt E (1997). Determination of microbial diversity in environmental samples: pitfalls of PCR-based rRNA analysis. *FEMS Microbiology Reviews*, Vol.21, No.3, (November 1997), pp. 213-229, ISSN 0168-6445

Xu Y, Wang G, Jin J, Liu J, Zhang Q & Liu X (2009). Bacterial communities in soybean rhizosphere in response to soil type, soybean genotype, and their growth stage. *Soil Biology and Biochemistry*, Vol.41, No.5, (May 2009), pp. 919-925, ISSN 0038-0717

Yang YH, Yao J, Hu S & Qi Y (2000). Effects of Agricultural Chemicals on DNA Sequence Diversity of Soil Microbial Community: A Study with RAPD Marker. *Microbial Ecology*, Vol.39, No.1, (January 2000), pp. 72-79, ISSN 0095-3628

Yergeau E, Filion M, Vujanovic V & St-Arnaud M (2005). A PCR-denaturing gradient gel electrophoresis approach to assess Fusarium diversity in asparagus. *Journal of Microbiological Methods*, Vol.60, No.2, (February 2005), pp. 143-154, ISSN 0167-7012

15

Qualitative and Quantitative Assessment of Sediments Pollution with Heavy Metals of Small Water Reservoirs

Bogusław Michalec
Agriculture University in Cracow
Poland

1. Introduction

Silting of reservoirs is a very complex process of entrainment, transport and deposition of sediment in reservoir basins, which has major impacts and several effects on river system and its environment (Batuca & Jordaan, 2000). The silting process is initiated when the river damming is completed and ends when the reservoir basin is completely filled with sediment. The silting process is naturally caused and anthropically influenced and the prevention of the original state of the reservoir is impossible. Silting of water reservoirs is one of the main factor limiting their proper operation. Erosion processes in catchment basins are the factor limiting the life-time of water reservoirs. Considerable amount of erosion products coming into the reservoirs is transported and supplied into reservoirs where it is deposited. Together with trapped small fraction of sediment, organic and inorganic compounds including heavy metals, are also deposited. Quantity of heavy metals concentration in water reservoir sediments is regarded as the environmental pollution index. Heavy metals are treated as elementary pollution commonly distributed in natural environment and their concentration corresponds with natural concentrations whose natural source in minerals or soils (O' Neil, 1993). However increase in concentration value of heavy metals in the environment grow at a considerable extent within the latest decades due to human activity (Preuss & Kollman, 1974; Prater, 1975).

Determination of sediment quantity ant their quality is especially vital in causes of small water reservoirs. This is due to the specificity of the silting process characterized by considerable intensity which can be expressed by means of the annual silting intensity ratio. According to Hartung [1959] the mean annual reservoir capacity loss in the case of the big reservoirs is 0.25%, of medium reservoir 0.5%, and of small ones 3,0%. According to Lara and Pemberton's criterion (1963) large water reservoirs are characterized by capacity exceeding 1 km^3, capacity of medium water reservoirs exceeds 0.1 km^3, whereas, small ones are objects of some thousand m^3 capacity. In Poland ("Program of small retention..." 2004) and in Romania (Batuca & Jordaan, 2000) small water reservoirs are distinguished as water bodies of capacity below 5 million m^3. In Great Britain, on the other hand, an object of small capacity less than 1 million m^3 which closes a catchment area below 25 km^2 is regard as a small water reservoir (White et al., 1996). According to World Commission on Reservoirs (Sawunyama, 2005) small reservoir capacity equals from 50 thousand m^3 to 1 million m^3.

With regard to quick reduction of small water reservoir capacity in a relatively short time there arises the necessity of sediment removal and its utilization. It is essential to determine the period after which the reservoir is silted in such a degree that removal of sediment accumulated in it will prove to be a necessity. This time may be determined basing upon the elaborated silting forecast by use of empirical and theoretical methods. The up till now applied Orth's empirical method from the year 1934 (Michalec, 2008), Gončarov's from 1962 (Wiśniewski & Kutrowski, 1973) and Łapszenkov's from 1957 (Batuca & Jordaan, 2000) where elaborated in results of investigations of silting of water reservoirs located in various geographical regions. Reliable calculations results of capacity reduction of a water reservoir can be obtained by use of these methods, maintaining similarity between hydrological conditions prevailing in the catchment area or hydraulic and hydrodynamic condition in the reservoir and conditions for which they were elaborated. Empirical methods serving at silting forecast concern mainly medium and large water reservoirs. These were elaborated in results of investigations carried out on those reservoirs but their application in silting forecast of small water reservoirs often leads to faulty results.

Small water reservoirs undergo quick silting. As given Dendy (1982) the mean annual silting ratio of twenty water reservoirs in the basin of the Rio Grande of original capacity from 0.2 do 1.2 million m^3 was from 0.6 to 4.5%. According to Soler-López (2001) water reservoirs in Puerto Rico of original capacity of 0.76 to 6.4 millions m^3 characterized by an average silting degree from 0.5 to 1.5%. According to Batuca and Jordaan (2000) the average annual silting degree of four small water reservoirs in Austria, whose original capacity were from 0.7 to 2.1 millions m^3 was from 0.5 to 2.5%. With regard to quick reduction of small reservoirs capacity these reservoirs are subjected desilting every ten or so or even tens of years. The mineral material removed from these reservoirs is often utilized for agricultural purposes or earth work carried out in the cachment area. The reservoir sediments may constitute of territories degraded by industry. Such utilization of chemically uncontrolled bottom sediments is connected with risk of causing increase in harmful substances content – including heavy metals in the soil environment. Determination of the quantity of sediment pollution, including concentration of heavy metals, is essential not only with regard to estimation of utilization possibilities of the removed sediment, but my also be helpful in evaluation of state of the environment. There is, however, some hardship in evaluation of sediment quality concerning not only small water reservoirs and this is lack of univocal criteria of determination of thresholds values as well as lack of advised methods referring determination of sediment pollution indices and their classification. Caeiro et al. (2005) assessing heavy metals in Sado Estuary sediment proved on the basis of analysis of eight from the presented sixteen indices that some among then gave equivalent information or supplied some additional enriching data concerning pollution or background enrichment indices and ecological risk indices. According to Caeiro et al. [2005] application of these indices requires elaboration of method of their standardization to make comparison of the obtained results possible. These criteria should be specified by directives of the European Parliament and should be related to the Water Framework Directive. It should be remembered according to Borja et al. (2004) that in European Water Framework Directive (WFD; Directive 2000/60/EC) the question "water" is referred to 373 occasions, but other matrices, such as sediment or biota (biomonitors), are mentioned explicitly only 7 and 4 times, respectively. Changes and supplements in the water directive were to be introduced by the Directive of the European Parliament and Council approved in 2007. According to point 1 of Article 3 in the „Environmental Quality Standards (EQS) Applicable to Sediment and/or Biota (2007)" were defined that: „In accordance with Article 1 of this Directive and Article 4 of Directive 2000/60/EC, Member States shall apply the EQS laid down in Annex I, Part A, to this Directive in bodies of surface water". Whereas, according to point 2a of this Article it was defined that one can: „establish and apply EQS other than those mentioned in point (a) for sediment and/or biota for specified substances. These EQS shall offer at least the same level of protection as the EQS for water set out in Annex I, Part A". Can, thus, in connection of a given harmful substances in water be the basis of the determination of the sediment pollution level? Should be the evaluation criteria of sediment quality be separately identified? If quality criteria were to be defined for sediment, then monitoring would be required to establish compliance with such criteria? Response to these questions was obtained by defining concentration of three heavy metals in the water flowing into the studied small water reservoirs and in their bottom sediments. Investigations results on silting of twelve small water reservoirs in South Poland (fig. 1) were presented in this paper.

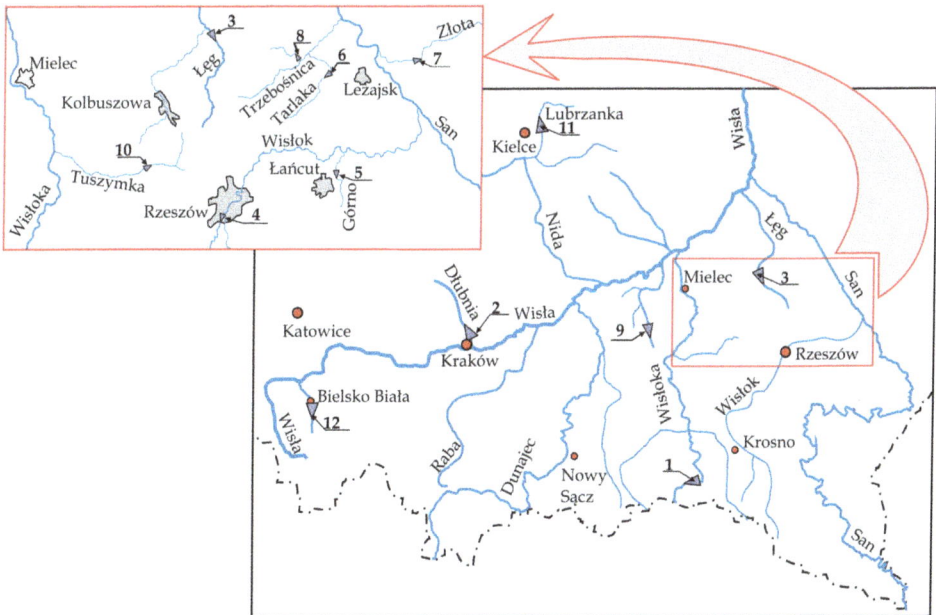

Fig. 1. Location of studied small water reservoirs: 1) Krempna, 2) Zesławice, 3) Maziarnia, 4) Rzeszów, 5) Głuchów, 6) Brzóza Królewska, 7) Ożanna, 8) Niedźwiadek, 9) Narożniki, 10) Cierpisz, 11) Cedzyna, 12) Wapienica

These results permitted elaboration of a method of determination of sediment quantity accumulated in small water reservoirs. This method bases upon indices determining silting intensity. The aims of this paper are: elaboration of a method of determination of the sediment quantity accumulated in small water reservoirs basing upon indices characterizing silting intensity and carrying out assessment of sediment pollution whit heavy metals applying Environmental Quality Standards (EQS) as well as other standard criteria of admissible concentrations of heavy metals in sediments.

2. Characteristics of research objects

Water reservoirs whose capacity, in agreement with the adopted in Poland criterion according to "Program of small retention…" (2004), is 5 millions m^3 were chosen for studies. This criterion is used for classification of small water reservoirs, among others in Romania (Batuca & Jordaan, 2000). Their average depth reaches from 0.8 m to 2.7 m, except the reservoirs at Wapienica whose average depth is above 6 m (table 1). Their surface of water table is from 1.5 ha to 160 ha. The length is also different from 0.34 km to 6.7 km.

Water reservoir / water-course	Water reservoir capacity [10^3 m³]	Surface of the water table [ha]	Mean depth [m]	Length of the reservoir [km]
Krempna-1 / Wisłok	112.0	3.20	3.50	0.40
Krempna-2 / Wisłok	119.1	3.20	3.72	0.40
Zesławice-1 and Zesławice-1 / Dłubnia	228.0	9.50	2.40	0.65
Maziarnia / Łęg	3 860.0	160.00	2.41	6.51
Rzeszów / Wisłok	1 800.0	68.20	2.64	6.70
Głuchów / Graniczny	22.6	1.50	1.51	0.64
Brzóza Królewska / Tarlaka	48.8	6.13	0.80	0.44
Ożanna / Złota	252.0	18.00	1.40	0.95
Niedźwiadek / Górno	124.5	8.10	1.54	0.55
Narożniki / Dęba	283.0	28.00	1.01	2.00
Cierpisz / Tuszymka	34.5	2.30	1.50	0.34
Cedzyna / Lubrzanka	1 554.0	64.00	2.43	2.20
Wapienica / Wapienica	1 100.0	17.50	6.29	1.00

Table 1. Chosen basic parameters of studied small water reservoirs

The basins of the studied water reservoirs are mainly under agricultural use. The general feature of the chosen reservoirs is their quick capacity reduction being a result of the trapped small fractions of sediment. These reservoirs belong mainly to water reservoirs of uncontrolled water management. From the twelve reservoirs only four of them are located on water flows being under hydrological observations. These are: Krempna on the river Wisłoka, Zesławice on the river Dłubnia, Maziarnia on the river Łęg, and Rzeszów on the river Wisłok. With regard to the executed desilting works and rebuilding of two water reservoirs i.e. Krempna and Zesławice additional denotation of these reservoirs were introduced in presentation of the research results. In consequence of desilting and rebuilding of the Krempna reservoir in 1986 i.e. after 15 years of operation that caused decrease in its capacity from 119 thousands m^3 to 112 thousands m^3 its denotation Krempna-1 and Krempna-2, before and after desilting respectively, were introduced. This distinction was necessary with regard to change hydraulic characteristics of water flow through the reservoir which influenced the results of analysis of its silting process. The Zesławice water reservoir was also desilted and rebuilt without introduction any changes of its capacity. But, the character and quantity of water flow changed as a consequence of division of the total water flow in the water branching point on the inflow in the reservoir when a part of the flow was directed to a new built side reservoir. Hence, denotation Zesławice-1 and Zesławice -2 respectively for the period before and after desilting were introduced.

3. Methods

Surveys of the deposited sediment volume of studied water reservoirs consisted in determination of the change of ordinates of the reservoirs bottom in cross-section corresponding with post-execution cross-section and in points beyond these cross-sections. These measurements were performed from a boat by use a rod probe. With regard to the considerable depth of the reservoirs Wapienica, Cedzyna and Maziarnia a Human Bird 1000 echo-sounder was used in measurements of depth. Disposing of post-execution projects of the studied water reservoirs changes of cross-section surface were determined. Surveys of the deposited sediment volume of the studied water reservoirs were performed by a team directed by the author in years 1996-2006 at various time intervals.

3.1 Methods of assessment of silting intensity of studied reservoirs

The obtained surveys results were supplied with archival data. Basing upon the calculated volume of the sediment deposited in the reservoir (V_{dep}) the silting ratio of each of the studied water reservoirs in particular years of operation was calculated. The silting ratio (S) was calculated as a quotient of volume of the sediment deposited in the reservoir (V_{dep}) and its original volume (V_{res}). The mean annual silting ratio (S_A) was also calculated as a quotient of silting ratio determined on the basis of silting measurements in a given years of operation and number of years of operation. The calculated annual silting ratio was compared with the value of the mean annual silting ratio calculated by use Hachiro's formula (Batuca & Jordaan, 2000) elaborated on the basis of investigations of 106 water reservoirs in Japan and 39 in USA and one in Taiwan. Hachiro elaborated the regressive relation of the mean silting ratio (S_A) and the capacity-inflow ratio (C-I) in form:

$$S_A = 0.214 \cdot (C\text{-}I)^{-0.473} \tag{1}$$

The capacity -inflow ratio (C-I) is the relation of the original reservoir capacity (V_{res}) and the sum of average annual runoff (Q) to the reservoir.

Another characteristic describing the silting process is the silting rate (W_z) defined as the volume of sediment deposits in water reservoir in given time interval referred to the cachment area. It is expressed in $m^3 \cdot km^{-2} \cdot year^{-1}$. The silting ratio calculated on the basis of the results of measurements of sediment volume deposited in the studied water reservoirs was compared with the value of this rate calculated by use of the Khosl's formula. This rate was elaborated on the basis of the results of measurements of water reservoirs silting in Europe, India, and USA which close small and medium catchment of area W<2500 km²:

$$W_z = 3180 \cdot W^{0.72} \tag{2}$$

Silting intensity (S_i) according to Šamov (Batuca & Jordaan, 2000) was also determined:

$$S_i = \frac{V_{res}}{V_{SS}} \tag{3}$$

Where (V_{res}) is the original water reservoir capacity, and (V_{SS}) is the mean annual volume of sediment flowing into the reservoir.

Simultaneously with the measurements of silting bottom sediments were sampled in cross-sections of the reservoirs in part close to the dam, in middle part and in the part of back waters. In each of the parts of the reservoir samples were collected from the bottom sediment surface (i.e. at the depth 0-15 cm – upper layer) and at the depth 40-55 cm under sediment surface (lower part). Samples were collected by use of a standard sampler Beeker, made by Eijkelkamp company, according to the methods elaborated by Madeyski (2002). Sediment samples collected in twelve water reservoirs were subjected to analysis of granulometric composition and content of organic matter. In the sampled bottom sediments of water reservoirs: Krempna-2, Zesławice-2, Cierpisz, Wilcza Wola, Niedźwiadek and Narożniki concentration of heavy metals i.e. lead, cadmium, and nickel as well as chromium, zinc, and cooper were determined. Annual means and maximal concentration of lead, cadmium and nickel in the water inflowing into reservoirs were also determined.

The granulometric composition of bottom sediments was determined by Cassagrande's method modified by Prószyński in agreement with the Polish standard PN-R-04032 from the year 1998 (Michalec, 2008). In this method 1.5 g of Na_2CO_3, which is a deflocculant causing decomposition of flocculi and particle aggregates into elementary particles is added to the solution of suspended mater. The granulometric composition was determined on the basis of granular analysis of particular grains of sediment. Specific density was pictometrically determined and bulk density by of sand replacement method. Content of organic matter (O_m) was determined according to the annealing method.

3.2 Methods of determination of heavy metal contents in sediments

Determination of heavy metal contents in sediments was carried out by means of the method Flame Atomic Absorption Spectrometry method (FAAS) by use of a spectrophotometer Solaar M6 of Unicam mark. Performance of analysis was preceded by preparation of samples in the mineralization process carried out by use of a mixture of acids $HClO_4$ and HNO_3 in proportion 1 to 3 and dissolution in heat of dry residue in HCl in proportion 1 to 1. Determination was performed in flame of a gas mixture air-acetylene with deuterium correction of the background (Tarnawski, 2009). Determination of lead, cadmium, and nickel content in water inflowing into the six studied reservoirs were performed by use of the Inductively Coupled Plasma Atomic Emission Spectroscopy (ICP-AES) in an apparatus Viacan Vista-MPX PU 7000. After collecting water samples and proving them into Teflon container HNO_3 and $30\%H_2O_2$ was added. The sample fixed in this way was transported to the laboratory were it was subjected to a two-stage pressure microwave mineralization and subsequently subjected to spectroscopic analysis ICP-AES. Water samples were collected within three years i.e. in the period 2003-2005 in water inflowing to the reservoirs Krempna-2 and Niedźwiadek, in the period 2004-2005 in water inflowing to the reservoir Zesławice-2, in the period 2001-2003 in water inflowing to the reservoirs Cierpisz and Maziarnia, and in the period 2003-2005 in water inflowing to the reservoir Narożniki. Water samples for determination of heavy metals content (Pb, Cd, Ni) were collected four times every year i.e. in spring in April, in summer in July, in autumn in November and in winter in January. When collecting water samples suspended mineral matter should be taken into regard since in pollution transport it plays a significant role. Majority of chemical reactions occurring in the water environment proceeds at the interphase border solid body – solution. Hence, majority of trace elements especially of heavy metals do not stay in dissolved form for a long time but is often accumulated on particles of suspension. Therefore, in agreement with Polish instructions given in Measurement Programs of Integrated Monitoring of Natural Environment (Kostrzewski et al., 2006) when determining concentration of heavy metals in surface waters the sample after fixing should be immediately filtered off. The filtering off process may be neglected in the case of clean surface waters. According to this instruction water samples collected in the tributaries of the studied water reservoirs were not filtered off since sampling took place mainly during period of medium and low water discharges when the content of solid substances in samples was small since concentrations of the suspended sediment were from some to ten or so mg·l⁻¹. The water samples collected during freshet when suspended sediment concentrations were above 100-150 mg·l⁻³ were subjected to filtering off by so called hard filters of diameter equaling 18,5 cm of the type 360 firm VEB Spezialpapierfabrik Niederschlag. During collecting water samples for determination of heavy metals content measurements of suspended sediment concentration were performed by use of a photooptic apparatus Portable Suspended Solids and Turbidity Monitor System 770.

3.3 Methods of determination of sediment transport

Apart from measurements of suspended sediment concentration made during water sampling in tributaries of the chosen six reservoirs series of suspended sediment concentration were also performed in cross-sections of the rivers which were under permanent hydrological observations carried by the Institute of Meteorology and Water Management. These were rivers flowing into the reservoirs Krempna, Zesławice, Maziarnia and Rzeszów. These measurements made possible determination of the amount of suspended sediment transport. Calculations of suspended sediment transport were made on the basis of series of data comprising average 24 hours' flows (Q) and corresponding with them suspended sediment concentration (Pp) determined at the station point of regular water sampling. The lacking data of suspended sediment concentration were complemented using the elaborated relations of suspended sediment concentration in function of water flow. The relation curves Pp=f(Q) for each of these seasons served for determination of suspended sediment transport in periods without measurements of suspended sediment concentration (Michalec, 2008). Heterogeneity of hydrological data concerning measurements of water flow in gauging station was studied applying a non-parametric rank-sum test.

The complemented hydrological data permitted to calculate the so called mean 24 hours' suspended sediment transport U_i [g·s⁻¹] which is the product of daily suspended sediment concentration P_p [g·m⁻³] and daily water discharges (Q) [m³·s⁻¹]. Subsequently a 24 hours', monthly, and yearly sediment transport in a given cross-section of the rivers was calculated. These calculations were performed on the basis of measurements in the point of regular water sampling. In calculation of suspended sediment transport a suspended sediment concentration in the whole cross-section should be considered. Aiming at it measurements of suspended sediment concentration in the whole cross-section of studied rivers were performed and the mean suspended sediment concentration (P_m) in the cross-section was determined. The calculated corrective coefficient "k" being the quotient of the mean suspended sediment concentration in the river cross-section (P_m) and the suspended sediment concentration in the point of the river cross-section (P_p), permitted calculation of intensity of the sediment transport.

The mass of sediment delivering to the other retention reservoirs, i.e. Głuchów, Brzóza Królewska, Ożanna, Niedźwiadek, Narożniki, Cierpisz, Cedzyna and Wapienica, located on small streams without hydrological data, was calculated by use of the DR-USLE method. The USLE equation (Wischmeier & Smith, 1965) in this method enables calculation of the annual mean mass of erosion products in catchment area. The amount of erosion products delivered to the rivers was determined by means of parameter DR (delivery ratio) according to Roehl (1962).

The calculated mass of sediment delivered into the water reservoirs permitted determination of silting intensity defined by Šamov (Batuca & Jordaan, 2000) as the quotient of the original capacity of reservoir (V_{res}) and the annual mean of the sediment volume entering the reservoir (V_{SS}). In order to determine on the basis of calculated sediment transport intensity, expressed in mass units, the sediment volume delivered to the reservoir the determined volumetric density of sediments was used.

3.4 Methods of assessment sediments pollution

Determined values of heavy metals content in sediments collected from six water reservoirs made determination of maximal, minimal, and mean concentrations of given chemical element possible. Evaluation of sediment quality as pollution with heavy metals was carried out according to the binding Regulation of the Minister of Environment concerning soil quality standards (Dz. U. Nr 165, poz. 1359) and following instructions and method:

- geochemical classification of water sediments (table 2) elaborated and applied by the Polish Geological Institute (PIG) and State Inspection of Environment Protection (SIEP).

According to the instruction PGI and SIEP contents of heavy metals are determined in a separated sediment fractions of grain diameter less than 0.2 mm since this fraction constitutes particles of cohesive soils reflects due to its adhesive properties the amount of heavy metals in the environment. However, according to Bojakowska and Sokołowska (1998) small grain fractions are characterized by a higher trace metals content as compared with coarser fraction of the same sample. Therefore, making use of smallest sediment fractions for investigations leads to overestimation of results. In spite of this provision the method of trace element determination in sediment was adopted. This classification was first attempt to standardization of the results of bottom sediment studies and in spite of provision and doubts it is applied for assessment of every year investigation results under monitoring. In later years i.e. in 2001, following Canadian and American classification, a IV class classification of water sediments was introduced (table 2), based on threshold values which consider the harmful effect of accumulated in sediments pollution on aquatic organisms (Bojakowska, 2001).

Component	Geochemical background	Classes						
		1998			2001			
		I	II	III	I	II	III	IV
Elements (mgkg⁻¹)								
Ar	5	10	20	50	7	30	70	>70
Cd	0.5	1	5	20	0.7	3.5	6	>6
Cr	5	20	100	500	50	100	400	>400
Cu	6	20	100	200	20	100	300	>300
Pb	10	50	200	500	30	100	200	>200
Hg	0.05	0.1	0.5	1.0	0.2	0.7	0.7	>0.7
Ni	5	30	50	100	16	40	50	>50
Zn	48	200	1000	2000	125	300	1000	>1000

Table 2. Qualitative classification of bottom sediments applied by the Polish Geological Institute and the State Inspection of Environment Protection

According to this classification of following classes were distinguished: I class – unpolluted sediments, no harmful effects of trace elements and toxic organic compounds on aquatic organisms are observed; II class – sediments moderately polluted, harmful effect of trace elements on aquatic organisms occurs occasionally; III class – sediments of medium pollution in which content of at least one harmful component exceeds the threshold content for sediments of II class, hence, sediments in which contamination is present in concentrations of which the harmful effect on aquatic organisms is frequently observed, but the content of harmful chemical elements is lower than the admitted content; and IV class – sediments heavily polluted in which at least for one of the components the admissible for III class is exceeded.

-   instruction elaborated by the Institute of Soil Science and Plant Cultivation (Kabata-Pendias et al., 1995). The ISSPC instructions distinguish six degrees of contamination (table 3) ascribing to each degree the way of agricultural utilization of soil, characteristics tillage and possible recultivation works: degree 0 – unpolluted soils of natural content of trace metals, degree I – soils of increased metal content, degree II – soils slightly polluted, degree III – soils of medium pollutions, degree IV – soils heavily polluted, degree V – soils extremely polluted.
-   Müller's method according to which the geoaccumulative index (I_{geo}). In this method the metal content is related to the natural content in sediments constituting the geochemical background of the given heavy metal (Müller, 1981). The index permits to distinguish seven classes: from unpolluted sediments – class 0 (I_{geo} <0) to very heavily polluted sediment – class VI (I_{geo} >5).

A sediment pollution index (SPI) was established according to Singht et al. (2002). Singh et al. (2002) introduced the concept of sediment pollution index (SPI). The SPI is a multi-metal approach for an assessment of sediment quality with respect to trace metal concentrations along with metal toxicity:

$$SPI = \Sigma \left(EF_m \cdot W_m \right) / \Sigma W_m \qquad (4)$$

where EF_m is ratio between the measured metal concentration (C_n) and the reconstructed background metal concentration (C_R) instead of the average metal concentration in shale. W_m is toxicity weight. Toxicity weight 1 was assigned for Cr and Zn, 2 for Cu and Ni, and 5 for Pb (Singh et al., 2002). The following classification is given for the SPI: 0–2 = natural sediment, 2–5 = low polluted sediment, 5–10 = moderately polluted sediment, 10–20 = highly polluted sediment, and >20 = dangerous sediment.

Evaluation of sediments was also made according to the criteria of admissible environmental levels of heavy metals concentrations adopted in Germany, Denmark, USA, and Canada (table 4) and presented in the paper Yau and Gray (2005) and also according to the criteria given by Sullivan (Chen & Lin, 2001) for dredged sediments.

Apart from the evaluation of sediments quality according to the above given criteria assessment was also performed according to Environmental Quality Standards of pollution with heavy metals: Pb, Cd and Ni of sediments and waters flowing into the six studied water reservoirs: Krempna-2, Zesławice-2, Cierpisz, Wilcza Wola, Niedźwiadek and Narożniki. This comparative assessment will permit to state whether the environmental pollution level determined on the basis of heavy metal concentration in water corresponds whit the level of the environmental pollution determined on the basis of heavy metals contamination in sediments.

Metal	Soil group	Degree of soil pollution [mg·kg⁻¹]					
		0	I	II	III	IV	V
Pb	a	20	70	100	500	2500	>2500
	b	40	100	250	1000	5000	>5000
	c	60	200	500	2000	7000	>7000
Zn	a	50	100	200	700	1500	> 1500
	b	70	150	300	1000	3000	>3000
	c	100	250	500	2000	5000	>5000
Cu	a	10	30	50	80	300	>300
	b	20	50	80	100	500	>500
	c	25	70	100	150	750	>750
Cd	a	0,3	1,5	2	3	5	>5
	b	0,5	2	3	5	10	>10
	c	1,0	3	5	10	20	>20
Ni	a	10	30	50	100	400	>400
	b	25	50	75	150	600	>600
	c	50	75	100	300	1000	>1000
Cr	a	20	40	80	150	300	>300
	b	30	60	150	300	500	>500
	c	50	80	200	500	1000	>1000

Table 3. Limits of heavy metals content in soils of various pollution degree (according to Kabata-Pendias et al., 1995), where: group a – soils containing 10-20% of washed off particles and of pH < 5.5; group b – soils containing 10-20% of washed off particles and of pH > 5.5 and soils containing more than 20% of washed off particles and of pH < 5.5; group c – other soils i.e. soils containing over 20% of washed off particles and of pH > 5.5

Background levels	Metal concentration [µg·g⁻¹]				
	Cd	Cu	Fe	Pb	Zn
Canadian background levels	1.1	25	31000	23	65
US background levels	–	20	28000	23	88
Minimum German background levels	0.15	10	–	12.5	50
Maximum German background levels	0.6	40	–	50	200
Dutch Target (DT) values	0.85	36	–	85	140
Dutch Intervention (DI) values	1.2	190	–	530	720

Table 4. Background concentration and quality objectives for heavy metals in sediments of freshwater ecosystems according to Woitke (Yau and Gray 2005)

4. Results and discussion

The volume of sediment trapped in the studied water reservoirs, determined on the basis of measurement results enables calculation of the silting ratio (S) in consequent years of operation (table 5, column 6).

Reservoir	Original capacity V_P [m³]	Year	Years of operation	Volume of deposited sediment [m³]	Silting ratio S [%]	Mean annual silting ratio S_A [%]	Capacity-inflow ratio C-I [%]
1	2	3	4	5	6	7	8
Krempna-1	119100	1986	15	35665	30.0	2.00	0.261
Krempna-2	112000	1996	9	27041	24.1	2.68	0.265
		1997	10	30464	27.2	2.72	0.255
		1998	11	34637	30.9	2.81	0.242
		1999	12	38002	33.9	2.83	0.231
		2000	13	40144	35.8	2.76	0.224
		2002	15	44200	39.5	2.63	0.212
		2003	16	44901	40.1	2.62	0.210
		2005	18	45810	40.9	2.27	0.207
Zesławice-1	228000	1968	2	26968	11.8	5.92	0.585
		1969	3	70425	30.9	10.30	0.458
		1970	4	75780	33.2	8.31	0.443
		1971	5	76251	33.4	6.69	0.441
		1974	8	86192	37.8	4.73	0.413
		1983	17	116091	50.9	3.00	0.326
Zesławice-2		1999	14	56162	24.6	1.76	0.786
		2005	20	75315	33.0	1.65	0.682
		2006	21	77232	33.9	1.61	0.673
Maziarnia	3860000	1999	10	504876	13.1	1.31	8.377
		2002	13	609600	15.8	1.21	8.116
		2003	14	625300	16.2	1.16	8.039
Rzeszów	1800000	1986	13	1188000	66.0	5.08	0.306
Głuchów	22570	2002	7	4126	18.3	2.61	0.104
Brzóza Królewska	48970	2002	17	4184	8.5	0.50	0.103
Ożanna	252000	1998	20	26000	10.3	0.52	0.634
		2003	25	30206	12.0	0.48	0.710
Niedźwiadek	124500	2003	5	3214	2.6	0.52	0.696
Narożniki	283800	2005	4	1646	0.6	0.15	2.317
Cierpisz	34500	1990	34	15000	43.5	1.28	5.770
		2001	11	6100	17.7	1.61	0.157
		2003	13	6745	19.6	1.50	0.229
Cedzyna	1550000	1999	26	145000	9.4	0.36	0.224
		2003	30	168500	10.9	0.36	4.032
Wapienica	1100000	1967	36	24250	2.2	0.06	3.964
		2003	71	46800	4.3	0.06	2.843

Table 5. Volume of deposited sediment and silting ratio of the studied reservoirs in individual years of operation

Considerable increase in silting ratio in consequent years of operation indicates it, and this is evidenced by results of calculations performed for the reservoirs: Krempna-2, Zesławice-1, Zesławice-2 and Maziarnia. It was stated that loads depositing in the reservoirs Krempna-1, Krempna-2 and Zesławice-1, Zesławice-2 increase their volumes very quickly. During the fifteen years of their operation, the 30 and 39.5% of the reservoirs Krempna-1 and Krempna-2 was silted up. During seventeen years operation before desilting and twenty one years operation after desilting, the silting ratio of the Zesławice reservoir is equal 50.9% and 33.9%. Directing a part of inflowing water and sediment into the building assistant reservoir for the time of the main reservoir desilting contributed to decrease in silting intensity of the main reservoir, i.e. reservoir Zesławice-2. The water reservoir Rzeszów is characterized by highest silting rate. During the thirteen years of its operation, 66% of its capacity was silted up.

Within the over 10 years of operation period the value of silting ratio, in the case of some studied reservoirs, is above 30-40% giving evidence of the silting process intensity. The silting ratio of small water reservoirs varies from 8 to 66%, already after a couple of years of operation. Reservoirs Brzóza Królewska, Niedźwiadek, Narożniki, and Wapienica, located in catchment areas covered in majority by forests and meadows are characterized by a lower silting ratio. Studied small water reservoirs are characterized by a considerably differentiated value of the mean annual silting degree ranging from 0.06 to 10.3% of the initial reservoir capacity. Whereas, the mean annual silting ratio (S_A) undergoes a relatively quick reduction what may be stated on the basis of its values calculated for the mentioned water reservoirs (table 5, column 7).

Reservoir	Years of operation	Annual mean silting ratio S_A [%] acc. to		Silting rate W_z [$m^3 \cdot km^{-2} \cdot year^{-1}$] acc. to	
		sediment volume surveys	Hachiro	sediment volume surveys	Khosl
1	2	3	4	5	6
Krempna-1	15	2.0	4.2	13.5	125881
Krempna-2	18	2.3	4.3	15.4	125881
Zesławice-1	17	3.0	2.3	31.3	153503
Zesławice-2	21	1.6	1.9	16.9	153503
Maziarnia	14	1.2	0.7	191.7	161036
Rzeszów	13	5.1	3.3	44.3	773602
Głuchów	7	2.6	5.0	47.9	19371
Brzóza Królewska	17	0.5	2.3	8.1	37162
Ożanna	25	0.5	1.1	8.9	109462
Niedźwiadek	5	0.5	1.3	34.3	26241
Narożniki	4	0.2	0.8	16.5	32281
Cierpisz	34	1.5	3.5	8.1	56576
Cedzyna	30	0.4	0.9	40.1	111594
Wapienica	71	0.1	1.1	11.9	57344

Table 6. Annual mean silting ratio of studied reservoirs in the year of last silting measurement and silting rate

The mean annual silting ratio, determined in the last year in which silting measurements were made, was compared with the mean annual silting ratio calculated according to Hachiro's formula (table 6). The value S_A calculated according to Hachiro's equation is from 0.9 to 19 times higher than that determined on the basis of sediment volume surveys. Also the silting ratio calculated according to Khosl's formula is higher than that established on the basis of measurements and being on the average 6000 higher.

4.1 Determination of silting intensity of small water reservoirs

The obtained results of calculated silting ratio (S) of studied reservoirs (table 5, column 6) indicate at considerable silting intensity of small water reservoirs, whose capacity is lower than 5 millions m³ and capacity-inflow ratio is lower than 10%.

Water reservoir	Average annual mass of delivered sediment R_u [t]	Bulk density ρ_0 [t·m⁻³]	Average annual volume of sediment delivered to the reservoir V_{Ru} [m³]	Silting intensity acc. to Šamov S_i [-]
1	2	3	4	5
Krempna-1	3990	1.23	3244	37
Krempna-2	3990	1.23	3244	35
Zesławice-1	16400	1.03	15922	14
Zesławice-2	16400	1.03	15922	14
Maziarnia	97310	1.65	58976	65
Rzeszów	206110	1.41	173202	10
Głuchów	872	1.25	697	32
Brzóza Królewska	308	1.14	270	181
Ożanna	1644	1.19	1382	182
Niedźwiadek	737	1.11	664	188
Narożniki	720	1.07	673	420
Cierpisz	744	1.32	563	61
Cedzyna	7092	1.12	6332	245
Wapienica	919	1.29	712	1544

Table 7. Average annual sediment transport and silting intensity of the water reservoirs established for the last silting measurement

Calculation of silting intensity according to the ratio defined by Šamov (Batuca & Jordaan 2000) required estimation of annual mean volume of sediment flowing into the reservoir. The amount of sediment transport delivered to the reservoirs Krempna, Zesławice, Maziarnia and Rzeszów, located on hydrologically controlled streams, was calculated on the basis of measurements of water flow and suspended sediment concentration taking into a count the corrective coefficient "k" in each measurement cross-section in streams. The corrective coefficients "k" calculated by use of program Statistica for Windows equal respectively: k=0.906 for the river Wisłok (reservoir Krempna), k=1.065 for the river Dłubnia (reservoir Zesławice), k=1.148 for the river Łęg (reservoir Maziarnia) and k=1.034 for the river Wisłok (reservoir Rzeszów). Regression equations were established with the confidence interval equal 95%. The mass of delivered suspended sediment to the other reservoirs was calculated by use of the DR-USLE method. In table 7 there are presented the results of calculation of mean annual mass of suspended sediment delivered to the studied reservoirs and mean bulk density of deposited sediments, established on the basis of six samples of bottom sediments collected in each reservoir (table 7). The relation of the mean annual silting ratio (S_A), calculated in the last year of operation of studied small water reservoirs, and determined according to Šamov (Batuca & Jordaan, 2000) silting intensity (S_i) was given in figure 2.

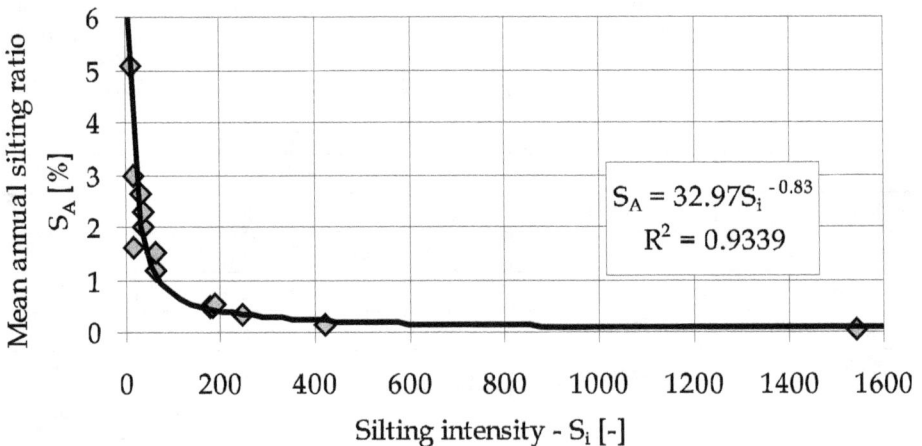

Fig. 2. Regression dependence of the mean annul silting ratio (S_A) defined on the basis of measurements and silting intensity (S_i) according to the Šamov's equation

This equation in form $S_A = 32.97 S_i^{-0.83}$ enables recognition of the mean annual silting ratio of small water reservoirs (fig. 2), whose capacity-inflow ratio is lower than 10%. Disposing of the volume of suspended sediment delivered to the reservoirs (V_{SS}) and its original capacity (V_{res}) it is possible, on the basis of silting intensity calculated by use of Šamov's formula (3), to determinate the annual silting ratio, and on its basis the annual mean volume of sediment deposits accumulated in a small water reservoir may be estimated. On the basis of annual mean volume of sediments accumulated in the reservoir and on the bulk density of sediments the mean annual mass of sediments, deposited in reservoir, may be determined. Disposing of the so determined mass of sediments in the small water reservoir and knowing the heavy metal concentration appointed in the sample the total mean annual pollution burden of sediment may be forecasted. The lack of possibility of determination of suspended sediment mass delivered to the reservoir or of sediment volumetric density makes application of the relation presented in figure 2 impossible.

The mean annual silting ratio, calculated on the basis of the last measurement of silting was related to capacity-inflow ratio (C-I) with the determined for its original capacity. The relation presented in figure 3 may serve for estimation of silting intensity of small water reservoirs.

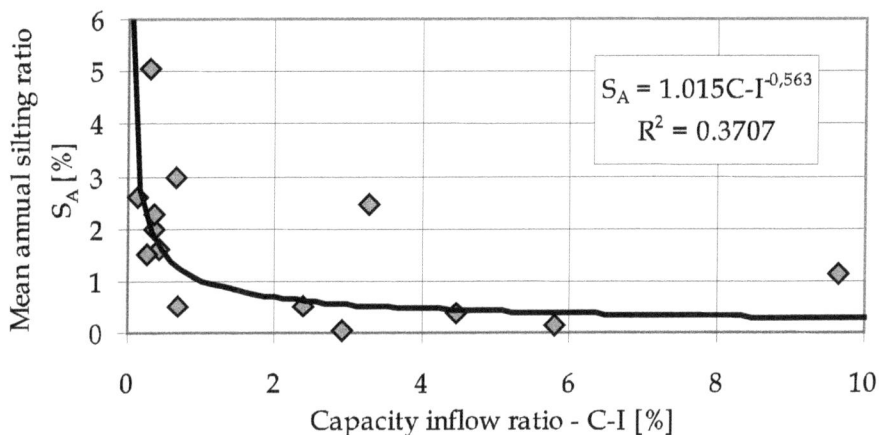

Fig. 3. Regression dependence of the mean annul silting ratio (S_A) defined on the basis of measurements and capacity-inflow ratio (C-I)

4.2 Characteristic of bottom sediment of studied small water reservoirs

The determined bulk density of sediments of lower layers in the inflow part of studied reservoirs ranges from 1.28 t·m^{-3} do 1.67 t·m^{-3} (table 8). These sediments are mostly constituted by sorted layers, forming clayey sands or clays. In the outflow parts of the reservoirs the surface layers of sediments are built from clayey or dusty mineral material of density from 0.68 t·m^{-3} to 1.19 t·m^{-3}. The average specific density equals 2.54-2.63 t·m^{-3}, whereas content organic matter in sediments does not exceed 10%. Fine grains of sediment deposited in the inflow part are characterized by bigger grain diameters as compared with the sediment trapped in the middle and in the outflow parts. Coarse grains of diameters above 2 mm are trapped above the inlet to the reservoirs. Variability of granulation in particular deposits layer was also stated. The upper layer of bottom sediment is constituted by a fines mineral material.

Bottom sediments collected from the water reservoirs: Krempna-2, Zesławice-2, Cierpisz, Wilcza Wola, Niedźwiadek and Narożniki were subjected - also apart from determination of physical properties – to evaluating of concentration of heavy metals: lead, cadmium and nickel as well as chromium, zinc and cooper. On the basis of six sediment samples collected in each water reservoirs the mean, minimal, and maximal concentration of heavy metals were determined in the analysed samples. Disposing of the mean concentration of the given metal its mass in sediments accumulated to the year in which last silting measurement was performed. The mass of sediments was calculated on the basis of sediment volume and mean volumetric density and subsequently the percentage content of metals in sediment was determined (table 9).

Water reservoir	Bulk density of sediment deposits ρ_0 [t·m^{-3}]							Content of organic matter O_m [%]
	1D	1G	2D	2G	3D	3G	m	
1	2	3	4	5	6	7	8	9
Krempna-2	1.28	1.31	1.32	1.16	1.17	1.16	1.23	0.82-7.22
Zesławice-2	1.47	0.78	1.16	0.80	1.30	0.68	1.03	3.35-7.36
Maziarnia	1.67	1.59	1.71	1.62	1.67	1.65	1.65	1.65-5.53
Rzeszów	1.62	1.51	1.47	1.22	1.42	1.19	1.41	4.86-7.87
Głuchów	1.41	1.27	1.33	1.24	1.14	1.09	1.25	3.22-6.89
Brzóza Królewska	1.51	1.08	1.38	0.91	1.11	0.87	1.14	1.69-4.98
Ożanna	1.42	1.07	1.38	1.12	1.23	0.94	1.19	2.41-8.12
Niedźwiadek	1.39	0.91	1.31	0.93	1.30	0.79	1.11	3.37-9.14
Narożniki	-	-	-	-	1.19	0.94	1.07	5.53-7.88
Cierpisz	1.65	1.23	1.53	1.03	1.44	1.06	1.32	2.89-4.19
Cedzyna	1.38	1.01	1.31	0.92	1.37	0.83	1.12	1.08-7.31
Wapienica	1.39	1.22	1.36	1.23	1.37	1.19	1.29	2.11-3.29

Table 8. Values of bulk density of sediment deposits and mean content of organic parts, where: 1 – outlet part, 2 – middle part, 3 – inlet part of the reservoir and G – upper layer, D – lower layer of deposits, m – mean value

Reservoir	Value	Heavy metals (μg·g^{-1})						Percentage content of heavy metals in sediments [%]					
		Pb	Cd	Ni	Cr	Zn	Cu	Pb	Cd	Ni	Cr	Zn	Cu
Krempna-2	Min.	9.8	0.1	42.3	8.4	49.1	19.2						
	Mean	17.5	0.3	52.9	10.4	76.5	50.8	1.75	1.03	5.29	1.04	7.56	5.08
	Max.	21.0	0.5	66.1	11.2	92.8	78.6						
Zesławice-2	Min.	10.5	0.1	14.7	16.9	34.9	6.9						
	Mean	20.0	0.5	16.2	18.5	72.9	10.7	2.00	0.05	1.62	1.85	7.29	1.07
	Max.	30.2	0.9	17.5	20.1	129.9	14.0						
Cierpisz	Min.	1.0	0.1	5.1	14.2	12.5	0.5						
	Mean	5.4	0.9	7.4	17.3	64.6	1.9	0.54	0.09	0.74	1.73	6.46	0.19
	Max.	10.5	2.8	12.6	19.8	208.5	6.0						
Maziarnia	Min.	8.7	0.1	8.4	10.1	23.9	4.5						
	Mean	9.5	0.3	9.3	11.5	24.6	5.2	0.95	0.03	0.93	1.15	2.46	0.52
	Max.	11.5	0.5	11.2	12.6	34.5	6.7						
Niedźwiadek	Min.	11.1	0.5	19.2	25.4	45.0	12.6						
	Mean	23.8	1.3	23.7	29.9	157.8	27.6	2.38	0.13	2.37	2.99	5.78	2.76
	Max.	34.3	1.9	25.4	33.2	230.0	33.8						
Narożniki	Min.	8.3	0.6	19.1	7.2	71.5	7.5						
	Mean	12.4	0.8	23.4	9.5	77.3	12.6	1.24	0.08	2.34	0.95	7.73	1.26
	Max.	15.2	1.2	29.6	11.0	83.8	15.3						

Table 9. Content of heavy metals Pb, Cd, Ni, Cr, Zn and Cu in sediments deposited in the small water reservoirs at Krempna-2, Zesławice-2, Maziarnia, Niedźwiadek and Narożniki

4.3 Assessment of sediment pollution with heavy metals

Water samples for determination of heavy metals content (Pb, Cd, Ni) were collected within three years four times every year. With regard to a short period, comparing only three years, no seasonal differentiation of concentration of heavy metals was stated. Table 10 shows the minimal, maximal, and mean annual concentration of the researched heavy metals.

Water reservoir	Value	Heavy metals in sediments ($\mu g \cdot l^{-1}$)			Heavy metals in water ($\mu g \cdot l^{-1}$)		
		Pb	Cd	Ni	Pb	Cd	Ni
Krempna-2	Min.	7.97	0.04	34.39	1.20	0.01	9.50
	Mean	14.19	0.28	43.03	2.40	0.03	12.30
	Max.	17.07	0.37	53.74	3.10	0.05	14.10
Zesławice-2	Min.	10.19	0.10	14.27	1.80	0.01	6.80
	Mean	19.44	0.47	15.73	3.10	0.04	8.90
	Max.	29.32	0.87	16.99	4.90	0.09	10.60
Cierpisz	Min.	0.76	0.08	3.86	0.10	0.01	1.30
	Mean	4.11	0.64	5.61	0.60	0.03	2.10
	Max.	7.95	2.12	9.55	1.40	0.05	2.80
Maziarnia	Min.	5.27	0.06	5.09	1.10	0.01	2.20
	Mean	5.77	0.16	5.64	1.50	0.03	3.70
	Max.	6.97	0.27	6.79	1.90	0.05	6.10
Niedźwiadek	Min.	10.00	0.41	17.30	1.30	0.02	6.30
	Mean	21.47	1.18	21.35	5.00	0.04	8.70
	Max.	30.90	1.71	22.88	6.10	0.08	11.20
Narożniki	Min.	7.76	0.56	21.20	0.90	0.03	5.90
	Mean	11.59	0.75	23.87	1.90	0.06	7.80
	Max.	14.21	1.12	28.32	2.40	0.07	9.30

Table 10. Content of heavy metals Pb, Cd and Ni in sediments and in water of water-course of six small water reservoirs

According to the proposition presented by Foernster (2007) determination of harmful substances content in water, including heavy metals, should be performed in agreement with one of two standards. According to the first standard evaluation of pollution level should be performed on the basis of the mean value established during at least one year. The mean value of substance concentration determined in such a way should be compared with threshold values of the Environmental Quality Standards. Whereas, according to the second standard the maximal recorded concentration of a given chemical element is compared with threshold values of the Environmental Quality Standards. In spite of the fact that chemical analyses of researched heavy metals were performed within the period of three years so with regard to low frequency of these measurements the pollution level of water was established according to the second standard proposed by Foernster (2007).

The recorded maximal concentrations of lead, cadmium and nickel (table 10) do not exceed the threshold values established in the Environmental Quality Standards according to which for the second water purity class the content of cadmium should be lower than 0.08 $\mu g \cdot l^{-1}$, whereas, the content of lead and nickel should be respectively not higher than 7.2 and 20 $\mu g \cdot l^{-1}$.

Application of the Environmental Quality Standards for evaluation of heavy metal pollution of bottom sediments of the studied water reservoirs required change of weight concentration of lead, cadmium, and nickel into bulk concentration expressed in µg $\cdot$l^{-1} (table 10). For this purpose the established mean bulk density of bottom sediments was used. The limits of admissible concentrations of Pb, Cd and Ni given in the Environmental Quality Standards were compares with maximal concentration in the sampled sediment. Contrary to the appraised water quality as pollution with lead, cadmium, and nickel in concerned the sediments of the studied water reservoirs contained metal concentrations exceeding limits of Environmental Quality Standards. In three water reservoirs i.e. in Krempna-2, Niedźwiadek and Narożniki, content of Pb, Cd and Ni in sediments is higher than admitted. In the other water reservoirs content of nickel did not exceed 20 µg $\cdot$l^{-1}. Content of cadmium in sediments in all studied reservoirs is higher than the admissible concentration equaling 0.08 µg $\cdot$l^{-1}. With regard to lead content which did not exceed 7.2 µg $\cdot$l^{-1}, only sediments in water reservoir Maziarnia may be, according to the concentration limits admitted by Environmental Quality Standards, regarded as unpolluted. A similar evaluation may be obtained in the sediments basing upon the mean lead, cadmium and nickel contamination.

Sediments accumulated in six water reservoirs were subjected to assessment of quality according to the instruction of PGI and SIEP, ISSPC instructions, and according to Müller's method with regard to mean concentrations of lead, cadmium, and zinc as well as chromium, zinc, and copper. The contamination degree of sediments was determined for mean concentrations of heavy metals according to the ISSPC instructions limits of heavy metals content corresponding with soils of group c (table 3) were adopted. As researches results of Michalcec and Tarnawski (2009) showed the pH of the studied reservoirs is within range 5.5-8.0. The results are given in table 11.

Reservoir	Method	Heavy metals (µg $\cdot$g^{-1})					
		Pb	Cd	Ni	Cr	Zn	Cu
Krempna-2	PGI and SIEP	I	I	III	I	I	II
	ISSPC	0	0	I	0	0	I
	Müller's	0,16 / I	-1,32 / 0	2,43 / II	-0,24 / 0	-0,69 / 0	1,8 / II
Zesławice-2	PGI and SIEP	I	II	II	I	I	I
	ISSPC	0	0	0	0	0	0
	Müller's	0,82 / I	3,58 / IV	1,43 / I	1,73 / II	-0,21 / 0	-0,10 / 0
Cierpisz	PGI and SIEP	I	II	I	I	I	I
	ISSPC	0	0	0	0	0	0
	Müller's	-0,84 / 0	-0,68 / 0	-1,21 / 0	1,60 / II	0,48 / I	-1,91 / 0
Maziarnia	PGI and SIEP	I	I	I	I	I	II
	ISSPC	0	0	0	0	0	0
	Müller's	-0,76 / 0	-1,67 / 0	-0,61 / 0	0,14 / I	-2,12 / 0	-1,75 / 0
Niedźwiadek	PGI and SIEP	I	II	II	I	II	I
	ISSPC	0	0	0	0	I	II
	Müller's	-0,13 / 0	1,34 / I	0,33 / I	-0,96 / 0	0,62 / I	0,17 / I
Narożniki	PGI and SIEP	I	I	II	I	I	I
	ISSPC	0	0	0	0	0	0
	Müller's	-1,47 / 0	-0,71 / 0	0,83 /I	-0,54 / 0	-0,27 / 0	0,49 / I

Table 11. Assessment of heavy metal pollution in bottom sediments

In order to establish geoaccumulative index (I_{geo}) according to Müller concentration of metals was determined sieved separated in a part of sediment grains smaller than 2 mm and the total concentration of trace metals were measured in <1 mm fraction of sediment, the background concentration in the shale sediment was adopted in agreement with Wedepohl (1995) as elemental concentrations in the continental crust.

Both according to the criteria PGI and SIEP, ISSPC and according to the Müller's method the studied reservoirs are characterized by highest pollution with nickel and chromium. Whereas, taking into consideration the obtained pollution classes (PGI and SIEP, Müller's method) and contamination degree (ISSPC) with particular heavy metals the heaviest polluted sediments are stated in water reservoirs Zesławice-2, Niedźwiadek and Krempna-2.

According to PGI and SIEP classification, assessed sediment contamination with trace metals may be classified as class I and II. They are characterized by toxic substances concentrations that do not have any ill effect on live organisms or this effect is occasional only. According to this assessment least polluted are sediments of the biggest reservoir i.e. Maziarnia as well as sediments of reservoirs Cierpisz and Narożniki. Sediments of these reservoirs are not contaminated with heavy metals (class I), except moderate contamination of sediments of the reservoir Maziarnia (class II) with copper, sediments of the reservoir Cierpisz contaminated with cadmium (class II) and sediments of the reservoir Narożniki with nickel (class II). With regard to the directives of ISSPC the sediments corresponds with soils of contamination degree from 0 to II. The lowest degree of pollution points at the possibility of unrestricted agricultural utilization of soils and in the case of sediments of this degree 0 there is an unlimited possibility of their agricultural utilization. With regard to a low content of lead, cadmium, nickel, chromium, zinc and copper such sediments are characterized of the reservoirs Maziarnia, Cierpisz, Narożniki and Zesławice-2. The other sediments were classified to soils poorly polluted (degree II) with regard to content of even one of the particular heavy metals indicating at the II degree of contamination. Out of this reason utilization of sediments from the water reservoirs Krempna-2 and Niedźwiadek is not admitted for recultivation or enrichment of agricultural soils used for some horticulture purposes, such as carrot, salad, spinach, since these plants assimilate heavy metals intensively.

This assessment confirmed by values of the sediment pollution index (SPI) according to according to Singht et al. (2002). These are for sediments of reservoirs Zesławice-2, Niedźwiadek and Krempna-2 respectively 5.12, 4.32 and 4.01; this indicates a moderately polluted sediment of the water reservoir Zesławice-2 water reservoir and low polluted sediment of the water reservoirs Niedźwiadek and Krempna-2. In the case of the other reservoirs i.e. Cierpisz, Narożniki and Maziarnia the obtained assessment indicated natural sediment since the obtained SPI values equal 1.49, 1.35 and 1.11 respectively.

According to the background levels in sediments adopted in Germany, Denmark, USA, and Canada (Yau & Gray, 2005), presented in table 4, the sediment quality may be appraised with regard to pollution with cadmium, cooper, iron, lead, and zinc. According to Canadian and USA levels of environmental pollution with lead only the sediments from the reservoir Niedźwiadek do not satisfy requirements with regard to its medium concentration. In this context the most restrict minimum German background levels of environmental pollution with lead of sediments both in the reservoir Niedźwiadek as well as in reservoirs Krempna-2 and Zesławice-2. Concentrations of lead in sediments of six studied reservoirs are lower than the environmental levels of the other criteria. Content of cadmium, according to the

Canadian, USA, and Danish criteria indicate that sediments from water reservoirs Ciepisz and Niedźwiadek exceed the admissible values. Whereas, according to the minimum German background levels mean concentrations of cadmium qualify the sediments as polluted. Sediments of reservoirs Krempna-2, Zesławice-2 and Niedźwiadek, similarly, as well as sediments from the reservoir Narożniki may be qualified as polluted sediments with regard to mean zinc concentration according to the environmental background levels of USA, Canada, and minimum German background levels. Trespassing the maximum German background levels of pollution and Dutch Target values with zinc was recorded only in the case of sediments from the reservoir Cierpisz and Niedźwiadek taking into regard only maximal concentrations of this chemical element. With regard to the mean zinc concentration only sediments from reservoirs Krempna-2, Zewławice-2, Niedźwiadek and Narożniki may be assessment according to minimum German background levels as polluted. Mean concentration of zinc in sediments from water reservoir Krempna-2 exceed additionally the Canadian and USA background levels and Dutch Target values. Taking into regard the highest number of exceeded concentration limits of Pb, Cd, Zn and Cu according to the above criteria the sediments of the reservoir Niedźwiadek are most intensively polluted. Sediments of the reservoirs Krempna-2, Zesławice-2, Cierpisz and Narożniki are characterized by a lower general pollution. Lowest pollution is found in the sediments of water reservoir Maziarnia.

The ISSPC directives are stricter in assessment of dredged sediments in comparison to the criteria given by Sullivan (Chen, Lin 2000). This follows from the fact that assessment of reservoir sediments according to the ISSPC directives takes into consideration not only quantities of trace elements pollution but also with regard to nature-agricultural use of sediments and accumulation heavy metals in cultivated plants. This caused establishment of lower admissible thresholds of heavy metal concentrations in sediments. According to criteria given by Sullivan (Chen, Lin 2000) dredged sediments should not contain nickel, chromium, zinc, and copper higher than 100 μg·g^{-1} and more than of 50 μg·g^{-1} lead. Neglecting cadmium, which was not included in Sullivan's criterion, sediments of the studied reservoirs satisfy the criterion given by Sullivan except zinc content. Only sediments of the water reservoir Niedźwiadek are characterized by moderate zinc concentration equaling over 100 μg·g^{-1}. Hence, it would be advisable to apply the ISSPC directives for assessment of possibilities of dredged sediments utilization. It is so important because due to a very intensive silting process small water reservoirs require desilting even after only a few years of operation. In the case of agricultural utilization of the dredged sediments substances present in it may cause contamination agricultural cultivated soils. Carrying on researches concerning determination of heavy metals content in sediments and recording their changes in time of small water reservoirs operation is extremely important with regard to environmental condition and human health protection.

5. Conclusions

The mean silting ratio (S_A) of small water reservoirs whose capacity does not exceed 5 millions m^3 is within range from 0.06% to 10.3%. The considerable differentiation of values of silting ratio results both from the way in which the catchment is used, its physiogeographic parameters as well as from extreme hydrological-meteorological phenomena or different hydraulic water flow conditions through the reservoir. A high silting ratio is characteristic of water reservoirs Krempna, Zesławice, Maziarnia, Rzeszów,

Głuchów and Cierpisz (table 5, column 6). The mean silting ratio of these reservoirs shows intensive silting. The silting intensity is influenced by the way of catchment area cultivation. Crop lands cover over 45% of catchment areas of the studied reservoirs. The considerable participation of arable land is conductive to surface erosion processes of mineral material which is carried away from the catchment.

Small water reservoirs whose capacity-inflow ratio is lower than 10% are characterized by a quick reduction of capacity in a relatively short time and their annual silting ratio is 2.5%. With regard to the intensity of the silting process their capacity is reduced within a dozen of years of operation on the average by about 40% (table 5). Assessment of the Hachiro's and Khosl's formulas for determination the mean annual silting ratio of the studied small water reservoirs proved to the impossible since the obtained results differed considerably for the mean annual silting ratio calculated on the basis of silting measurements.

The elaborated relation in form $S_A = 32.97S_i^{-0.83}$ permits calculation of the mean annual silting ratio of the small water reservoirs whose capacity-inflow ratio is lower than 10%. For this purpose it is necessary to determine the mean annual volume of carried suspended sediment entering the reservoir (V_{SS}) and this makes calculation of silting intensity (S_i) possible using Šamov's formula (3). In case of the volume of the carried sediment entering the reservoirs is not available the mean annual silting ratio can be assessed by use of the relation $S_A = 1.015C\text{-}I^{-0.563}$. Admittedly on the basis of the determination coefficient equaling 0.3707 poor agreement accordance of the regression model to the data at disposal was found, the correlation coefficient 0.61 indicates at average strength of the correlation. Hence, the calculation results obtained by use of this formula should be treated only as approximate values. Extension of researches over a bigger number of small water reservoirs located in various catchment areas could enable verification of the proposed equations and possibility their modification.

In the studied water reservoirs a fine fractions of sediment smaller than 1 mm is accumulated. Mean bulk density of bottom sediments is from 1.03 to 1.65 t·m^{-3}, and the content of the organic matter does not exceed 10%. The content of heavy metals in bottom sediments of six small water reservoirs does not exceed on the average a few percent of sediment mass deposited in the reservoirs. Appraising pollution of sediments with trace metals according to the Environmental Quality Standards it was stated that concentration of metals in deposits exceed the EQS limits. According to the assessment of environmental pollution elaborated on the basis of assessment of lead, cadmium, and nickel concentrations in waters flowing into the studied water reservoirs which do not exceed the threshold value defined in EQS no threat of environmental pollution of the reservoirs ecosystem may be stated. The obtained researches results shows that correct environmental assessment of rivers, streams and dam reservoirs based on surface water monitoring results cannot be guaranteed. Identical level of heavy metal pollution in sediment and in biota cannot be serving at assessment of environment pollution on the basis of the concentration of heavy metals in surface water. Hence, fulfillment of notation of EQS Article 2 concerning environmental quality, saying that Member States shall arrange for the long term trend analysis of concentrations of priority substances, that tend to accumulate in sediment and/or biota on the basis of monitoring of water status carried out. They shall take measures aimed at ensuring, that such concentrations do not significantly increase in sediment and/or relevant biota. With regard to lack of extensive and reliable information in the European Community on concentration of priority substances in fauna and flora and in sediments one cannot confine oneself to establishing standards of environmental quality basing on surface water assessment. It is assumed that Member States are supposed to establish more in detail Environmental Quality Standards or its supplement. Thus, is important to perform a comparison of standards of particular states serving for assessment of pollution not only of sediments but also of fauna and flora. These constitute important matrices for monitoring of some substances for assessment of long term effect of human activity and control if the Member States ensure no increase in pollution level of biota as well as of sediments. The exceeded assessment of chosen criteria of admissible levels of environmental concentrations of heavy metals adopted in Germany, Denmark, USA, and Canada showed that according to these criteria great differentiation of quality of sediments accumulated in the studied water reservoirs was recorded. Assessing sediment quality to the background levels in sediments it was stated, that the most restrictive criteria are the German ones with regard to minimum background levels. This is in particularly noticeable in the case of cadmium concentration. Sediments of each of the analysed reservoirs do not satisfy the criterion of admissible concentration of this metal. Both the maximal and medium concentrations of cadmium in sediments are higher than admissible ones in background concentration levels adopted in Germany, Denmark, USA, and Canada. Whereas, according to PGI and SIEP sediments of three reservoirs, i.e. Zesławice-2, Cierpisz and Niedźwiadek, are sediments moderately polluted (class II) with regard to cadmium concentration. Sediments of the other reservoirs according to this criterion are not polluted with cadmium and are classified as class I. This indicates at adoption of unified criteria of determination of contamination level of bottom sediments, that may enable gaining comparable assessment of long term effects oh human activity. Criteria of sediment assessment should not be treated neglecting the influence sediments on live organisms. The classification elaborated and applied in Poland by the Polish Geological Institute and the Inspection of Environment Protection State are an example of introduction of such criteria.

Elaboration of more accurate normatives of Environmental Quality Standards, based on heavy metals content in surface waters and sediments, requires the elaboration of more precise comparison methods and tools for the assessment of heavy metals pollution levels and contaminants constant monitoring. Currently, more reliable surface water quality assessments are still demanded.

6. References

Batuca, G. D; Jordaan, M. J. Jr. (2000). Silting and desilting of reservoirs, *A.A.Balkema*, ISBN 90-5410-477-5, Rotterdam, Netherlands

Bojakowska, I. (2001). Criteria for evaluation of water sediments pollution. *Polish Geological Review*, No.49(3), pp. 213-219

Bojakowska, I.; Sokołowska G. (1998). Geochemical purity classes of bottom sediments, *Polish Geological Review*, No.46, pp. 49–54

Borja, A.; Valencia, V.; Franco, J.; Muxika, I.; Bald, J.; Belzunce, M. J.; Solaun, O. (2004). The water framework directive: water alone, or in association with sediment and biota, in determining quality standards? *Marine Pollution Biuletine*, No.1-2, pp. 8-11

Caeiro, S.; Costa M. H.; Ramos, T. B.; Fernandes, F; Silveira, N; Coimbra, A.; Medeiros, G; Painho, M. (2005). Assessing heavy metal contamination in Sado Estuary sediment: An index analysis approach. *Ecological Indicators*, No.5, pp. 151-169

Chen, S. Y.; Lin, J. G. (2001). Bioleaching of heavy metals from sediment: significance pH. *Chemiosphere*, No.44, pp. 1093-1102

Dendy, F. E. (1982). Distribution of sediment deposits in small reservoirs. *Transactions of the American Society of Agricultural Engineers*, vol.25, No.1, pp. 100-104

Dz.U. Nr 165, poz. 1359. Regulation of the Minister of Environment dated 9 September 2002, http://dokumenty.e-prawnik.pl/akty-prawne/dziennik-ustaw/2002/165/1359 (in Polish)

Environmental Quality Standards (EQS) Applicable to Sediment and/or Biota. (2007). Science and Policy. Council of the European Union, Brussels, Appendix, pp. 1-20, http://dx.doi.org/10.1065/jss2007.07.240.1

Foernster, U. (2007). Environmental Quality Standards (EQS) Applicable to Sediment and/or Biota. *Journal of Soils and Sediments*, Vol.7, No.4, pp. 270, http://www.springerlink.com/content/52r5g23782841u39/

Hartung, F. (1959). Ursache und Verhuetung der Staumraumverlandung bei Talsperren, *Wasserwirtschaft*, Nr.1, pp. 3-13

Kabata-Pendias, A.; Piotrowska, M.; Motowicka-Terelak, T.; Maliszewska-Kordybach, B.; Filipiak, K.; Krakowiak, A.; Pietruch, Cz. (1995). Basis of assessment of chemical contamination of soils - heavy metals, sulfur and PAH. IUNG, *Environmental Monitoring Library*, Warsaw, pp 1–34

Kostrzewski, A.; Kruszyk R.; Kolander R. (2006). Measurement Programs of Integrated Monitoring of Natural Environment, 8.11. Measurement Program H1: surface waters – rivers. *Environmental Monitoring Library*, Warsaw, http://www.staff.amu.edu.pl/~zmsp/org/8_11.htm

Lara, J. M.; Pemberton, E.L. (1963). Initial weight of deposited sediments, Proceedings of Federal Interagency Sedimentation Conference. USDA-ARS, Miscellaneous Publication 82, in: *Proceedings of the Federal Inter-Agency Sedimentation Conference 1963*, U.S. Department of Agriculture Miscellaneous Publication, No.970, pp. 818–845

Madeyski, M. (2002). Quality of bottom sediments in some small water reservoirs and fish ponds. *Proc. 21-th Conference of the Danube Countries on Hydrological Forecasting and Hydrological Bases on Water Management*, Bucharest, Romania, pp. 85-89

Michalec, B. (2008). Appraisal of silting intensity of small water reservoirs in the Upper Vistula river basin. *Scientific Fascicles* No. 451, Fascicle 328, Agricultural University Cracow, pp. 193 (in Polish)

Michalec, B.; Tarnawski, M., (2009). Appraisal of bottom sediment pollution with trace metal of water reservoirs located in South Poland, *Archives of Environmental Protection*, Vol.35(3), pp. 73-85

Müller, G. (1981). Die Schwermetallbelastung der Sedimenten des Neckars und seiner Nebenflüsse. *Chemiker-Zeitung*, Nr 6, pp. 157-168

O' Neil, P. (1993). Environmental Chemistry, *Chapman and Hall*, London. pp. 193

Ozga-Zieliński, B. (1987). The study of statistical heterogeneity of measuring strings. Water Management, Nr 10, Warszawa, 226-228 (in Polish)

Prater, B.E. (1975). The metal content and characteristics of Steelwork effluents discharging to the Tees estuary, *Water Pollution Control*, No.74, pp. 63-78

Preuss, E.; Kollman, H. (1974) Metallgehalte in Klarschlammen, *Naturwissenschafter*, No.61, pp. 270-274

Program of small retention of the Małopolski District, (2004). Project of Marshal Office of Małopolski District and Land Melioration and Water Units Board of Małopolski Provincie in Cracow, Cracow, ersion for CD, pp. 47 (in Polish)

Roehl, J. (1962). Sediment source area, delivery rations and influencing morphological factors. *IAHS Publication*, No.59, pp. 202-213

Sawunyama, T. (2005). Estimation of small reservoirs storage capacities in Limpopo river basin using GIS and remotely sensed surface areas. Dep. of Civil Engineering, University of Zimbabwe, pp. 68

Singh, M.; Müller, G.; Singh, I. B. (2002). Heavy metals in freshly deposited stream sediments of rivers associated with urbanization of the Ganga Plain, India. *Water, Air, and Soil Pollution*, No.141, pp. 35–54

Soler-López, L. R. (2001). Sedimentation survey results of the principal water supply reservoirs of Puerto Rico. *Proc. of 6th Caribbean Islands Water Res. Congr.* Puerto Rico (unpaginated CD), http://pr.water.usgs.gov/public/reports/soler.html

Tarnawski, M. (2009). Appraise of chemical quality of bottom sediments of chosen small water reservoirs. *Environmental Protection and Natural Resources*, No.38, pp. 372-379

Wedepohl, K. H. (1995). The composition of the continental crust. *Geochimica et Cosmochimica Acta*, No.59, pp. 1217–1239

White, P.; Labadz, J. C.; Buchter, D. P. (1996). Sediment yield estimation from reservoir studies: An appraisal of variability in the southern Pennines of the UK. In: *Erosion and sediment yield: global and regional perspectives*, IAHS publication No. 236, pp. 163-173

Wischmeier, H. W.; Smith, D. D. (1965). Predicting rainfall erosion losses-aquide from cropland east of the Rocky Mountains. USDA, *Agriculture Handbook*, No. 282, p. 47

Wiśniewski, B.; Kutrowski, M. (1973). Special constructions in water managements. *Water reservoirs. Predicting silting rate. Manual.* Water Management Study and Design Office „Hydroprojekt", Warsaw, pp. 55 (in Polish)

Yau, H; Gray, N. F. (2005). Riverine sediment metal concentration of the Avoca-Avonmore catchment, south-east Ireland: a baseline assessment. *Biology and Environment*: Proceedings of the Royal Irish Academy, Vol.105B, No.2, pp. 95-106

16

Earthworm Biomarkers as Tools for Soil Pollution Assessment

Maria Giulia Lionetto, Antonio Calisi and Trifone Schettino
University of Salento - Dept. of Biological and Environmental Sciences and Technologies
Italy

1. Introduction

Soil pollution has enormously increased during the last decades due to the intensive use of biocides and fertilizers in agriculture, industrial activities, urban waste and atmospheric deposition. Its occurrence is related to the degree of industrialization and intensity of chemical usage. Soil pollution causes decrease in soil fertility, alteration of soil structure, disturbance of the balance between flora and fauna residing in the soil, contamination of the crops, and contamination of groundwater, constituting a threat for living organisms.

The most diffusive chemicals occurring in soil are heavy metals, pesticides, petroleum hydrocarbons, polychlorobiphenyl (PCBs), dibenzo-p-dioxins/dibenzofurans (PCDD/Fs). Heavy metals from anthropogenic sources are widely spread in the environment and most of them finally reach the surface soil layers. Heavy metals can enter the soil from different sources, such as pesticides, fertilizers, organic and inorganic amendants, mining, wastes and sludge residues (Capri & Trevisan, 2002). In contrast to harmful organic compounds, heavy metals do not decompose and do not disappear from soil even if their release to the environment can be restricted (Brusseau, 1997). Therefore, the effects of heavy metal contamination on soil organisms and decomposition processes persist for many years. Pesticides are widely used in agriculture for counteracting insects, fungi, rodents or other animals living in or on the crops. They are either directly applied to soil to control soil borne pests or deposited on soil as run off from foliar applications and their concentrations are high enough to affect the soil macro-organisms (Bezchlebova et al., 2007). The pesticides most widely used in the past have been organochlorine pesticides, characterized by high hydrophobicity and persistence. Currently, they have been replaced by less persistent compounds. Organophosphates have become the most widely used pesticides today. They are used for pest control on crops in agriculture and on livestock, for other commercial purposes, and for domestic use. Due to their water solubility, the organophosphate residues in agricultural practices are capable of infiltrating through soil into surface water. As a consequence of their wide diffusion they have been detected in food, ground and drinking water, and natural surface waters (Dogheim et al., 1996; Garrido et al., 2000). Soil pollution by petroleum hydrocarbons usually originates from spills or leaks of storage tanks during fuel supply and discharge operations. Petroleum hydrocarbons include aliphatic and aromatic compounds; some of them are known or suspected human carcinogens, and are classified as priority pollutants. PCBs are persistent soil contaminants due to their hydrophobicity and resistance to biodegradation (Weber et al., 2008). They can be released into the environment from poorly maintained hazardous waste sites that contain PCBs, illegal or improper dumping of PCB wastes, such as transformer fluids, leaks or releases from electrical transformers containing PCBs, and disposal of PCB-containing consumer products into municipal or other landfills not designed to handle hazardous waste. PCBs are also currently released into the environment by municipal and industrial incinerators from the burning of organic wastes. PCDD/Fs are chemically very stable and are highly hydrophobic compounds. PCDD/Fs have a high affinity to organic matter and have limited mobility unless transported in association with particulate organic matter. However, these compounds are bioaccumulative, can be found in the terrestrial food chain, and have been reported to impact the biota at the higher trophic levels. Recently, the attention of the scientific community focused on emerging contaminants in the soil, such as pharmaceuticals, endocrine disruptors, personal care products, surfactants, flame retardants. They are currently not included in routine monitoring programmes, but may be candidates for future regulation depending on research on their toxicity, potential effects on the environment and occurrence in the environmental compartments.

Due to the increasing concern about chemical contamination of soil there is an increasing interest in the scientific community and international agencies for soil pollution monitoring and assessment. The traditional approach to soil pollution assessment, based on the analysis of the concentrations of pollutants in the soil and comparison with specific threshold values, does not provide indication of deleterious effects of contaminants on the biota. It neglects several essential aspects such as toxicity of chemicals not included in the selection of contaminants to be analyzed, interactive effects (synergism and antagonisms) of pollutants on biota and bioavailability. Bioavailability refers to the fraction of a contaminant that is taken up by an organism from the environmental media (i.e., through both passive and active routes), and directly influences toxicity (Smith et al., 2010). Bioavailability of pollutants in soil to terrestrial invertebrates and plants can be influenced by some characteristics of the soil such as pH, cation exchange capacity, and organic matter content (Bradham et al., 2006; Spurgeon et al., 2006; Criel et al., 2008). The influence of a single factor on pollutant bioavailability is usually site-, chemical-, and soil-specific. For this reason, it is difficult to model pollutant bioavailability based on total concentration and soil characteristics alone. The best integrators of these complex effects are the exposed organisms themselves.

For these reasons, new biological approaches to soil monitoring, such as the measurement of biochemical and cellular responses to pollutants (i.e. biomarkers) on organisms living in the soil (bioindicators), have become of major importance for the assessment of the quality of this environmental compartment (Kammenga et al. 2000). Soil invertebrates may represents good sentinel organisms of soil chemical pollution because they are in direct contact with soil pore water or food exposure, in contrast to many vertebrates that are indirectly exposed through the food chain (Kammenga et al., 2000). Among soil invertebrates earthworms are relevant organisms for soil formation and organic matter breakdown in most terrestrial environments. Because of their particular interactions with soil, earthworms are significantly affected by pollution originated on intensive use of biocides in agriculture, industrial activities, and atmospheric deposition. Hence, earthworms as been proved as valuable bioindicators of soil pollution (Lanno et al., 2004).

The aim of the chapter is to review the use of molecular and cellular biomarkers in earthworms as early and sensitive indicators of soil pollution and stress.

2. Earthworm as bioindicator organisms of soil pollution

Earthworms are very important organisms for soil formation and organic matter breakdown in most terrestrial environments and traditionally they have been considered to be convenient indicators of land use impact and soil fertility. They contribute to pedogenesis and soil profile, affect the physical, chemical and microbiological properties of soil (Barlett et al., 2010) and contribute to improve soil fertility. In particular earthworms may increase mineralization and humification of organic matter by food consumption, respiration, and gut passage (Lavelle & Spain 2001). They may indirectly stimulate microbial mass and activity as well as the mobilization of nutrients by increasing the surface area of organic particles and by their casting activity (Emmerling & Paulsch 2001). Moreover, their burrowing activities significantly contribute to increase water infiltration and soil aeration.

The role of earthworms in the decomposition of organic matter and subsequent cycling of nutrients has raised the interest of their use as indicator organisms for the biological impact of soil pollutants. This in turn has lead to a large body of work on earthworm ecotoxicology (Spurgeon et al., 2003). In addition, earthworms manipulation is relatively simple; this facilitates the measurement of different life-cycle parameters e.g. growth and reproduction, as well as accumulation and excretion of pollutants, and biochemical responses. Thus, earthworms are suitable organisms for soil ecotoxicological research.

Because of their particular interactions with soil, they are significantly affected by pollutants reaching the soil system. Earthworms can be exposed to contaminants in various ways. Firstly, living in the soil they are in direct contact with soil pore water and therefore with pollutants therein dissolved. The earthworm skin is extremely permeable to water (Wallwork, 1983) and it represents a main route for contaminant uptake (Jajer et al., 2003; Vijver et al., 2005). Secondly, these organisms ingest large amounts of soil, therefore they are continuously exposed to contaminants adsorbed to solid particles through their alimentary tract (Morgan et al., 2004). Chemicals uptake via the dermal route can be directly related to the pore water concentration. Vijver et al. (2005) measured the contribution of each pathway for heavy metals uptake in the earthworm *Lumbricus rubellus*. The authors concluded that the dermal route is the main uptake route for metals and that pore water uptake via ingestion does not significantly contribute to metal accumulation.

Earthworms are able to accumulate various organic and inorganic contaminants (Morrison et al., 2000) present in the soil. Previous studies under both field and laboratory conditions demonstrated that earthworms bioaccumulate certain metals (such as Cd, Cu, Zn and Pb) from soils. They accumulate efficiently and tolerate high tissue metal concentrations using a variety of sequestration mechanisms (Peijnenburg, 2002; Andre et al., 2009). Earthworms appear to have well-developed trafficking and storage pathways for heavy metals, particularly for essential trace metals such as Cu and Zn (Morgan & Morgan, 1999). Metals are primarily accumulated within the posterior alimentary canal of the earthworm when inhabiting heavy metal contaminated sites. This part of the body includes the intestine and the related chloragogenous tissue that separates the absorptive epithelia from the coelomic cavity (Andre et al., 2009). The chloragogenous tissue is composed of pedunculated cells and its main functions are synthesis of haemoglobin, homeostasis of cation composition in the blood and coelomic fluid, maintenance of a balanced pH level, storage of nutrients and waste, and uptake and detoxification of toxic cations (Jamieson, 1992). Therefore, chloragogenous tissue represents a major metal sink (Morgan et al., 2002).

Bioaccumulation is species-specific (Heikens et al., 2001; Hendrickx et al., 2004; Nahmani et al., 2007) and is influenced by the physicochemical properties of the pollutants and of the environmental scenario (Vijver et al., 2005). In fact it depends on factors such as metal speciation and concentration (Heikens et al., 2001; Hobbelen et al., 2006; Spurgeon et al., 2006; Nahmani et al., 2007), soil type and characteristics (Hendrickx et al., 2004; Kizilkaya, 2005; Hobbelen et al., 2006; Spurgeon et al., 2006), temperature (Olchawa et al. 2006), and exposure duration (Nahmani et al. 2007). Bioaccumulation in earthworms can be expresses as Biota to- Soil Accumulation Factor (BSAF) (Cortet et al., 1999), calculated by the following formula: BSAF = metal content in earthworm/total metal content in soil. In Tab.1 the BSAF values calculated for several soil pollutants in two earthworms species is reported.

	Cd	Zn	Cu	Pb	DDTs	PCBs	PBDEs
A. caliginosa	6.18–17.02	1.95–7.91	0.27–0.89	0.08–0.38	-	-	-
L. rubellus	3.64–6.34	1.5–6.35	0.29–0.87	0.04–0.13	1.48–1.70	1.09–2.76	1.99–5.67
Ref.	Dai et al. 2004				Vermeulen et al., 2010		

Table 1. Biota-to-soil accumulation factors (BSAF) in earthworms

Earthworms serve as a major food source for numerous animals, in particular, amphibians, reptiles, birds and mammals. Therefore, bioaccumulation of chemical contaminants by earthworms implies the risk of the transfer of pollutants to higher trophic levels (Marino et al., 1992).

The importance of earthworms in testing the adverse effects of chemicals on soil organisms has been recognised by several environmental organisations and, as a result, a set of standard test guidelines are available. For official guidelines soil toxicity assessment must be addressed by at least two-three different assays, thus combining plant assays and earthworm assays. During the early 1980s, acute toxicity tests for earthworms, based on mortality and grow rate, were developed jointly by the European Union (EU) and by the Organization for Economic Cooperation and Development (OECD) (OECD, 1984). They were later implemented with chronic toxicity tests based on reproduction rate measurement (OECD, 2004).

Several standardized tests for assessing the toxicity of contaminated soil to earthworms exist (OECD 1984, 2004), but these methods remain relatively costly and time-consuming. For instance, the acute earthworm toxicity test requires 14 days (OECD 1984), and since it only measures lethality it may be insufficient for predicting long-term population fitness following chronic exposure to contaminated soil (Whitfield et al., 2011). Chronic test, based on the inhibition of earthworm reproduction (OECD 2004), provides a more ecologically relevant sub-lethal endpoint than lethality, but it requires a longer exposure period for accurate assessment (50 days). In recent years there is a growing interest for increasing the knowledge on molecular and cellular responses of earthworms to pollutants as biomarkers of pollutant exposure and effect to be used in soil monitoring and assessment programmes (Beliaeff & Burgeott, 2002; Handy et al., 2003). With respect to standard toxicity tests the biomarker approach can offer more information about the organism's stress response to individual toxicants and mixtures (Kammenga et al., 2000; Scott-Fordsmand & Weeks, 2000; Hankard et al., 2004; Svendsen et al., 2004) and a tool in monitoring field populations where the assessment of conventional endpoints is difficult to be applied.

Most studies on earthworm biomarkers has been conducted on *Eisenia spp.* while other earthworm species remain less investigated. *Eisenia fetida* is the standard testing organism used in terrestrial ecotoxicology due to its rapid life cycle and simple rearing in the laboratory. The ecological relevance of *Eisenia spp.* as bioindicator organism in soil monitoring based on biomarker approach has been recently questioned (Sanchez-Hernandez, 2006) because they are epigeic species, forming no permanent surface burrows on the soil surface, feeding on decaying organic matter, while in most cases contaminants occur at soil depths where these earthworms are not found. *Eisenia fetida* is a North European litter-dwelling species inhabiting the soil surface, living primarily in sites rich in organic matter (Jänsch et al., 2005) such as compost heaps, manure piles, or sewage sludge, and thus unlikely to be present naturally in agricultural soils or contaminated land sites (Spurgeon et al., 2002). The study of anecic species (i.e. *Lumbricus terrestris*), that forms temporary deep burrows and comes to the surface to feed, and endogeic species (i.e. *Aporrectodea caliginosa*), which build complex lateral burrow systems through all layers of the upper soil rarely coming to the surface, could be more ecologically relevant. In fact these species can be exposed to pollutants present not only in the soil surface but also in the soil deeper layer, providing an integrated response to soil pollution.

3. Biomarkers: Definition

A biomarker is defined as a "biochemical, cellular, physiological or behavioural variations that can be measured in tissue or body fluid samples, or at the level of whole organisms, to provide evidence of exposure and/or effects from one or more contaminants" (Depledge, 1994). The effects of contaminants at lower levels of biological organization (e.g. biochemical, cellular, physiological) in general occur more rapidly than those at higher levels (e.g., ecological effects) and therefore may provide a more sensitive early warning of toxicological effects within populations. Potentially, any alterations in any of the molecular, cellular, biochemical, and physiological processes occurring within an organism following pollutant exposure could be used as biomarkers.

Biomarkers, in general, may be classified into biomarkers of exposure, and biomarkers of effect (Chambers et al., 2002). The former indicate that an organism has experienced exposure to a pollutant, and offer an early signal for exposure to micropollutants. They can provide qualitative and quantitative estimates of exposure to various compounds. However, the change in these biomarkers may not be predictive of the degree of adverse effects either on the organism or in the population. Biomarkers of effect are associated specifically with the toxicant's mechanism of action and are sufficiently well characterized to relate the degree of biomarker modification to the degree of adverse effects (Chambers et al., 2002). Biomarker of exposure and biomarkers of effect can differently contribute to environmental assessment. Biomarkers of exposure may have the potential to offer an alternative to some chemical analyses or to measure effects of short-lived chemicals as well as giving a more biologically relevant indication of exposure (Hagger et al., 2006). On the other hand, biomarkers of effect can shed light on qualitative aspects of hazard identification by both demonstrating that a hazard is occurring and elucidating the mechanisms responsible (Chambers et al. 2002). These biomarkers can provide insights into both the causal factors of the hazard and its ecological consequences, depending on the degree of specificity of the biomarker with the pressure and the degree to which its expression to higher order effects is understood. The specificity of biomarkers to pollutants ranges from highly specific biomarkers to not specific biomarkers such as immune system impairment or DNA damage. Therefore, the use of a suite of biomarkers with a range of specificity is an important aspect of environmental monitoring based on the biomarkers. Although all biomarker types can provide useful information about exposure or effects of pollutants on living organisms, certain criteria have been developed to address the selection of the most useful and relevant to use for pollutant impact assessment in environmental monitoring. These include whether a biomarker is sensitive, whether it responds in a dose- or time dependent manner to the toxicant (Walker et al., 1998), including its transient and biochemical memory (how long after exposure the response lasts), whether the variability in biomarker response due to natural variation is known (i.e., season, temperature, sex, weight, age) (Hagger et al., 2006). To provide an accurate toxicity assessment, biomarkers should responds to a pollutant in a dose-dependent manner over a concentration range of pollutant that is environmentally meaningful. The validity of any biomarker depends on its ability to precisely separate anthropogenic stressors from the influence of natural variability. Finally, biomarkers that can be linked to important biological processes and for which a correlation between the observed responses and deleterious effects at the individual or population/community level is established are considered of relevance in environmental assessment.

4. Biomarkers in earthworms

Overall, the use of biomarkers in earthworm is becoming increasingly important for the evaluation of effects of contaminants on soil organisms. Acetylcholinesterase, metallothionein, biotransformation enzymes and antioxidant defences are among the most used biomarkers due to their crucial role in the neurocholinergic transmission and in cell homeostasis preventing toxic action of chemicals (Sanchez-Hernandez, 2006; Novais et al 2011). However, the research of novel biomarkers in earthworms is receiving increasing attention for its potentiality in soil pollution monitoring and assessment.

4.1 Metallothioneins

Metallothioneins (MTs) are low-molecular-weight cystein-rich metal-binding proteins that are involved in homeostasis of essential metals like Cu and Zn and detoxification of non essential metals such as Ag, Cd and Hg (Costello et al., 2004; Amiard et al., 2006). In addition to their function as metal chelators, MTs act as free radical scavengers (Min, 2007). They contain 25%–30% cysteine, but few aromatic or histidine residues. Vertebrate and invertebrate metallothioneins contain two unique metal-thiolate clusters determined by the presence along the sequence of the protein of metal chelating Cys-X-Cys sequences, where X can be any amino acid other than cysteine. The MT protein is dumbbell-shaped, and the polypeptide backbone is wrapped around the metal thiolate core, forming the scaffold for two domains, designated α and β, separated by a short linker region. Induction of MTs by metal exposure has been detected in a wide variety of organisms including earthworms (Stürzenbaum et al., 2001). For example a significant induction of MT proteins was observed in different earthworm species such as *Lumbricus rubellus*, *Eisenia fetida*, *Eisenia andrei* exposed to cadmium (Calisi et al., 2009; Demuynck et al., 2006; Ndayibagira et al., 2007; Brulle et al., 2007), or in *Lumbricus mauritii* exposed to Pb and Zn contaminated soil (Maity et al., 2011) and in *Lumbriucus terrestris* exposed to cadmium, copper and mercury (Calisi et al., 2011a). It is known that earthworms share a high tolerance to heavy metal exposure

(Stürzenbaum et al., 1998) also thanks to the fundamental contribution of these metal-binding proteins. In *Lumbricus terrestris* Calisi et al (2011b) found the major concentration of MT in the postclitellar portion of the animal body with respect to the preclitellar part. This result is in agreement with the hymmunoistochemical localization of MT in the intestine and chloragogenous tissue previously reported by Stürzenbaum et al. in *Lumbricus rubellus* (2001).

Although the amino acid sequences of more than 50 invertebrate MT and MT-like proteins have already been determined, little is known about the biochemical properties of earthworm MTs. So far, only 5 MT genes of earthworms have been cloned from *Lumbricus castaneus, Eisenia fetida, Lumbricus rubellus,* and *Lumbricus terrestris* (Gruber et al., 2000; Liang et al., 2009) whose expression is differentially regulated by different heavy metals (Sturzenbaum et al., 1998, 2001). Metallothionein induction is one of the mostly utilized biomarker in earthworms and is applied as early biomarker of exposure to heavy metals in soil monitoring.

4.2 Acetylcholinesterase

Acetylcholinesterase (AChE) is a key enzyme in the nervous system, terminating nerve impulses by catalyzing the hydrolysis of neurotransmitter acetylcholine. AChE is the target site of inhibition by organophosphorus and carbamate pesticides. In particular, organophosphorus pesticides inhibit the enzyme activity by covalently phosphorylating the serine residue within the active site group. They irreversibly inhibit AChE, resulting in excessive accumulation of acetylcholine, leading to hyperactivities and consequently impairment of neural and muscle system. Acetylcholinesterase represents the main cholinesterase in earthworms (Rault et al., 2007). Its activity has been identified and biochemically characterised only in a few earthworm species (Caselli et al., 2006). According to Rault et al. (2007) and Calisi et al. (2011b), the highest concentration of AChE activity was found in the pre-clitellar part of the animal and suggests a main role of this enzyme in functioning of the dorsal brain localized near the prostomium. Rao et al. (2003) and Rao & Kavitha (2004) found a time-dependent AChE inhibition in *Eisenia fetida* exposed to chlorpyrifos and azodrin, two organophosphate pesticides, in the standardized paper contact test. Calisi et al. (2009) reported a significant inhibition (about 45 %) of AChE activity in *Eisenia fetida* after two weeks of exposure to the carbamate methiocarb added into the soil at the maximal concentrations recommended in vineyards (EEC, 2001). Moreover, Calisi et al (2011b) observed that in *Lumbricus terrestris* high percentage inhibition of AChE activity by pesticide exposure was not paralleled by a corresponding high mortality value (higher than) as observed in birds and mammals (see Tab.2), where AChE inhibition higher than 50% of normal is referred to be irreversible and regarded as being in the lethal range (for a review see Lionetto et al., 2010).

ACHE inhibition (%)	Mortality (%)	Species	Ref.
70	30	Earthworm *(Lumbricus terrestris)*	Calisi et al., 2011b
70	100	Mammals and birds	Walker, 1998

Table 2. AChE inhibition and mortality in earthworm and higher vertebrates

The lower sensitivity of animal survival to AChE inhibition compared to vertebrates was also recently documented in other earthworm species (Rault et al., 2008) and suggests that the toxic action of pesticide on earthworms can involve also other molecular or cellular target beyond AChE.

Recently, the potential of some metallic ions, such as Hg^{2+}, Cd^{2+}, Cu^{2+} and Pb^{2+}, to depress the activity of AChE of fish and invertebrates, *in vitro* and or *in vivo* conditions has been demonstrated in several studies (for a review see Lionetto et al., 2010). On the contrary in *Lumbricus terrestris* (Calisi et al., 2011b) and *Eisenia fetida* (Calisi et al., 2009) AChE activity was unaffected by copper sulphate exposure.

AChE inhibition in earthworms is presently regarded as giving early warning of adverse effects of pesticides (Booth & O'Halloran, 2001), and consistently included among the batteries of biomarkers employed for early assessments of pollutant impact on wildlife in terrestrial ecosystems. However, concerning AChE in earthworms only a few pesticides in use have been tested against relatively few earthworm species both in laboratory tests and under field conditions (Rao et al., 2003; Calisi et al., 2009; Scott-Fordsmand &Weeks, 2000; Rao & Kavitha, 2004; Gambi et al., 2007). As pointed out by Scott-Fordsmand & Weeks (2000) and by Sanchez-Hernandez (2006) the potential use of AChE in earthworms as biomarker of pesticide exposure has not been sufficiently explored.

4.3 Biotransformation enzymes

In their habitat earthworms can be exposed to a variety of plant alkaloids, PAHs and pesticides known to be inducers of detoxification responsible enzymes. In eukaryotes, detoxification of organic compounds usually occurs in two phases. Phase I detoxification processes involve the cytochrome P450 enzyme system and results in the introduction of a functional group, such as hydroxyl or sulphonyl, to non-polar compounds. In some cases the metabolites of phase I reactions are more toxic than the parent compound. Phase II detoxification enzymes, such as glutathione S-transferase (GST), attach a large polar, water-soluble moiety to the products of phase I metabolism to promote excretion and elimination of the toxicant.

The presence of cytochrome P450 was demonstrated in *Lumbrucus terrestris* by Liimaitainen & Hänninen (1982), but only the occurrence of the monooxygenase activity benzoxy-resorufin-Odealkylase (BenzROD) but not of other phase I enzymes was proven (Berghout et al., 1991). However, induction of CYP1A in earthworms has demonstrated to be quite difficult to be measured because of interference from endogenous pigments (Liimatainen & Hänninen, 1982) and the identification of non-inducible forms of cytochrome P450 (Milligan et al., 1986). Achazi et al. (1998), by utilizing ethoxy-, pentoxy- and benzoxyresorufin as substrates for monooxygenase activity, demonstrated pentoxy-resorufin-Odealkylase (PentROD) and BenzROD activities in *Eisenia fetida* microsomes, but exposure of the animals for up to four weeks to 100 mg fluoranthene or benzo[a]pyrene kg-1 soil (dry weight) did not induce significant changes in the activity of these monooxygenases. The same authors demonstrated the presence of etoxy-resorufin-Odealkylase (EROD) and PentROD activities in *Eisenia crypticus* but failed to demonstrate an induction of these activities following xenobiotic exposure. On the other hand short-term exposure to benzo[a]pyrene by feeding reduced the EROD activity significantly by 45%, but did not affect PentROD activity. After long-term (8 weeks) exposure to benzo[a]pyrene in the agar–agar medium EROD activity was not changed but PentROD was decreased to zero (Achazi et al., 1998).

Glutathione transferases (GSTs) form a ubiquitous superfamily of multi-functional dimeric enzymes (w50 kDa) with roles in phase-II detoxification. GSTs neutralise a broad range of xenobiotics and endogenous metabolic by-products via enzymatic glutathione conjugation, glutathione-dependent peroxidase activity or isomerisation reactions (Hayes et al., 2005). Several studies have demonstrated the sensitivity of earthworm GST to metals and pesticide exposure (Aly & Schröder, 2008; Maity et al., 2008; Lukkari et al., 2004; Saint-Denis et al., 2001; Booth et al., 2000). Recently, transcriptome approaches in the earthworm *Lumbricus rubellus* highlight GSTs as responders to several classes of pollutants including inorganic (cadmium, copper), organic (fluoranthene) and agrochemicals (atrazine) (Bundy et al., 2008; Owen et al., 2008). LaCourse et al. (2009) demonstrated *Lumbricus rubellus* to possess a range of GSTs related to previously known GSTs from other taxa including nematodes and humans, with evidence of tissue-specific isoforms, activity, location, the ability to detoxify products of cellular toxicity and potential response to pollution. This study combined sub-proteomics, bioinformatics and biochemical assay to characterise the *Lumbricus rubellus* GST complement as pre-requisite to initialise assessment of the applicability of GST as a biomarker.

4.4 Antioxidant enzymes

The exposure to either organic or inorganic pollutants is known to induce oxidative stress in the cells. A by-products of the metabolism of xenobiotics is the production of free radicals, on the other hand exposure to metals leads to the generation of reactive oxygen species (ROS) such as hydrogen peroxide (H_2O_2), superoxide (O_2^-) and hydroxyl ($OH\cdot$) radicals (Dazy et al., 2009). In order to scavenge ROS and avoid oxidative damage on biological macromolecules (lipids, proteins or DNA), cells protect themselves using enzymes and small molecular-weight antioxidants, such as glutathione (Valavanidis et al., 2006). Superoxide dismutase, catalase, glutathione peroxidase and glutathione reductase are important enzymatic antioxidants in the response to oxidative stress: superoxide dismutase metabolizes the superoxide anion (O_2^-) into molecular oxygen and H_2O_2, which is then deactivated by catalase, thus preventing oxidative damage. The glutathione reductase enzyme also plays an important role in cellular protection by reducing glutathione in the oxidized form (GSSG) to GSH (reduced and active form). Several studies (Liu et al., 2010) indicate that exposure to either organic or inorganic pollutants are able to induce a stress response in the antioxidant enzymes, suggesting their potential application as general biomarkers for assessing effects of pollutants in terrestrial ecosystems at early stages and with low concentrations. However, the dose-dependent and time-dependent response of antioxidant enzymes to pollutant exposure is sometimes complex and a better understanding of their behaviour in stress condition is needed for their application in monitoring and assessment programmes. For example Liu et al. (2010) demonstrated that exposure to toluene, ethylbenzene and xylene in earthworms (*Eisenia fetida*) induced a bell-shaped change in superoxide dismutase and catalase activities with a tendency of inducement firstly and then inhibition with increasing concentrations of the pollutants. Moreover, Wu et al (2011) found superoxide dismutase to be induced during the early period of phenanthrene exposure while with longer exposure times its activity decreased.

4.5 Cellular biomarkers on coelomocytes

Earthworm coelomic fluid is particularly interesting from a toxicological perspective for the development of novel cellular biomarkers. It can transport pollutants throughout the exposed organism and its cells (coelomocytes) are involved in the internal defence system

(Cooper et al., 2002; Reinhart & Dollahan, 2003; Engelmann et al., 2004). The coelomocyte population is comprised of amoebocytes originating from mesenchymal lining of the coelom (Hamed et al., 2002) and eleocytes (chloragocytes) sloughed into the coelomic fluid from the chloragogen tissue surrounding the intestine and blood vessels (Affar et al., 1998). Thank to the important role played by coelomocytes in the animal physiology, any impairment of their functioning can alter the health of the entire organism. Five cell types were observed by Calisi et al. (2009) in *Eisenia fetida* celomic fluid, corresponding to the previously described coelomocyte cell types (Valembois et al., 1985): leukocytes type I (basophilic) and II (acidophilic), granulocytes, neutrophils, and eleocytes.

The most recent cellular biomarkers standardized on earthworm coelomocytes are summarized in Tab.3.

Biomarker	Type	Analytical technique	Species	Ref.
Eleocyte riboflavin concentration	General biomarker of exposure	Flow cytometry	*Dendrodrilus rubidus*	Plytycz et al., 2007
Lysosomal membrane stability	General biomarker of exposure and effect	Neutral red retention assay	*Lumbricus spp. Eisenia spp. Aporrectodea caliginosa*	(for review see Sanchez-Hernandez, 2006)
Granulocyte morphometric alteration	General biomarker of exposure and effect	Diff Quick® stain	*Eisenia fetida Lumbricus terrestris*	Calisi et al 2009 Calisi et al. 2011b
Gene expression	Specific biomarkers of exposure	Real-Time PCR	*Eisenia fetida*	Brulle et al., 2010

Table 3. Cellular biomarkers on earthworm coelomocytes

Eleocytes are characterized by the presence of granules (chloragosomes), showing a high autofluorescence derived from riboflavin stored in (Cholewa et al., 2006, Plytycz et al., 2007) and from other fluorophores, putatively including lipofuscins (Cygal et al., 2007, Plytycz et al., 2009). Riboflavin storage was detected in all earthworm species studied, either in chloragocytes localised in chloragogen tissue of *Lumbricus spp.* and *Aporrectodea spp.* or in freely floating eleocytes (Plytycz et al., 2006). This suggests that riboflavin plays an important role in immunity of lumbricid worms, as it does in vertebrates (e.g. Verdrengh & Tarkowski, 2005). The amount of riboflavin in eleocytes is species-specific (Plytycz et al., 2006) and changes in response to environmental factors, including metal pollution, in a metal- and species-specific manner (e.g. Kwadrans et al., 2008, Plytycz et al., 2009). It was proposed as general biomarkers of exposure to environmental pollutants.

The most investigated coelomocyte alteration is represented by lysosomal membrane stability used as an indicator of chemical exposure and associated biological effects (Svendsen et al., 1996; Maboeta et al., 2002; Svendsen et al. 2004). Responses of the lysosomal system are generally thought to provide a first answer to pollutant exposure in a wide variety of animals including earthworms, since injurious lysosomal reactions frequently precede cell and tissue pathology (Moore et al., 2006). The neutral red retention assay (NRRA) has been successfully applied for the *in vivo* evaluation of lysosomal membrane stability in earthworm coelomocytes (Svendsen et al., 1996; Weeks & Svendsen, 1996; Scott-Fordsmand et al., 1998; Svendsen et al., 2004, Gastaldi et al., 2007). The quantification of this biomarker is based on the time at which 50% of the cells, previously incubated with neutral red, show sign of lysosomal leaking (the cytosol becoming red and the cells rounded), as evaluated by microscopic observations. Several studies have been demonstrated that lysosomal membrane destabilization is a useful predictor of adverse effect on lifecycle parameters such as survival, reproduction, and growth (Sanchez-Hetnandez 2006).

Recently Calisi et al. (2009, 2011b) demonstrated pollutant-induced morphometric alterations in both *Eisenia fetida* and *Lumbricus terrestris* granulocytes with possible applications as sensitive, simple, and quick biomarker for monitoring and assessment applications (Calisi et al., 2009; Calisi et al., 2011b). Granulocyte morphometric alterations were determined by image analysis on Diff-Quick® stained cells (Calisi et al., 2009; Calisi et al., 2011b). The rapid alcohol-fixed Diff-Quick stain is widely utilised in clinical and veterinary applications for immediate interpretation of histological samples. It was successfully applied to earthworm coelomocyte staining (Calisi et al., 2009). Granulocytes appeared as large cells with broad pseudopodial processes; they were filled with numerous acidophilic granules, presumably corresponding to the lysosomal compartment. They are the cell type mainly involved in phagocytosis (Engelmann et al., 2002; Cooper & Roch, 2003). A considerable enlargement of granulocytes was observed in copper sulphate exposed earthworms with respect to control group. The enlargement was quantified by measuring the area of 2D digitalised granulocyte images. The same effect was observed also when the animals were exposed to xenobiotics, such as the pesticide carbamate methiocarb. Either copper sulphate or methiocarb exerted the same effect on granulocyte dimension. In general, cell swelling can result from the impairment of mechanisms regulating intracellular osmolarity, such as alteration in protein catabolism and/or amino acid and ion transport across cell membrane and it is often an indication of cell damage or metabolic alterations. Heavy metals and pesticides are known to interfere with a wide range of metabolic functions and membrane transport mechanisms (Lionetto et al., 1998; Scott Fordsman & Weeks, 2000; Sanchez-Hernandez, 2006). This could result in an increase of intracellular osmolyte content, followed by osmotic influx of water and cellular swelling. Therefore, granulocyte enlargement could be the resulting integrated effect of the impairment of several cellular functions by different classes of toxic chemicals and can be related to manifestations of sublethal injury due to pollutant exposure. Moreover, in either copper sulphate or methiocarb exposed animals the increase in the granulocyte dimension was accompanied by cell rounding with loss of pseudopods. This effect could be ascribed to toxic chemical-induced reduction of the microfilament and microspine number. This result can be assigned at alteration on actin or tubulin cytoskeletal components by either copper or methiocarb. The cytoskeleton has been demonstrated to be an intracellular target of heavy metals and xenobiotic such as pesticides and polycyclic aromatic hydrocarbon (Gomez-Mendikute & Cajaraville, 2003). The cytoskeleton has also been shown to play a role in cell volume regulation (Pedersen et al., 2001). Therefore, a possible pollutant induced alteration of cytoskeletal components could contribute to the observed morphometric alterations of earthworm granulocytes.

A pollutant induced increase in the cell size was previously documented in the granulocytes of *Mytilus galloprovincialis* (Calisi et al., 2008) following cadmium exposure. In earthworms granulocyte enlargement was similar in the two species investigated, suggesting the potential application of this response in several earthworm species. Due to the important immunological role of granulocytes, which mediate many of the innate immune responses in earthworms (Cooper et al., 2002), the observed adverse effects of pollutants on these cells may increase the susceptibility of animals to diseases and reduce their survival ability. In fact, the immune system is extremely vulnerable to injury by chemical pollutants. Major changes in the immune system can be expressed in considerable morbility and even mortality of the organisms involved. Therefore, early subtle alterations in some of the components of the immune system can be used as early indicators of altered organism health. Pollutant induced granulocyte enlargement in *Lumbricus terrestris* was consistent with alterations at the organism level, such as mortality and reduced reproduction rate, suggesting a possible link to organism health impairment. Compared to the other biological responses to pollutant, granulocyte enlargement showed high percentage variation, very similar to the values of specific biomarkers (such as MT induction in copper exposure and AChE inhibition in methiocarb exposure). This result pointed out the high sensitivity of the granulocyte enlargement with respect to other general standardized biomarkers, such as lysosomal membrane stability, and indicated its possible applications as a sensitive, simple, and quick general biomarker for monitoring and assessment applications (Calisi et al., 2009; Calisi et al. 2011b) to be included in a multibiomarker strategy. It demonstrates several of the necessary characteristics for successful application as an effective biomarker in monitoring and assessment programs. This includes an evaluation of pollutant-induced stress at the cellular level in an easy, sensitive, and inexpensive way. Moreover, it provides a sensitive generalized response to pollutants that can integrate the combined effect of multiple contaminants present in the soil.

Earthworm coelomocytes have been recently exploited for trascrittomic studies to identify genes whose expression varies during metal exposure (Brulle et al., 2010). Brulle et al. (2008) identified and assayed (by Real-Time PCR (RTPCR)) 3 transcripts that were significantly elevated in coelomocytes when *Eisenia fetida* was exposed to a metalliferous field soil from the vicinity of a Pb-smelter. These were Cd-MT, and two hitherto unstudied earthworm immunity biomarkers (lysenin, and a transcript identified as coactosin-like protein, CLP). The lysenin is a haemolytic protein, produced in coelomocytes. CLP is a member of the ADF/cofilin group of actin-binding proteins which support the activity of the 5-lipoxygenase (5-LO), an enzyme of central importance in cellular leukotriene synthesis, which are key mediators of inflammatory disorders in vertebrates.

4.6 Genotoxicity biomarkers

Coelomocyte are also interesting for ecotoxicological research and application being the cells of choice for the assessment of the genotoxic effect of pollutants on earthworms. Many pollutants in soil either metals or POPs can alter both the structure and integrity of DNA. Since DNA damage may result in severe consequences for individuals and species, it is considered as an important indicator to be used in the assessment of earthworm health (Reinecke & Reinecke 2004). However, so far there have been only a few studies which used earthworms for assessing the genotoxicity of field-contaminated soils (Button et al., 2010; Espinosa-Reyes et al., 2010; Klobučar et al., 2011; Quiao et al., 2007). The single cell gel electrophoresis (or comet assay) and micronucleus test are two most extensively used methods in the detection of genotoxicity of chemicals in the environment. Compared to other assays, they are sensitive, rapid and easy to handle. Comet assay measures DNA damage in single cells, as single- and double-strand breaks, alkali-labile sites, oxidative DNA base damage (Cotelle & Ferard, 1999). The comet assay technique involves embedding cells in agarose gel on microscope slides and lysing with detergent and high salt. Slides are then soaked in an alkaline solution to allow cleavage of DNA at alkali labile sites. During electrophoresis under alkaline conditions, cells with damaged DNA display increased migration of DNA from the nucleus towards the anode. Broken DNA migrates further in the electric field, and the cell then resembles a 'comet' with a brightly fluorescent head and a tail region which increases as damage increases. The degree of migration is related to DNA damage (Lee & Steinert 2003). The Comet assay presents various advantages, because of its sensitivity for detecting low levels of DNA damage in single cells and the relative ease of application (Tice et al., 2000). The Comet assay has been demonstrated to be effective in determining DNA damage levels in the coelomocytes of earthworms exposed to genotoxic compounds, both in vivo and in vitro, in several studies (Reinecke & Reinecke, 2004; Fourie et al., 2007; Di Marzio 2005; Bonnard et al., 2009). Dose-dependent DNA damage in earthworm coelomocytes has been demonstrated *in vivo* for chromium (Manerikar et al., 2008), cadmium (Fourie et al., 2007) nickel (Reinecke & Reinecke, 2004; Bigorgne et al., 2010) and arsenic (Button et al 2010).

Besides comet assay the micronucleus test has emerged as one one of the most powerful methods for assessing chromosome damage (both chromosome loss and chromosome breakage) accumulated during lifespan of the cell in vertebrates and invertebrates. A micronucleus is formed during cell division. It may arise from a whole lagging chromosome or an acentric chromosome fragment detaching from a chromosome after breakage which do not integrate in the daughter nuclei. Sforzini et al (2010) provided the first step of validation of this test on earthworm (*Eisenia andrei*) cells.

4.7 Haemoglobin oxidation

Changes in haematology are reported to be early warning signals of the toxic effects of pollutants in vertebrates (Bowerman et al. 2000; Dauwe et al. 2006; Rogival et al. 2006), but they are poorly explored in invertebrates and for comparison in earthworms. Earthworms have a closed circulatory system The blood contains haemoglobin which is a large extracellular hemoprotein flowing in a closed circulatory system. In spite of the fundamental role of this respiratory pigment in earthworm physiology, little is known about its sensitivity to environmental pollutants. Recently Calisi et al (2011a) demonstrated heavy metal (cadmium, copper, mercury) exposure to significantly induce changes in either Hb concentration or its oxidation state in the earthworm Lumbricus terrestris. Exposure to heavy metals (10^{-5}-10^{-3} M for Cd, 10^{-4}-10^{-3} M for Hg, and 10^{-4}-10^{-2} M for Cu) was found to increase blood Hb concentration. The observed effects were seen at concentrations in the order of 65 (for Cu) to 200 (for Hg) mg/l, below the LC50 value for heavy metal exposure previously observed in earthworms (Neuhauser et al., 1985). Further studies are needed to demonstrate if the observed effect is due to a metal induced increased expression of Hb protein and/or to a reduced degradation of the molecule. In addition to changes in the Hb concentration, heavy metals showed a dramatic effect on the oxidation state of the respiratory pigment. A strong dose-dependent increase of blood methemoglobin (MetHb) percentage was observed following 48 h exposure with the highest Hb oxidation sensitivity to mercury, followed by cadmium and copper. The role of trace metals in the generation of free radical mediated oxidative stress is known. This could account for the Hb oxidation observed in the earthworms during metal exposure. In addition, a direct action of metals on the earthworm haeme group cannot be excluded. In fact, copper is a known direct-acting methemoglobin-producing agent in humans directly converting Fe(II) to Fe(III) in a two-stage reaction (Smith & Reed 1993, French et al. 1995). Compared to other biological responses to heavy metals, such as the known metallothionein induction, MetHb increase showed a higher sensitivity. In fact, the lowest concentration able to significantly increase MetHb concentration was 10^{-8} M for mercury, 10^{-7} M for cadmium and 10^{-6} M for copper while 10^{-5} M was the concentration of each metal able to significantly induce metallothionein increase in the same species and in the same exposure conditions. Moreover, it is interesting to observe that MetHb formation was very suitable for routine application in monitoring assessment in terms of measurable biological response. In fact, it showed a very high percentage variation following heavy metal exposure, being about ten fold higher compared to Mt induction. Future studies will be addressed to evaluate if the observed response is specific for heavy metal exposure or represents a biomarker of general health of earthworms in polluted sites. In any case it demonstrated to be a suitable biomarker of exposure/effect to be included in a multibiomarker strategy in earthworm in soil monitoring assessment.

5. Earthworm biomarker relevance for soil pollution monitoring and assessment

Earthworm biomarkers represent useful tools in soil monitoring and assessment as an early warning of adverse ecological effects (Sanchez-Hernandez, 2006; Rodriguez-Castellanos & Sanchez-Hernandez, 2007). As indicated by Sanchez-Hernandez (2006) four types of approaches can be performed in soil pollution monitoring : 1) biomarker analysis on native earthworm populations; 2) use of transplanted organisms in *in situ* exposure bioassays; 3) exposure of a selected earthworm population to the environmental medium (soil) in laboratory standardized conditions; 4) simulated field studies. The use of natural population offers the advantage of an ecologically more relevant approach to environmental monitoring and assessment. In addition the usefulness of native organisms arises mainly when studying pollutant long-term effects that may be emphasized in organisms from natural populations. In fact it is difficult to extrapolate effects from spiked soils to field soils, when these are already polluted for a long time. The bioavailability of pollutants in comparable soil types polluted in the field or spiked in the lab is different (Smolders et al., 2003). Soil characteristics, e.g. pH, organic matter and clay content (Peijnenburg et al., 2002) also play an important role in determining the bioavailability of pollutants. Using native earthworm populations for biomarker analysis integrates the bioavailability of pollutants, exposure pathways and temporal aspect of exposure (Spurgeon et al., 2002; Sanchez-Hernandez, 2006). However, so far only few studies have explored the potentiality of the biomarker approach to native earthworms, if compared with studies on aquatic environments. For example Laszczyca et al. (2004) found spatial and temporal variation of AChE and antioxidant enzymes in three natural earthworm population (*Aporrectodea caliginosa, Lumbricus terrestris and Eisenia fetida*) collected from meadow sites along a 32 km long transect from a Zn/Pb ore mine and a smelter metallurgic complex. Lukkari et al (2004b) documented an increase in the response of three biomarkers (metallothionein, cytochrome P4501A, glutathione transferase) along a 4 km long transect from an area contaminated by a steel smelter. The response of the three biomarkers was positively correlated with decreasing distance from the steel smelter. Moreover, Svendensen et al (2003) demonstrated the validity of using NRRA in biological impact assessment along gradients of contamination. The authors collected earthworms (*Lumbricus castaneus*) at the site of a large industrial plastics fire in Thetford, UK along a 200 m transect leading from the factory perimeter fence. NRRA response was positively correlated with decreasing distance from the factory. While metal residues in soil and earthworms were found to be highly elevated close to the factory perimeter and to rapidly drop to background levels within the first 50 m of the transect, the NRRA values were significantly different from the NRRA determined in control animals also in the sourrounding forest along the transect. This results shows that NRRA determination represents a more sensitive indicator of pollutant exposure along a contamination gradient with respect to the analytical metal residue determination. Most of the available studies have been carried out on earthworm population inhabiting areas contaminated with heavy metals (Button et al., 2010). However, in some cases the employment of native earthworms to determine soil toxicity (particularly in studies of long-term exposure) is complicated by the fact that some populations appear to have developed a resistance to metals in soil (Spurgeon & Hopkin, 2000).

In the case of transplanted organisms earthworms can be collected from a population at one location and translocated to the monitoring sites. This approach provides the advantage of ensuring comparable biological samples, reducing the variability of results usually encountered in field sampling programmes. Using caged organisms in biomonitoring studies, as well as in related research, makes it easier to standardize the results and to compare control organisms to animals collected from potentially polluted sites.

The application of the biomarker approach to a selected earthworm population exposed to the test soil in laboratory standardized conditions offers a complementary approach to standard toxicity tests (i.e., mortality and reproduction rates) to investigate the effects of contaminant toxicity on living organisms at earlier stages and lower concentrations (Lukkari et al. 2004; Gastaldi et al. 2007; Schreck et al. 2008). Appropriate biomarkers may be applied in standardized bioassays to provide evidence of the cause-effect relationship between soil contaminants and toxic effects in the individuals. Biomarkers can give a contribution in acute bioassays as a measurement of the bioavailable and bioactive fraction of contaminants and in chronic bioassays as sublethal endpoints (Sanchez-Hernandez, 2006). For example, many studies have reported that the lysosomal destabilization linearly correlate with the bioavailable fraction of heavy metals in the soil (Sanchez-gHernandez, 2006). In addition, NRRA was demonstrated to be more sensitive to Cu exposure than reproduction rate (Scott-Fordsmand et al., 2000). The need to develop biomarkers for earthworms to supplement standard toxicity tests has been widely discussed by Van Gestel & Weeks (2004) and Sanchez-Hernandez (2006).

Simulated field studies offer the opportunity to study the effects of pollutants on the biomarker responses in the organisms under the influence of multiple environmental variables (for review see Sanchez-Hernandez, 2006). They can be carried out in microcosm or mesocosm. Soil microcosm experiments are carried out in laboratory scale, under standardized ambient conditions, while mesocosm experiments are carried out in field scale and are functionally closer to the real environmental scenarios.

For the potential of the biomarker approaches to be realized in soil monitoring and assessment a crucial aspect need to be clearly addressed. Ideally, the degree to which the magnitude of the biological response relates to the dose of exposure should be known, enabling severity of the exposure to be clearly assessed. In general, the extent to which a molecular or cellular response occurs is generally related to the dose of chemical received. Nevertheless, exposure to low doses may produce no effects because of the presence of a threshold level of effect, variable for each responses. In other cases where the threshold level is exceeded, protective mechanisms may mask the effects, such as the induction of metallothionein. In addition contaminated soil typically contains a complex mixture of contaminants that often interact and the organisms are exposed to multiple chemicals and multiple stresses which can confound a simple dose-response curve. It is important, however, to note that although it is useful to understand the toxic responses of contaminants and how they alter with dose and over time it is harder to quantify these responses especially in field conditions in terms of dose-response. It is well understood that no one biomarker has been validated as unique tool of detecting specific pollutant exposure and effects. The biological response of an organisms to pollutant exposure can be various because of the variety of pollutants that may be present in the environment. Thus, a suite of biomarkers is required to be effectively applicable in a biomonitoring programme. By using a suite of biomarkers future attempts should be made to try and develop a quantitative biomarker index that could simplify the complex biological alterations measured by multiple biomarkers into a single, predefined quality class.

6. Conclusions

During the last years, earthworm biomarkers have become increasingly relevant for the evaluation of contaminants effects on soil organisms However, the application of the biomarker approach to soil pollution monitoring, compared to aquatic environment monitoring, is recent and some aspects need to be further evaluated. First, it is necessary to identify and characterise appropriate sentinel earthworm species to be used as field collected organisms, in order to provide a quick assessment of soil pollution. Second, earthworm biomarkers studies have been mostly conducted for heavy metals. Thus, developing biomarkers of exposure/effects to a wider range of chemicals of concern for soil pollution constitutes a major demand. Third, there is a growing interest for increasing the knowledge of biological responses of earthworms to pollutants in order to standardize a suite of sensitive and reliable biomarkers for the detection of the pollutant induced stress syndrome in soil organisms. Earthworm physiological fluids, such as coelomic fluids and blood, offer an interesting field for exploitation of novel sensitive non destructive biomarkers, including cytological, biochemical and trascriptomic parameters. Thus, granulocyte morphometric alteration has been recently demonstrated as a suitable general biomarker of effect that could be included in a multibiomarker strategy. It provides a sensitive generalized response to pollutants that can integrate the combined effect of multiple contaminants present in the soil. Finally, earthworm biomarkers have been scarcely investigated under field conditions. The most studies on earthworm biomarkers have been carried out in laboratory condition, but only few studies are available which used native earthworm populations for assessment of polluted soils. Hence, there has been a growing interest in field studying earthworm biomarkers and validating their effectiveness in the field conditions as an early warning of adverse ecological effects. This represents an attractive field of research in the light of the growing interest in the use of earthworm biomarkers as valuable tools for soil pollution monitoring and assessment.

7. References

Achazi, R.K., Flenner, C., Livingstone, D.R., Peters, L.D., Schaub, K., & Scheiwe, E. (1998). Cytochrome P450 and dependent activities in unexposed and PAH-exposed terrestrial annelids. *Comparative Biochemistry and Physiology Part C,* Vol. 121, No. 1-3, (November, 1998), pp. 339–350, ISSN 1095-6433

Affar, E.B, Dufour, M., Poirier, G.G. & Nadeau, D. (1998). Isolation, purification and partial characterization of chloragocytes from the earthworm species *Lumbricus terrestris. Molecular and Cellular Biochemestry,* Vol. 185, No. 1-2, (August, 1998), pp. 123-133, ISSN 0270-7306

Aly, M.A., Schröder, P. (2008). Effect of herbicides on glutathione S-transferases in the earthworm, *Eisenia fetida. Environmental Science and Pollution Research International,* Vol. 15, No. 2, (March 2008), pp. 143–149, ISSN 0944-1344

Amiard, J.C., Amiard-Triquet, C., Barka, S., Pellerin J. & Rainbow, P.S. (2006). Metallothioneins in aquatic invertebrates: their role in metal detoxification and their use as biomarkers. *Aquatic Toxicology,* Vol. 76, No. 2, (February 2006), pp. 160–202, ISSN 0166-445X

Andre, J., Charnock, J., Sturzenbaum, S.R., Kille, P., Morgan, A.J. & Hodson, M.E. (2009). Metal speciation in field populations of earthworms with multi-generational exposure to metalliferous soils: cell fractionation and high energy synchrotron analysis. *Environmental Science and Technology,* Vol. 43, No. 17, (September 2009), pp. 6822–6829, ISSN 0013-936X

Barlett, M.D., Briones, M.J.I., Neilson, R., Schmidt, O., Spurgeon, D. & Creamer, R.E. (2010). A critical review of current methods in earthworm ecology: from individuals to populations. *European Journal of Soil Biology,* Vol. 46, No. 2, (March 2010), pp. 67-73, ISSN 1164-5563

Beliaeff, B., Burgeott, T. (2002). Integrated biomarker response: a useful tool for ecological risk assessment. *Environmental Toxicology and Chemistry,* Vol. 21, No. 6, (June 2002) pp. 1316-1322, ISSN 0730-7268

Berghout, A.G.R.V., Wenzel, E., Buld, J. & Netter, K.J. (1991). Isolation, partial purification, and characterisation of the cytochrome P-450-dependent monooxygenase system from the midgut of the earthworm *Lumbricus terrestris. Comparative Biochemistry and Physiology Part C,* Vol. 100, No. 3, (March 1991), pp. 389–396, ISSN 1532-0456

Bezchlebova, J., Cernohlavkova, J., Ivana Sochova, J.L., Kobeticova, K. & Hofman, J. (2007). Effects of toxaphene on soil organisms. *Ecotoxicology and Environmental Safety,* Vol. 68, No. 3, (June 2007), pp. 326–334, ISSN 0147-6513

Bigorgne, E., Cossu-Leguille, C., Bonnard, M. & Nahmani, J. (2010). Genotoxic effects of nickel, trivalent and hexavalent chromium on the *Eisenia fetida* earthworm. *Chemosphere,* Vol. 80, No. 9, (August 2010), pp. 1109-1112, ISSN 0045-6535

Bonnard, M., Eom, I.C., Morel, J.L. & Vasseur, P. (2009). Genotoxic and reproductive effects of an industrially contaminated soil on the earthworm *Eisenia Fetida. Environmental and Molecular Mutagenesis,* Vol. 50, No. 1, (January 2009), pp. 60–67, ISSN 0893-6692

Booth, L.H., Heppelthwaite, V. & Mc Glinchy, A. (2000). The effect of environmental parameters on growth, cholinesterase activity and glutathione S-transferase activity in the earthworm *Aporectodea caliginosa. Biomarkers,* Vol. 5, No. 1, (January 2000), pp. 46–55, ISSN 1354-750X.

Booth, L.H., O'Halloran, K. (2001). A comparison of biomarker responses in the earthworm *Aporrectodea caliginosa* to the organophosphorus insecticides diazinon and chlorpyrifos. *Environmental Toxicology and Chemistry*, Vol. 20, No. 11, (November 2011), pp. 2494–2502, ISSN 0730-7268

Bowerman, W.W., Stickle, J.E., Sikarskie, J.G. & Giesy, J.P. (2000). Hematology and serum chemistries of nestling bald eagles (*Haliaeetus leucocephalus*) in the lower peninsula of MI, USA. *Chemosphere*, Vol. 41, No. 10, (November 2000), pp. 1575–1579, ISSN 0045-6535

Bradham, K.D., Dayton, E.A., Basta, N.T., Schroder, J., Payton, M. & Lanno, R.P. (2006). Effect of soil properties on lead bioavailability and toxicity to earthworms. *Environmental Toxicology and Chemistry*, Vol. 25, No. 3, (March 2006) 769–775, ISSN 0730-7268

Brulle, F., Mitta, G., Leroux, R., Lemière, S., Leprêtre, A. & Vandenbulcke, F. (2007). The strong induction of metallothionein gene following cadmium exposure transiently affects the expression of many genes in *Eisenia fetida*: a trade-off mechanism?. *Comparative Biochemistry and Physiology Part C*, Vol. 144, No. 4, (April 2007), pp. 334–341, ISSN 1532-0456

Brulle, F., Cocquerelle, C., Mitta, G., Castric, V., Douay, F., Leprêtre, A. & Vandenbulcke, F. (2008). Identification and expression profile of gene transcripts differentially expressed during metallic exposure in *Eisenia fetida* coelomocytes. *Developmental and Comparative Immunology*, Vol. 32, No. 12, (July 2008), pp. 1441-1453, ISSN 0145-305X

Brulle F., Morgan A.J., Cocquerelle C. & Vandelbulcke F. (2010). Transcriptomic underpinning of toxicant-mediated physiological function alterations in three terrestrial invertebrate taxa: A review. *Environmental Pollution*, Vol. 158, No. 9, (September 2010), pp. 2793-2808, ISSN 0269-7491

Brusseau, M.L. (1997). Transport and fate of toxicants in soils. In: *Soil Ecotoxicology*, J., Tarradellas, G., Bitton, & D., Rossel, (Eds.), pp. 33–53, Lewis Publishers, ISBN 978-1566701341, New York, USA

Bundy, J.G., Sidhu, J.K., Rana, F., Spurgeon, D.J., Svendsen, C., Wren, J.F., Stürzenbaum, S.R., Morgan, A.J. & Kille, P. (2008). 'Systems toxicology' approach identifies coordinated metabolic responses to copper in a terrestrial non-model invertebrate, the earthworm *Lumbricus rubellus*. *BMC Biology*, Vol. 6, No. 6 (June 2008), pp. 1-25, ISSN 1741-7007

Button, M., Jenkin, J.R.T., Bowman, K.J., Harrington, C.F., Brewer, T.S., Jones, D.D. & Watts M.J. (2010). DNA damage in earthworms from highly contaminated soils: Assessing resistance to arsenic toxicity by use of the Comet assay. *Mutation Research*, Vol. 696, No. 2, (February 2010), pp. 95–100, ISSN 1383-5718

Calisi, A., Lionetto, M.G., Caricato, R., Giordano, M.E. & Schettino T. (2008). Morphometrical alterations in *Mytilus galloprovincialis* granulocytes: a new potential biomarker. *Environmental Toxicology and Chemistry*, Vol. 27, No. 6, (June 2008), pp. 1435-1441, ISSN 0730-7268

Calisi, A., Lionetto, M.G. & Schettino, T. (2009). Pollutant-induced alterations of granulocyte morphology in the earthworm *Eisenia foetida*. *Ecotoxicology and Environmental Safety*, Vol. 72, No.5, (July 2009), pp. 1369-1377, ISSN 0147-6513

Calisi, A., Lionetto, M.G., Sanchez-Hernandez, J.C. & Schettino, T. (2011a). Effect of heavy metal exposure on blood haemoglobin concentration and methemoglobin percentage in *Lumbricus terrestris*. *Ecotoxicology*, Vol. 20, No.4, (Jun 2011), pp. 847-854, ISSN 0963-9292

Calisi, A., Lionetto, M.G. & Schettino, T. (2011b). Biomarker response in the earthworm *Lumbricus terrestris* exposed to chemical pollutants. *Science of the Total Environment*, in press, Jul 23 2011, Available from: <http://www.sciencedirect.com/science/article/pii/S0048969711007145>, ISSN 0048-9697

Capri, E., Trevisan, M. (2002). *I metalli pesanti di origine agricola nei suoli e nelle acque sotterranee*, Pitagora Editrice, ISBN 9788837112622, Bologna, Italy

Caselli, F.; Gastaldi, L.; Gambi, N. & Fabbri, E. (2006). In vitro characterization of cholinesterases in the earthworm *Eisenia Andrei*. Comparative Biochemistry and Physiology Part C, Vol. 143, No. 4, (August 2006), pp. 416-421, ISSN 1532-0456

Chambers, J.E., Boone, J.S., Carr, R.L., Chambers, H.W. & Straus, D.L. (2002). Biomarkers as predictors in health and ecological risk assessment. *Human and Ecological Risk Assessment*, Vol. 8, No. 1, (February 2002), pp. 165–176, ISSN 1080-7039

Cholewa, J., Feeney, G.P., O'Reilly, M., Sturzenbaum, S.R., Morgan, A.J. & Plytycz, B. (2006). Autofluorescence in eleocytes of some earthworm species. *Folia Histochemical et Cytobiolica*, Vol. 44, No. 1, (January 2006), pp. 65–71, ISSN 0239-8508

Cooper, E.L., Kauschke, E. & Cossarizza, A. (2002) Digging for innate immunity since Darwin and Metchnikoff. *BioEssays*, Vol. 24, No.4, (April 2002), pp. 319–333, ISSN 0265-9247

Cooper, E.L., Roch, P. (2003). Earthworm immunity: a model of immune competence. *Pedobiologia*. Vol. 47, No. 5-6, (June 2003), pp. 1–13, ISSN 0031-4056

Cortet, J., Gomot-De Vauflery, A., Poinsot-Balaguer, N., Gomot, L., Texier, C. & Cluzeau D. (1999). The use of invertebrate soil fauna in monitoring pollutant effects. *European Journal of Soil Biology*, Vol 35, No.3, (March 1999), pp. 115-134, ISSN1164-5563

Costello, L.C., Guan, Z., Franklin, R.B. & Feng, P. 2004. Metallothionein can function as a chaperone for zinc uptake transport into prostate and liver mitochondria. *Journal of Biological Inorganic Chemistry*, Vol. 98,No. 4, (April 2004), pp. 664–666, ISSN 0949 - 8257

Cotelle, S., Ferard, J.F. (1999). Comet assay in genetic ecotoxicology: a review. *Environmental and Molecular Mutagenesis*, Vol. 34, No. 4, (April 1999), pp. 246–255, ISSN 0893-6692

Criel, P., Lock, K., Van Eeckhout, H., Oorts, K., Smolders, E. & Janssen, C.R. (2008). Influence of soil properties on copper toxicity for two soil invertebrates. *Environmental Toxicology and Chemistry*, Vol. 27, No. 8, (August 2008), pp. 1748–1755, ISSN 0730-7268

Cygal, M., Lis, U., Kruk, J. & Plytycz, B. (2007). Coelomocytes and fluorophores of the earthworm *Dendrobaena veneta* raised at different ambient temperatures. *Acta Biologica Cracovensia Series Zoologica*, Vol. 49, pp. 5–11, ISSN 0001-530X

Dai, J., Becquer, T., Rouiller, J.H., Reversat, G., Bernhard-Reversat, F., Nahmani, J. & Lavelle, P. (2004). Heavy metal accumulation by two earthworm species and its relationship to total and DTPA-extractable metals in soils. *Soil Biology and Biochemistry*, Vol. 36, No. 1, (January 2004), pp. 91-98 , ISSN 0038-0717

Dauwe, T., Janssens, E. & Eens, M. (2006). Effects of heavy metal exposure on the condition and health of adult great tits (*Parus major*). *Environmetal Pollution*, Vol. 140, No. 1, (March 2006), pp. 71–78, ISSN 0269-7491

Dazy, M., Masfaraud, J.F. & Ferard, J.F. (2009). Induction of oxidative stress biomarkers associated with heavy metal stress in Fontinalis antipyretica Hedw. *Chemosphere*, Vol. 75, No. 3, (April 2009), pp. 297-302, ISSN 0045-6535

Demuynck, S., Grumiaux, F.,Mottier,V., Schikorski, D., Lemière, S. & Leprêtre, A. (2006). Metallothionein response following cadmium exposure in the oligochaete *Eisenia fetida*. *Comparative Biochemistry and Physiology Part C*, Vol. 144, No. 1, (September 2006), pp. 34–46, ISSN 1532-0456

Depledge, M.H. (1994). The rational basis for the use of biomarkers as ecotoxicological tools. In *Nondestructive Biomarkers in Vertebrates*, M.C., Fossi, C., Leonzio, (Eds.), pp. 271–295, Lewis Publisher, ISBN 978-0873716482 Boca Raton, USA

Di Marzio, W.D., Saenz, M.E., Lemière, S. & Vasseur, P. (2005). Improved single-cell gel electrophoresis assay for detecting DNA damage in *Eisenia foetida*. *Environmental and Molecular Mutagenesis*, Vol. 46, No. 4, (December 2005), pp. 246–252, ISSN 0893-6692

Dogheim, S., Mohamed, E.Z., Alla, S.A.G., El-Saied, S., Emel, S.Y., Mohsen, A.M. & Fahmy, S.M. (1996) Monitoring of pesticide residues in human milk, soil, water, and food samples collected from Kafr El-Zayat Governorate. *The Journal of AOAC International*, Vol. 79, No. 1, (January 1996), pp. 111-116, ISSN 1060- 3271

EEC (Council Regulation), (2001). Commission Directive 2001/58. *Official Journal of the European Union*, No. 212, (July 2001), pp. 24–33, ISSN 1725-2555

Emmerling, C., Paulsch, D. (2001). Improvement of earthworm (Lumbricidae) community and activity in mine soils from open-cast coal mining by the application of different organic waste materials. *Pedobiologia*, Vol. 45, No. 5, (September 2001), pp. 396–407, ISSN 0031-4056

Engelmann, P., Pal, J., Berki, T., Cooper, E.L. & Nemeth, P. (2002). Earthworm leukocytes react with different mammalian antigen-specific monoclonal antibodies. *Zoology*. Vol. 105, No. 3, (March 2002), pp. 257–265, ISSN 0944-2006

Engelmann, P., Molnar, L., Palinkas, L., Cooper, E.L. & Nemeth, P. (2004). Earthworm leukocyte populations specifically harbor lysosomal enzymes that may respond to bacterial challenge. *Cell and Tissue Research*, Vol. 316, No. 3, (June 2004), pp. 391–401, ISSN 0302-766X

Espinosa-Reyes, G., Ilizaliturri, C.A., González-Mille, D.J., Costilla, R., Díaz-Barriga, F., Cuevas, M.D.C., Martínez, M.A. & Mejía-Saavedra, J. (2010). DNA damage in earthworms (*Eisenia spp.*) as an indicator of environmental stress in the industrial zone of Coatzacoalcos, Veracruz, Mexico. *Journal of Environmental Science and Health, Part A*, Vol. 45, No. 1, (January 2010), pp. 49-55, ISSN 1093-4529

Fourie, F., Reinecke, S.A. & Reinecke, A.J. (2007). The determination of earthworm species sensitivity differences to cadmium genotoxicity using the comet assay. *Ecotoxicology and Environmental Safety*, Vol. 67, No. 3, (July 2007), pp. 361–368, ISSN 0147-6513

French, C.L., Yaun, S.S., Baldwin, L.A., Leonard, D.A., Zhao, X.Q. & Calabrese, E.J. (1995). Potency ranking of methemoglobin-forming agents. *Journal of Applied Toxicology*, Vol. 15, No. 3, (May-June 1995), pp. 167-174, ISSN 0260-437X

Gambi, N., Pasteris, A. & Fabbri, E. (2007). Acetylcholinesterase activity in the earthworm *Eisenia Andrei* at different conditions of carbaryl exposure. *Comparative Biochemistry and Physiology Part C*, Vol. 145, No. 4, (May 2007), pp. 678–685, ISSN 1532-0456

Garrido, T., Fraile, J., Ninerola, J.M., Figueras, M., Ginebreda, A., Olivella, L. & Ginebreda, A., (2000). Survey of ground water pesticide pollution in rural areas of Catalonia (Spain). *International Journal of Environmental and Analytical Chemistry*, Vol. 78, No. 1, (January 2000), pp. 51-65, ISSN 0306-7319

Gastaldi, L., Ranzato, E., Caprì, F., Hankard, P., Pérès, G., Canesi, L., Viarengo, A. & Pons G. (2007). Application of a biomarker battery for the evaluation of the sublethal effects of pollutants in the earthworm *Eisenia andrei*. *Comparative Biochemistry and Physiology, Part C*, Vol. 146, No. 3 (September 2007), pp. 398–405, ISSN 1532-0456

Gomez-Mendikute, A., Cajaraville, M.P. (2003). Comparative effects of cadmium, copper, paraquat and benzo(a)pyrene on the actin cytoskeleton and production of reactive oxygen species (ROS) in mussel haemocytes. *Toxicology in Vitro*, Vol. 17, No. 5-6, (October-December 2003), pp. 539- 546, ISSN 0887-2333

Gruber, C., Stürzenbaum, S.R., Gehrig, P., Sack, R., Hunziker, P., Berger, B. & Dallinger, R. (2000). Isolation and characterization of a self-sufficient one-domain protein (Cd)-Metallothionein from *Eisenia foetida*. *European Journal of Biochemistry*, Vol. 267, No. 2 (January 2000), pp. 573–582, ISSN 0014-2956

Hagger, J.A., Jones, M.B., Leonard, D.L.P., Owen, R. & Galloway, T.S. (2006). Biomarkers and Integrated Environmental Risk Assessment: Are There More Questions Than Answers?. *Integrated Environmental Assessment and Management*, Vol. 2 No. 4, (October 2006), pp. 312–329, ISSN 1551-3777

Hamed, S.S., Kauschke, E. & Cooper, E.L. (2002). Cytochemical properties of earthworm coelomocytes enriched by Percoll. In: *A New Model for Analyzing Antimicrobial Peptides with Biomedical Applications*, A., Beschin, M., Bilej, & E.L., Cooper, (Eds.), pp. 29-37, IOS Press, ISBN 1-58603-237-2 , Tokyo, Japan

Handy, R.D., Galloway, T.S. & Depledge, M.H. (2003). A proposal for the use of biomarkers for the assessment of chronic pollution and in regulatory toxicology. *Ecotoxicology*, Vol. 12, No. 1-4, (February-August 2003), pp. 331-343, ISSN 0963-9292

Hankard, P.K., Svendsen, C., Wright, J., Wienberg, C., Fishwick, S.K., Spurgeon, D.J. & Weeks, J.M. (2004). Biological assessment of contaminated land using earthworm biomarkers in support of chemical analysis. *Science of the Total Environment*, Vol. 330, No. 1-3, (September 2004), pp. 9-20, ISSN 0048-9697

Hayes, J.D., Flanagan, J.U. & Jowsey, I.R., (2005). Glutathione transferases. *Annual Review of Pharmacology and Toxicology*, Vol. 45, (February 2005), pp. 51–88. , ISSN 0362-1642

Heikens, A., Peijnenburg, W.J.G.M. & Hendriks, A.J. (2001) Bioaccumulation of heavy metals in terrestrial invertebrates. *Environmental Pollution*, Vol. 113, No. 3, (August 2001), pp. 385–393, ISSN 0269-7491

Hendrickx, F., Maelfait, J.P., Bogaert, N., Tojal, C., Du Laing, G., Tack, F.M.G. & Verloo, M.G. (2004). The importance of biological factors affecting trace metal concentration as revealed from accumulation patterns in co-occurring terrestrial invertebrates. *Environmental Pollution*, Vol. 127, No. 3, (February 2004), pp. 335–341, ISSN 0269-7491

Hobbelen, P.H.F., Koolhaas, J.E. & Van Gestel, C.A.M. (2006) Bioaccumulation of heavy metals in the earthworms *Lumbricus rubellus* and *Aporrectodea caliginosa* in relation to total and available metal concentrations in field soils. *Environmental Pollution*, Vol. 144, No. 2, (November 2006), pp. 639–646, ISSN 0269-7491

Homa, J., Stürzenbaum, S.R., Morgan, A.J. & Plytycz, B. (2007). Disrupted homeostasis in coelomocytes of Eisenia fetida and *Allolobophora chlorotica* exposed dermally to heavy metals. *European Journal of Soil Biology*, Vol. 43, Supp. 1, (November 2007), pp. S273-S280, ISSN 1164-5563

Jajer, T., Fleuren, R.H.L.J., Hogendoorn, E.A. & de Korte, G. (2003). Elucidating the Routes of Exposure for Organic Chemicals in the Earthworm, *Eisenia andrei* (Oligochaeta). *Environmental Sciences and Technologies*, Vol. 37, No. 15, (August 2003), pp. 3399–3404, ISSN 0013-936X

Jamieson, B.G.M. (1992) Oligochaeta. In: *Microscopic Anatomy of Invertebrates*, F.W., Harrison, S.L., Gardiner, (Eds.), pp. 217–322, Wiley-Liss Inc., ISBN 978-0471561170, New York, USA.

Jänsch, S., Amorim, M.J. & Römbke, J. (2005). Identification of the ecological requirements of important terrestrial ecotoxicological test species. *Environmental Reviews*, Vol. 13, No. 2, (February 2005), pp. 51-83, ISSN 1181-8700

Kammenga, J.E., Dallinger, R., Donker, M.H., Köhler, H.R., Simonsen, V., Triebskorn, R. & Weeks, J.M. (2000). Biomarkers in terrestrial invertebrates for ecotoxicological soil risk assessment. *Review of Environmental Contamination and Toxicology*, Vol. 164, (June 2000), pp. 93-147, ISSN 0179-5953

Kizilkaya, R. (2005) The role of different organic wastes on zinc bioaccumulation by earthworm *Lumbricus terrestris* L. (Oligochaeta) in successive Zn added soil. *Ecological Engineering*, Vol. 25, No. 4, (November 2005), pp. 322–331 , ISSN 0925-8574

Klobučar, G.I.V., Koziol, B., Markowicz, M., Kruk, J. & Plytycz, B. (2006). Riboflavin as a source of autofluorescence in *Eisenia fetida* coelomocytes. *Photochemical and Photobiology*, Vol. 82, No. 2, (March-April 2006), pp. 570–573, ISSN 0031-8655

Kwadrans, A., Litwa, J., Woloszczakiewicz, S., Ksiezarczyk, E., Klimek, M., Duchnowski, M., Kruk, J. & Plytycz, B. (2008). Changes in coelomocytes of the earthworm, *Dendrobaena veneta*, exposed to cadmium, copper, lead or nickel- contaminated soil. *Acta Biologica Cracovensia Series Zoologica*. Vol. 50, pp. 57–62. , ISSN 0001-530X

LaCourse, E.J., RiboflavinHernandez-Viadel, M., Jefferies, J.R., Svendsen, C., Spurgeon, D.J., Barrett, J., Morgan, A.J., Kille, P. & Brophy, P.M. (2009). Glutathione transferase (GST) as a candidate molecular-based biomarker for soil toxin exposure in the earthworm *Lumbricus rubellus*. *Environmental Pollution*, Vol. 157, No. 8-9, (August-September 2009), pp. 2459–2469, ISSN 0269-7491

Lanno, R., Wells, J., Conder, J., Bradham, K. & Basta, N. (2004). The biovailability of chemicals in soil for earthworms. *Ecotoxicology and Environmental Safety*. Vol. 57, No. 1, (January 2004), pp. 39-47, ISSN 0147-6513

Lavelle, P., Spain, A. (2001). *Soil ecology*, Kluwer Scientific Publications, ISBN 978-0-7923-7123-6, Amsterdam, Netherlands

Lee, R. F., Steinert, S. (2003). Use of the single cell gel electrophoresis/comet assay for detecting DNA damage in aquatic (marine and freshwater) animals. *Mutation Research*,Vol. 544, No. 1, (September 2003), pp. 43–64, ISSN 1383-5718

Liang, S.H., Jeng, Y.P., Chiu, Y.W., Chen, J.H., Shieh, B.S., Chen, C.Y. & Chen, C.C. (2009). Cloning, expression, and characterization of cadmium-induced metallothionein-2 from the earthworms Metaphire posthuma and Polypheretima elongate. *Comparative Biochemistry and Physiology, Part C*, Vol. 149, No. 3, (April 2009), pp. 349-357, ISSN 1532-0456

Liimatainen, A., Hanninen, O. (1982). Occurrence of cytochrome P450 in the earthworm Lumbricus terrestris. In: *Cytochrome P450: Biochemistry, Biophysics, and Environmental Implication*, E., Hietanen, M., Larteinen, & O., Hanninen, (Eds.), pp. 255-257, Elsevier Biomedical Press, ISBN 0444804595, Amsterdam, Netherlands

Lionetto, M.G., Vilella, S., Trischitta, F., Cappello, M.S. & Schettino, T. (1998). Effects of CdCl₂ on electrophysiological parameters in the intestine of teleost fish *Anguilla anguilla*. *Aquatic Toxicology*, Vol. 4, No. 3, (March 1998), pp. 251-264, ISSN 0166-445X

Lionetto, M.G., Caricato, R., Calisi, A. & Schettino T. (2010). Acetylcholinesterase inhibition as a relevant biomarker in environmental biomonitoring: new insights and perspectives. In *Ecotoxicology around the globe*, J.E. Visser, (Ed.), pp. 87-115, Nova Science Publishers, ISBN 9781617611261 , New York, USA

Liu, Y., Zhou, Q., Xie, X., Lin, D. & Dong, L. (2010). Oxidative stress and DNA damage in the earthworm *Eisenia fetida* induced by toluene, ethylbenzene and xylene. *Ecotoxicology*, Vol. 19, No. 8, (November 2010), pp.1551-1559, ISSN 0963-9292

Lukkari, T., Taavitsainen, M., Soimasuo, M., Oikari, A. & Haimi, J. (2004a). Biomarker responses of the earthworm *Aporrectodea tuberculata* to copper and zinc exposure: differences between population with and without earlier metal exposure. *Environmental Pollution*, Vol. 129, No. 3, (June 2004), pp. 377-386, ISSN 0269-7491

Lukkari, T., Taavitsainen, M., Väisänen, A., haimi, J. (2004b). Effects of heavy metals on earthworms along contamination gradients in organic rich soil. Ecotoxicology Environmental Safety, Vol. 59, N.3, pp. 340-348, ISSN 0147-6513

Maboeta, M.S., Reinecke, S.A. & Reinecke, A.J. (2002). The relationship between lysosomal biomarker and population responses in a field population of *Microchaetus* sp. (Oligochaeta) exposed to the fungicide copper oxychloride. *Ecotoxicology and Environmental Safety*, Vol. 52, No. 3, (July 2002), pp. 280-287, ISSN 0147-6513

Maity, S., Roy, S., Chaudhury, S., & Bhattacharya, S. (2008). Antioxidant responses of the earthworm *Lampito mauritii* exposed to Pb and Zn contaminated soil. *Environmental Pollution*, Vol. 151, No. 1, (January 2008), pp. 1-7, ISSN 0269-7491

Maity, S., Roy, S., Bhattacharya, S., & Chaudhury, S. (2011). Metallothionein responses in the earthworm *Lampito mauritii* (Kinberg) following lead and zinc exposure: A promising tool for monitoring metal contamination. *European Journal of Soil Biology*, Vol. 47, No. 1, (January-February 2011), pp. 69-71, ISSN 1164-5563

Manerikar, R.S., Apte, A.A. & Ghole, V.S. (2008). In vitro and in vivo genotoxicity assessment of Cr(VI) using Comet assay in earthworm coelomocytes. *Environmental Toxicology and Pharmacology*, Vol. 25, No. 1, (January 2008), pp. 63-68, ISSN 1382-6689

Marino, F., Ligero, A. & Cosin, D.J.D. (1992). Heavy metals and earthworms on the border of a road next to Santiago. *Soil Biology and Biochemistry*, Vol. 24, No. 12, (December 1992), pp. 1705-1709, ISSN 0038-0717.

Milligan, L., Babish, J. & Neuhauser, F. (1986). Non-inducibility of cytochrome P450 in earthworm *Dendrobaena veneta*. *Comparative Biochemistry and Physiology Part C*, Vol. 85, pp. 85-87, ISSN 1532-0456

Min K.S. (2007). The physiological significance of metallothionein in oxidative stress . *Journal of the Pharmaceutical Society of Japan*, Vol. 127   No. 4, (April 2007), pp. 695-702, ISSN 0031-6903

Moore, M.N., Icarus, A.J. & McVeigh, A. (2006). Environmental prognostics: an integrated model supporting lysosomal stress responses as predictive biomarkers of animal health status. *Marine Environmental Research*, Vol. 61, No. 3, (April 2006), pp. 278–304. , ISSN 0141-1136

Morgan, A.J., Turner, M.P. & Morgan, J.E. (2002). Morphological plasticity in metal-sequestering earthworm chloragocytes: Morphometric electron microscopy provides a biomarker of exposure in field populations . *Environmental Toxicology and Chemistry*, Vol. 21, No. 3, (March 2002), pp. 610-618 , ISSN 0730-7268

Morgan, A.J., Stürzenbaun, S.R., Winters, C., Grime, G.W., Aziz, N.A.A. & Kille, P. (2004). Differential metallothionein expression in earthworm (*Lumbricus rubellus*) tissues. *Ecotoxicology and Environmental Safety*, Vol. 57, No. 1, (January 2004), pp. 11-19, ISSN 0147-6513

Morgan, J.E.,. Morgan, A.J. (1999). The accumulation of metals (Cd, Cu, Pb, Zn and Ca) by two ecologically contrasting earthworm species (*Lumbricus rubellus* and *Aporrectodea caliginosa*): Implications for ecotoxicological testing. *Applied Soil Ecology*, Vol. 13, No. 1, (September 1999), pp. 9-20, ISSN 0929-1393

Morrison, D.E., Robertson, B.K. & Alexander, M. (2000). Bioavailability to earthworms of aged DDT, DDE, DDD, and dieldrin in soil. *Environmental Sciences and Technologies*, Vol. 34, No. 4, (January 2000), pp. 709–713, ISSN 0013-936X

Nahmani, J., Hodson, M.E. & Black, S. (2007). A review of studies performed to assess metal uptake by earthworms. *Environmental Pollution*, Vol. 145, No. 2, (January 2007), pp. 402–424, ISSN 0269-7491

Ndayibagira, A., Sunahara, G.I. & Robidoux, P.Y. (2007). Rapid isocratic HPLC quantification of metallothionein-like proteins as biomarkers for cadmium exposure in the earthworm *Eisenia andrei*. *Soil Biology and. Biochemistry*, Vol. 39, No. 1, (January 2007), pp. 194–201, ISSN 0038-0717.

Neuhauser, E.F., Loehr, R.C., Milligan, D.L. & Malecki, M.R. (1985). Toxicity of metals to the earthworm *Eisenia fetida*. *Biology and Fertility of Soils*. Vol. 1, No. 3, (March 1985), pp. 149–52. , ISSN 0178-2762

Novais S.C., Gomes S.I.L., Gravato C., Guilhermino L., De Coen W., Soares A.M.V.M., Amorim M.J.B. (2011). Reproduction and biochemical responses in *Enchytraeus albidus* (Oligochaeta) to zinc or cadmium exposures. *Environmental Pollution*, Vol.159 No 7, (July 2011), pp. 1836-1843, ISSN 0269-7491

OECD (1984). Guidelines for testing of chemicals: earthworm acute toxicity test. No. 207, Paris, France

OECD. (2004). Guideline for testing of chemicals: earthworms reproduction test. No. 222, Paris, France

Olchawa, E., Bzowska, M., Stuerzenbaum, S.R., Morgan, A.J. & Plytycz, B. (2006). Heavy metals affect the coelomocyte-bacteria balance in earthworms: environmental interactions between abiotic and biotic stressors. *Environmental Pollution*, Vol. 142, No. 2, (July 2006), pp. 373–381, ISSN 0269-7491

Owen, J., Hedley, B.A., Svendsen, C., Wren, J., Jonker, M.J., Hankard, P.K., Lister, L.J., Stürzenbaum, S.R., Morgan, A.J., Spurgeon, D.J., Blaxter, M.L. & Kille, P. (2008).

Transcriptome profiling of developmental and xenobiotic responses in a keystone soil animal, the oligochaete annelid *Lumbricus rubellus*. *BMC Genomics* , Vol. 9, (June 2008), pp. 266, ISSN 1471-2164

Pedersen, S.F., Hoffmann, E.K. & Mills, J.W. (2001). The cytoskeleton and cell volume regulation. Comparative Biochemistry and Physiology part A, Vol. 130, No. 3, (October 2001), pp. 385-399. , ISSN 1095-6433

Peijnenburg, W.J.G.M. (2002). Bioavailability of metals to soil invertebrates. In: *Bioavailability of Metals in Terrestrial Ecosystems: Importance of Partitioning for Bioavailability to Invertebrates, Microbes, and Plants*, H.E., Allen, (Ed.), pp. 89–112, Society of Environmental Toxicology and Chemistry (SETAC), ISBN 978-1880611463, Pensacola, USA

Plytycz, B., Homa, J., Koziol, B., Rozanowska, M. & Morgan, A.J. (2006). Riboflavin content in autofluorescent earthworm coelomocytes is species-specific. *Folia Histochemica et Cytobiologica*, Vol. 44, No. 4, (April 2006), pp. 275–280, ISSN 0239-8508

Plytycz, B., Klimek, M., Homa, J., Tylko, G. & Kolaczkowska, E. (2007). Flow cytometric measurement of neutral red accumulation in earthworm coelo- mocytes: novel assay for studies on heavy metal exposure. *European Journal of Soil Biology*, Vol. 43, Supp. 1, (November 2007), pp. S116–S120, ISSN 1164-5563

Plytycz, B., Lis-Molenda, U., Cygal, M., Kielbasa, E., Grebosz, A., Duchnowski, M., Andre, J. &Morgan, A.J., (2009). Riboflavin content of coelomocytes in earthworm (*Dendrodrilus rubidus*) field populations as a molecular biomarker of soil metal pollution. *Environmental Pollution*, Vol. 157, No. 11, (November 2009), pp. 3042–3050, ISSN 0269-7491

Quiao, M., Chen, Y., Wang, C.X., Wang, Z. & Zhu, Y.G. (2007). DNA damage and repair process in earthworm after in-vivo and in vitro exposure to soils irrigated by wastewaters. *Environmental Pollution*, Vol. 148, No. 1, (July 2007), pp. 141–147, ISSN 0269-7491

Rao, J.V., Pavan, Y.S. & Madhavendra, S.S. (2003). Toxic effects of chlorpyrifos on morphology and acetylcholinesterase activity in the earthworm, *Eisenia foetida*. *Ecotoxicology and Environmental Safety*, Vol. 54, No. 3, (March 2003), pp. 296-301 , ISSN 0147-6513

Rao, J.V., Kavitha, P. (2004). Toxicity of azodrin on the morphology and acetylcholinesterase activity of the earthworm *Eisenia fetida*. *Environmental Research*, Vol. 96, No. 3, (November 2004), pp. 323–327, ISSN 0013-9351

Rault, M., Mazzia, C. & Capowiez, Y. (2007). Tissue distribution and characterization of cholinesterase activity in six earthworm species. *Comparative Biochemistry and Physiology Part B Biochemistry & Molecular Biology*, Vol. 147, No. 2, (June 2007), pp. 340-346, ISSN 1096-4959

Rault M, Collange B, Mazzia C, Capowiez Y. (2008). Dynamics of acetylcholinesterase activity recovery in two earthworm species following exposure to ethyl-parathion. *Soil Biology and Biochemistry*, Vol. 40, No. 12, (December 2008), pp. 3086–3091, ISSN 0038-0717

Reinecke, S.A., Reinecke, A.J. (2004). The Comet assay as biomarker of heavy metal genotoxicity in earthworms, *Archives of Environmental Contamination and Toxicology*, Vol. 46, No. 2, (February 2004), pp. 208–215, ISSN 0090-4341

Reinhart, M., Dollahan, N. (2003). Responses of coelomocytes from *Lumbricus terrestris* to native and non-native eukaryotic parasites. *Pedobiologia*, Vol. 47, No. 5-6, (May 2003), pp. 710–716, ISSN 0031-4056

Rodriguez-Castellanos, L., Sanchez-Hernandez, J.C. (2007). Earthworm biomarkers of pesticide contamination: current status and perspectives. *Journal of Pesticide Science*, Vol. 32, No. 4, (April 2007), pp. 360-371, ISSN 1348-589X

Rogival, D., Scheirs, J., De Coen, W., Verhagen, R. & Blust, R. (2006). Metal blood levels and haematological characteristics inwoodmice (*Apodemus sylvaticus L.*) along a metal pollution gradient. *Environmental Toxicology and Chemistry*, Vol. 25, No. 1, (January 2006), pp. 149–157, ISSN 0730-7268

Saint-Denis, M., Narbonne, J.F., Arnaud, C. & Ribera, D. (2001). Biochemical responses of the earthworm *Eisenia fetida andrei* exposed to contaminated artificial soil, effects of lead acetate. *Soil Biology and Biochemistry*, Vol. 33, No. 3, (March 2001), pp. 395–404, ISSN 0038-0717

Sanchez-Hernandez, J.C. (2006). Earthworms biomarkers in ecological risk assessment. *Reviews of Environmental Contamination and Toxicology*, Vol. 188, pp. 85-126, ISSN 0179-5953

Schreck, E., Geret, F., Gontier, L.M. & Treilhou, M. (2008). Neurotoxic effect and metabolic responses induced by a mixture of six pesticides on the earthworm *Aporrectodea caliginosa nocturna*. *Chemosphere*, Vol. 71, No. 10, (May 2008), pp. 1832-1839, ISSN 0045-6535

Scott-Fordsmand, J.J., Weeks, J.M. (1998). Review of selected biomarkers in earthworms. In: *Advances in Earthworm Ecotoxicology*, M. Holmstrup, J., Bembridge, S., Sheppard, et al. (Eds.), pp. 173–198, SETAC Press, ISBN 978-1880611258, Boca Raton, USA

Scott-Fordsmand, J.J., Weeks, J.M. (2000). Biomarkers in earthworms. *Reviews of Environmental Contamination and Toxicology*, Vol. 165, pp. 117-159, ISSN 0179-5953

Scott-Fordsmand, J.J., Weeks, J.M. & Hopkin, S.P. (2000). Importance of contamination history for understanding toxicity of copper to earthworm *Eisenia fetida* (Oligochaeta: anellida), using neutral red retention assay. *Environmental Toxicology and Chemistry*, Vol. 19, N.7, pp. 1774-1780, ISSN 0730-7268

Sforzini, S., Saggese, I., Oliveri, L., Viarengo, A. & Bolognesi, C. (2010). Use of the Comet and micronucleus assays for in vivo genotoxicity assessment in the coelomocytes of the earthworm *Eisenia andrei*. *Comparative Biochemistry and Physiology, Part A*, Vol. 157, Supp. 1, (September 2010), pp S13, ISSN 1095-6433

Smith B.A., Greenberg B. & Stephenson G.L. (2010). Comparison of biological and chemical measures of metal bioavailability in field soils: Test of a novel simulated earthworm gut extraction. *Chemosphere*, Vol. 81, No. 6, (October 2010), pp. 755–766, ISSN0045-6535

Smith, R.C., Reed, V.D. (1993). Reversal of copper(II)-induced methemoglobin formation by thiols. *Journal of Inorganic Biochemistry*, Vol. 52, No. 3, (November 1993), pp. 173-182, ISSN 0162-0134

Smolders, E., McGrath, S.P., Lombi, E., Karman, C.C., Bernhard, R., Cools, R., Van Den Brande, K., Van Os B. & Walrave, N. (2003). Comparison of toxicity of zinc for microbial processes between laboratory-contamined and polluted field soils. *Environmental Toxicology and Chemistry*, Vol. 22, No. 11, (November 2003), pp. 2592-2598, ISSN 0730-7268

Spurgeon, D.J., Hopkin, S.P., (2000). The development of genetically inherited resistance to zinc in laboratory-selected generations of the earthworm *Eisenia fetida*. *Environmental Pollution*, Vol. 109, No. 2, (August), pp. 193–201, ISSN 0269-7491

Spurgeon, D.J., Svendsen, C., Hankard, P.K., Weeks, J.M., Kille, P. & Fishwick, S.K. (2002). Review of Sublethal Ecotoxicological Tests for Measuring Harm in Terrestrial Ecosystems. P5-063/Technical Report, pp 108, Environment Agency, ISBN 1 85705 682 5, Bristol, United Kingdom

Spurgeon, D.J., Weeks, J.M. & Van Gestel, C.A.M. (2003). A summary of eleven years progress in earthworm ecotoxicology: The 7th international symposium on earthworm ecology, Cardiff, Wales, 2002. *Pedobiologia*, Vol. 47, No. 5-6, (May 2003), pp. 588– 606, ISSN 0031-4056

Spurgeon, D.J., Lofts, S., Hankard, P.K., Toal, M., McLellan, D., Fishwick, S. & Svendsen, C. (2006). Effect of pH on metal speciation and resulting metal uptake and toxicity for earthworms. *Environmental Toxicology and Chemistry*, Vol. 25, No. 3, (March 2006), pp. 788–796, ISSN 0730-7268

Stürzenbaum, S.R., Kille, P. & Morgan, A.J. (1998). The identification, cloning and characterization of earthworm metallothionein. *FEBS Letters*, Vol. 431, No. 3, (July 1998), pp. 437–442, ISSN 0014-5793

Stürzenbaum, S.R, Winters, C., Galay, M., Morgan, A.J. & Kille, P. (2001). Metal ion trafficking in earthworms. Identification of a cadmium-specific metallothionein. *Journal of Biological Chemistry*, Vol. 276, No. 36, (September 2001), pp. 34013-34018, ISSN 0021-9258

Svendsen, C., Meharg, A.A., Freestone, P. & Weeks, J.M. (1996).Use of an earthworm lysosomal biomarker for the ecological assessment of pollution from an industrial plastics fire. *Applied Soil Ecology*, Vol. 3, No. 2, (February 1996), pp. 99-107, ISSN 0929-1393

Svendsen, C., Meharg, A.A., Freestone, P& Weeks, J.M. (2003). Use of an earthworm lysosomal biomarker for the ecological assessment of pollution from an industrial plastics fire. *Applied Soil Ecology*, Vol. 3, N.2, pp.99-107, ISSN 0929-1393.

Svendsen, C., Spurgeon, D.J., Hankard, P.K. & Weeks, J.M. (2004) A review of lysosomal membrane stability measured by neutral red retention: is it a workable earthworm biomarker?. *Ecotoxicology and Environmental Safety*, Vol. 57, No. 1, (January 2004), pp. 20–29, ISSN 0147-6513

Tice, R.R., Agurell, E., Anderson, D., Burlinson, B., Hartmann, A., Kobayashi, H., Miyamae, Y., Rojas, E., Ryu, J.C. & Sasaki, Y.F. (2000). Single cell gel/comet assay: guidelines for in vitro and in vivo genetic toxicology testing. *Environmental and Molecular Mutagenesis*, Vol. 35, No. 3, (March 2000), pp. 206–221, ISSN 0893-6692

Valavanidis, A., Vlahogianni, T., Dassenakis, M. & Scoullos, M. (2006). Molecular biomarkers of oxidative stress in aquatic organisms in relation to toxic environmental pollutants. *Ecotoxicology and Environmental Safety*, Vol: 64, No. 2, (June 2006), pp. 178-189, ISSN 0147-6513

Valembois, P., Lassegues, M., Roch, P. & Vaillier, J. (1985). Scanning electron microscopic study the involvement of coelomic cells in earthworm antibacterial defence. *Cell and Tissue Research*, Vol. 240, No. 2, (February 1985), pp. 479 484, ISSN 0302-766X

Van Gestel, C.A.M., Weeks, J.M. (2004). Future recommendations of the third international workshop on earthworm ecotoxicology, Aarhus, Denmark (August 2001).

Ecotoxicology and Environmental Safety, Vol. 57, No. 1, (January 2004), pp. 100-105, ISSN 0147-6513

Verdrengh, M., Tarkowski, A. (2005). Riboflavin in innate and acquired immune responses. *Inflammation Research*, Vol. 54, No. 9, (September), pp. 390–393, ISSN 1023-3830

Vermeulen, F., Covaci, A., D'Havé, H., Van den Brink, N., Blust, R., De Coen, W. & Bervoets, L. (2010). Accumulation of background levels of persistent organochlorine and organobromine pollutants through the soil–earthworm–hedgehog food chain. *Environment International*, Vol. 36, No. 7, (October 2010), pp. 721-727 , ISSN 0160-4120

Vijver, M.G., Vink, J.P.M., Miermans, C.J.H. & Van Gestel, C.A.M. (2003). Oral sealing using glue: a new method to distinguish between intestinal and dermal uptake of metals in earthworms. *Soil Biology and Biochemistry*, Vol. 35, No. 1, (January 2003), pp. 125-132, ISSN 0038-0717

Walker, C.H. (1998). Biomarker strategies to evaluate the environmental effects of chemicals. *Environmental Health Pespectives*, Vol. 106, Suppement 2, (April1998), pp.613-520, ISSN 0091-6765

Wallwork, J.A. (1983). *Annelids: The First Coelomates. Earthworms Biology*. Edward Arnold Publisher, ISBN 0713128844, London, United Kingdom

Weber, R., Gaus, C., Tysklind, M., Johnston, P., Forter, M., Hollert, H., Heinisch, E., Holoubek, I., Lloyd-Smith, M., Masunaga, S., Moccarelli, P., Santillo, D., Seike, N., Symons, R., Torres, J.P., Verta, M., Varbelow, G., Vijgen, J, Watson, A., Costner, P., Woelz, J., Wycisk, P. & Zennegg, M. (2008). Dioxin- and POP-contaminated sites – contemporary and future relevance and challenger. *Environmental and Sciences Pollution Research*, Vol. 15, No. 5, (July 2008) pp. 363–393, ISSN 0944-1344

Weeks, J.M., Svendsen, C. (1996). Neutral red retention by lysosome from earthworm (*Lumbricus rubellus*) coelomocytes: a simple biomarker of exposure to soil copper. *Environmental Toxicology and Chemistry*, Vol. 15, No. 10, (October 1996), pp. 1801–1805, ISSN 0730-7268

Whitfield Åslund, M. L., Simpson, A. J. & Simpson M. J. (2011). 1H NMR metabolomics of earthworm responses to polychlorinated biphenyl (PCB) exposure in soil. *Ecotoxicology*, Vol. 20, No. 4, (June 2011), pp.836–846, ISSN 0963-9292

Wu, S., Wu, E., Qiu, L., Zhong, W. & Chen, J. (2011). Effects of phenanthrene on the mortality, growth, and anti-oxidant system of earthworms (*Eisenia fetida*) under laboratory conditions. *Chemosphere*, Vol. 83, No. 4, (April 2011), pp. 429–434, ISSN 0045-6535

Permissions

The contributors of this book come from diverse backgrounds, making this book a truly international effort. This book will bring forth new frontiers with its revolutionizing research information and detailed analysis of the nascent developments around the world.

We would like to thank M.C. Hernández-Soriano, for lending her expertise to make the book truly unique. She has played a crucial role in the development of this book. Without her invaluable contribution this book wouldn't have been possible. She has made vital efforts to compile up to date information on the varied aspects of this subject to make this book a valuable addition to the collection of many professionals and students.

This book was conceptualized with the vision of imparting up-to-date information and advanced data in this field. To ensure the same, a matchless editorial board was set up. Every individual on the board went through rigorous rounds of assessment to prove their worth. After which they invested a large part of their time researching and compiling the most relevant data for our readers. Conferences and sessions were held from time to time between the editorial board and the contributing authors to present the data in the most comprehensible form. The editorial team has worked tirelessly to provide valuable and valid information to help people across the globe.

Every chapter published in this book has been scrutinized by our experts. Their significance has been extensively debated. The topics covered herein carry significant findings which will fuel the growth of the discipline. They may even be implemented as practical applications or may be referred to as a beginning point for another development. Chapters in this book were first published by InTech; hereby published with permission under the Creative Commons Attribution License or equivalent.

The editorial board has been involved in producing this book since its inception. They have spent rigorous hours researching and exploring the diverse topics which have resulted in the successful publishing of this book. They have passed on their knowledge of decades through this book. To expedite this challenging task, the publisher supported the team at every step. A small team of assistant editors was also appointed to further simplify the editing procedure and attain best results for the readers.

Our editorial team has been hand-picked from every corner of the world. Their multi-ethnicity adds dynamic inputs to the discussions which result in innovative outcomes. These outcomes are then further discussed with the researchers and contributors who give their valuable feedback and opinion regarding the same. The feedback is then collaborated with the researches and they are edited in a comprehensive manner to aid the understanding of the subject.

Apart from the editorial board, the designing team has also invested a significant amount of their time in understanding the subject and creating the most relevant covers. They scrutinized every image to scout for the most suitable representation of the subject and create an appropriate cover for the book.

The publishing team has been involved in this book since its early stages. They were actively engaged in every process, be it collecting the data, connecting with the contributors or procuring relevant information. The team has been an ardent support to the editorial, designing and production team. Their endless efforts to recruit the best for this project, has resulted in the accomplishment of this book. They are a veteran in the field of academics and their pool of knowledge is as vast as their experience in printing. Their expertise and guidance has proved useful at every step. Their uncompromising quality standards have made this book an exceptional effort. Their encouragement from time to time has been an inspiration for everyone.

The publisher and the editorial board hope that this book will prove to be a valuable piece of knowledge for researchers, students, practitioners and scholars across the globe.

List of Contributors

Maria C. Hernández-Soriano
Department of Soil Science, College of Agriculture and Life Sciences, North Carolina State University, Raleigh NC, USA

Shinya Funakawa, Hiroshi Yoshida, Tetsuhiro Watanabe and Soh Sugihara
Graduate School of Agriculture, Kyoto University, Japan

Method Kilasara
Faculty of Agriculture, Sokoine Agricultural University, Tanzania

Takashi Kosaki
Graduate School of Urban Environmental Sciences, Tokyo Metropolitan University, Japan

Manoj K. Jha
North Carolina A&T State University, USA

Andrea Rubenacker, Paola Campitelli, Manuel Velasco and Silvia Ceppi
Departamento de Recursos Naturales, Facultad de Ciencias Agropecuarias, Universidad Nacional de Córdoba, Córdoba, Argentina

Benjamin O. Botwe and Elvis Nyarko
University of Ghana, Department of Oceanography & Fisheries, Ghana

William J. Ntow
University of California, Department of Plant Sciences, USA

Bart Minten
International Food Policy Research Institute, Addis Ababa, Ethiopia

Claude Randrianarisoa
United States Agency for International Development (USAID), Madagascar

Shinjiro Sato
Department of Environmental Engineering for Symbiosis, Soka University, Tokyo, Japan

Kelly T. Morgan
Southwest Florida Research and Education Center, University of Florida, Florida, USA

Song Xiaozong, Jiang Lihua, Lin Haitao, Xu Yu, Gao Xinhao, Zheng Fuli, Tan Deshui, Wang Mei, Shi Jing, Shen Yuwen and Liu Zhaohui
Institute of Agricultural Resources and Environment, Shandong Academy of Agricultural Sciences Jinan, 250100, China

Yasuyuki Fukumoto
Institute of Livestock and Grassland Science, National Agriculture and Food Research Organization (NARO), Japan

Celia Maria Maganhotto de Souza Silva and Elisabeth Francisconi Fay
Embrapa Environment, Brazil

Masaru Ogasawara
Weed Science Center (WSC), Utsunomiya University, 350 Mine-machi, Utsunomiya, Japan

Francisco Martín, Elena Arco, Ana Romero and Carlos Dorronsoro
Soil Science Department, Faculty of Science, University of Granada, Campus Fuentenueva s/n, Granada, Spain

Mariano Simón
Soil Science Department, EPS CITE IIB, University of Almería, Carretera Sacramento s/n, Almería, Spain

Timothy J. Gish, William P. Kustas, Lynn G. McKee and Andrew Russ
USDA-ARS, Hydrology and Remote Sensing Laboratory, Beltsville, Maryland, USA

John H. Prueger and Jerry L. Hatfield
USDA-ARS, National Laboratory for Agriculture and the Environment, Ames, Iowa, USA

Yuko Takada Hoshino
National Institute for Agro-Environmental Sciences, Japan

Bogusław Michalec
Agriculture University in Cracow, Poland

Maria Giulia Lionetto, Antonio Calisi and Trifone Schettino
University of Salento - Dept. of Biological and Environmental Sciences and Technologies, Italy